Computational and Experimental Studies in Alzheimer's Disease

This reference book compiles the recent advances in computational and experimental modelling to screen and manage Alzheimer's disease. It covers basic etiopathology and various in vitro and in vivo strategies of disease intervention. The book discusses how computer-aided drug design approaches reduce costs and increase biological test efficiency. It reviews the screening for anti-Alzheimer drugs and biomarker analysis of disease inhibitors. The book also explores mechanistic aspects of neurodegeneration and the use of natural products as therapeutics for Alzheimer's disease.

Key features:

- Elaborates on the computational modelling of protein target inhibitors as anti-Alzheimer's agents
- Explains the role of phytomolecules and natural products in Alzheimer's therapy
- Reviews preclinical ways to assess drugs focusing on Alzheimer's disease
- Covers biomarker analysis for Alzheimer's disease
- Discusses the onset and progression of Alzheimer's disease

The book is meant for professionals, researchers, and students of neuroscience, psychology, and computational neurosciences.

Computational and Experimental Studies in Alzheimer's Disease

Edited by
Kunal Bhattacharya and
Atanu Bhattacharjee

CRC Press
Taylor & Francis Group
Boca Raton London New York

CRC Press is an imprint of the
Taylor & Francis Group, an **informa** business

Designed cover image: © Shutterstock

First edition published 2024
by CRC Press
2385 NW Executive Center Drive, Suite 320, Boca Raton FL 33431

and by CRC Press
4 Park Square, Milton Park, Abingdon, Oxon, OX14 4RN

CRC Press is an imprint of Taylor & Francis Group, LLC

ISBN: 978-1-032-53007-9 (hbk)
ISBN: 978-1-032-53446-6 (pbk)
ISBN: 978-1-003-41206-9 (ebk)

DOI: 10.1201/9781003412069

Typeset in Times
by Apex CoVantage, LLC

Contents

Foreword

I am highly delighted to put my views and endorse the forthcoming publication of the book titled *Computational and Experimental Studies in Alzheimer's Disease*. This comprehensive reference book represents a compilation of scholarly articles in the field of neuroscience, as it focuses on latest advances in computational and experimental modeling for screening and managing Alzheimer's disease. Authored by distinguished experts, this book is an invaluable source of knowledge for professionals, researchers, and students in neuroscience, psychology, and computational neurosciences etc.

I appreciate the timeliness efforts of the authors and significance of this compilation. The book not only delves into the basic etiopathology of Alzheimer's disease but also explores various in vitro and in vivo strategies for disease intervention. It provides valuable insights into the use of computational modeling in identifying protein target inhibitors as potential anti-Alzheimer agents. Furthermore, it discusses the crucial role of phytomolecules and natural products in Alzheimer's therapy, an area that holds great promise for effective treatments. One of the key strengths of this reference book is its coverage of biomarker analysis for Alzheimer's disease. Identifying reliable biomarkers is pivotal in early diagnosis and monitoring, and the insights offered in this book can potentially impact clinical practice and future research efforts significantly. The insight in this field is essential for developing targeted interventions and improving our understanding of the disease process.

In conclusion, it is worth mentioning that the book offers a comprehensive and up-to-date resource for those deeply committed to addressing the challenges of Alzheimer's disease. I wholeheartedly recommend this book to professionals, researchers, and students in neuroscience, psychology, and computational neurosciences, as well as anyone interested in making a meaningful contribution to the fight against Alzheimer's disease.

Prof (Dr) S. P. Singh
Vice-Chancellor
The Assam Royal Global University
Guwahati-781035, Assam, India

Preface

Alzheimer's disease (AD) has become a major issue in global health in recent decades, impacting millions of people all over the world. Memory loss, confusion, and other cognitive difficulties, as well as severe impairment in everyday functioning, are hallmarks of this progressive neurodegenerative condition. The prevalence of AD has increased rapidly with the aging population and rising life expectancy, inflicting a significant burden on health care systems, families, and society at large.

The goal of the book "Computational and Experimental Studies in Alzheimer's Disease" is to further advance our understanding of this disease by combining computational and experimental methods. To provide a thorough review of the present status of study in this difficult domain, this book compiles a wide variety of themes, approaches, and views.

In the first chapter, we set the stage by discussing the neuropathology, neurochemistry, and epidemiology of Alzheimer's disease. It explores the nuances and gaps in our knowledge of the disease, covering topics like risk factors and the influence of age, genetics, lifestyle, and comorbidities. Alterations in neurochemistry and the emergence of neuropathological markers like -amyloid plaques and neurofibrillary tangles are also discussed in this chapter.

Preclinical procedures for assessing anti-AD medications are discussed in Chapter 2, which mainly covers in vivo and in vitro testing. It delves into using ex vivo models and cell-based tests in addition to transgenic and nontransgenic animal models. This chapter emphasizes the value of preclinical research in determining if a therapy intervention is safe and effective before proceeding with clinical trials.

In Chapter 3, we look at how computational methods can be used to investigate the feasibility of AD treatments derived from plants. Methods for identifying natural chemicals that interact with important targets implicated in AD pathogenesis are discussed, including the use of molecular docking, virtual screening, and molecular dynamics simulations. The potential of computational approaches to promoting efficient treatments for AD is highlighted in this chapter.

The fourth chapter examines recent advances in the application of computer-aided drug design (CADD) to AD. It emphasizes the use of high-performance computing and advanced algorithms to identify and modulate key protein targets implicated in the pathology of Alzheimer's disease. The combination of computational models, virtual screening, and experimental methods opens up new avenues for the discovery of effective Alzheimer's disease (AD) treatments.

Analysis of biomarkers for Alzheimer's disease is the subject of Chapter 5. The possibility of different biomarkers for use in early diagnosis, prognosis, and therapy of AD is discussed. The chapter highlights the potential of biomarkers in generating new treatments and facilitating drug development for AD, and emphasizes their usefulness in clinical practice and research.

In Chapter 6, we delve into the role of phytochemicals and the underlying mechanisms of neurodegeneration in AD. It explores the therapeutic potential of plant-based substances like curcumin and resveratrol in slowing the progression of Alzheimer's disease. In this chapter, we learn more about the potential benefits of phytochemicals and how they can be used to attack AD at several points in the disease's development.

As anti-AD agents, Chapter 7 concentrates on computational modelling of BACE-1, AChE, and phosphodiesterase inhibitors. It presents the results of molecular dynamics simulations and binding free energy calculations and discusses the identification of potent antagonists using in silico techniques. This chapter provides insightful information regarding potential lead molecules and their interactions with key targets in Alzheimer's disease.

Therapeutic applications of natural products for AD utilizing computational approaches are explored in Chapter 8. It emphasizes natural compounds' potential as a source of structurally varied molecules with extensive biological activity. This chapter delves into how computational methods can be used to discover new natural chemicals and their underlying mechanisms of action, paving the path for novel AD treatments.

Potential AD inhibitors are analysed in Chapter 9 using computational modelling of several target proteins. The article talks about how virtual screening, molecular docking, and molecular dynamics simulations can be used to find new lead compounds. This chapter provides information about how to create new treatments for Alzheimer's disease.

In summary, "Computational and Experimental Studies in Alzheimer's Disease" integrates computational approaches and experimental investigations to present an interdisciplinary approach to understanding AD. This book aims to contribute to the growing body of knowledge and inspire further research toward developing effective treatments for this life-threatening condition by exploring various aspects of AD pathology, preclinical evaluation, computational modelling, and therapeutic applications.

Editors Kunal Bhattacharya
Atanu Bhattacharjee

About the Editors

Kunal Bhattacharya is currently working as an assistant professor at the Pratiksha Institute of Pharmaceutical Sciences, Assam, India. His area of specialization includes computer-aided drug design, bioinformatics, and analytical chemistry. Kunal Bhattacharya obtained his M.Pharm in pharmaceutical chemistry from Nitte University, Mangalore, India. He is currently pursuing his Ph.D. in pharmaceutical sciences in neurodegenerative diseases from The Assam Royal Global University, Assam, India. He has worked extensively on multiepitope vaccine designing and developing various essential drug molecules. He has reviewed various research articles and book chapters from reputed publishers. He has published various Research and Review articles in various reputed journals of academic importance. He has also contributed chapters to numerous books in the fields of mental health and pharmaceutical science. He has received prestigious summer research fellowships in India: CSIR in 2020 and IASc-INSA-NASI in 2021.

Atanu Bhattacharjee is currently working as a professor in the School of Pharmacy, The Assam Royal Global University, Assam, India. He is considered a pioneer in the fields of pharmacognosy and phytochemistry. Dr Bhattacharjee achieved his Ph.D. in Pharmaceutical Sciences in the areas of phytochemistry and neuro-pharmacology from Nitte University, Mangalore, India. He worked extensively on various in silico, in vitro and in vivo experiments to encounter neuro-degenerative disorders such as Alzheimer's disease with folklore medicine from the Indian diaspora and to determine safety profiles and standardize medicinal plants. Many research and review articles published in internationally acclaimed journals are under his credentials. In 2021, he received the prestigious "Young Scientist Award" from the VDGOOD Professional Association in India for his significant contribution to scientific research. His extensive work on natural products to encounter SARS COV-2 has been recognized globally as he received a patent from South Africa in 2022 on "Bioinformatics platform for identification of quercetin derivatives as immunity booster molecules against SARS COV-2". He was the Honorable Academic Editor of the *Asian Journal of Medicine and Health*, SCIENCEDOMAIN International, India till 2021. He is also an editorial board member and reviewer of many internationally reputed journals. Dr Bhattacharjee is also a member of the Alzheimer's Association.

About the Editors

Contributors

Noor-ul-Ain
Department of Biotechnology, Faculty of Science & Technology
University of Central Punjab, Lahore, Pakistan

Alakesh Bharali
Department of Pharmaceutical Science, GIPS
Girijananda Chowdhury University, Assam, India

Hans Raj Bhat
Department of Pharmaceutical Sciences
Dibrugarh University, India

Kunal Bhattacharya
Pratiksha Institute of Pharmaceutical Science
Assam India

Nilutpal Sharma Bora
NETES Institute of Pharmaceutical Science
NEMCARE Group of Institutions, Assam, India

Raja Chakraborty
Assam Don Bosco University Institute of Pharmacy
Assam Don Bosco University, Assam India

Nongmaithem Randhoni Chanu
Pratiksha Institute of Pharmaceutical Science
Assam, India

Raushan Kumar Chaudhary
Department of Pharmacy Practice, NGSM Institute of Pharmaceutical Sciences (NGSMIPS)
Nitte (Deemed to be University), Karnataka, India

G.N. Darwhekar
Acropolis Institute of Pharmaceutical Education and Research
Indore, India

Aparoop Das
Department of Pharmaceutical Sciences
Dibrugarh University, India

Dibyajyoti Das
Pratiksha Institute of Pharmaceutical Science
Assam, India

Bhargab Deka
Department of Pharmaceutical Science, GIPS
Girijananda Chowdhury University, Assam, India

Satyendra Deka
Pratiksha Institute of Pharmaceutical Science
Assam, India

Biswajit Devanath
Pratiksha Institute of Pharmaceutical Science
Assam, India

Chanam Melody Devi
NETES Institute of Pharmaceutical Science
NEMCARE Group of Institutions, Assam, India

Jagadeesh Dodakallanavar
KLE College of Pharmacy, Belagavi, KLE Academy of Higher Education and Research
Karnataka, India

Ankita Dubey
Acropolis Institute of Pharmaceutical Education and Research
Indore, India

Darasaguppe R. Harish
ICMR-National Institute of Traditional Medicine
Belagavi, 590010, Karnataka, India

Hafiz Muhammad Hurraira
Department of Biotechnology, Faculty of Science & Technology
University of Central Punjab, Lahore, Pakistan

Shraddha M. Jadhav
Department of Zoology, Savitrbai Phule Pune University, Maharashtra, India

Praneetha Jain
Department of Pharmacy Practice, NGSM Institute of Pharmaceutical Sciences (NGSMIPS)
Nitte (Deemed to be University), Karnataka, India

Ramen Kalita
Pratiksha Institute of Pharmaceutical Science
Assam, India

Tutumoni Kalita
Department of Pharmaceutical Science, GIPS
Girijananda Chowdhury University, Assam, India

Rakhi Khabiya
Acropolis Institute of Pharmaceutical Education and Research
Indore, India

Pukar Khanal
Department of Pharmacology, KLE College of Pharmacy Belagavi
KLE Academy of Higher Education and Research (KAHER), India

Karan Kumar
School of Energy Science and Engineering, Indian Institute of Technology
Assam, India

Rupesh Kushwah
Government Shyam Sundar Agrawal PG College
Madhya Pradesh, India

Maleeha Manzoor
Department of Basic & Applied Chemistry, Faculty of Science & Technology
University of Central Punjab, Lahore, Pakistan

Uday Venkat Mateti
Department of Pharmacy Practice, NGSM Institute of Pharmaceutical Sciences (NGSMIPS)
Nitte (Deemed to be University), Karnataka, India

Banshongdor H. Mawlieh
NETES Institute of Pharmaceutical Science
NEMCARE Group of Institutions, Assam, India

B. Nagasudha
Department of Pharmaceutical Chemistry, CES College of Pharmacy
Andhra Pradesh, India

Muhammad Naveed
Department of Biotechnology, Faculty of Science & Technology
University of Central Punjab, Lahore, Pakistan

B M Patil
Department of Pharmacology, KLE College of Pharmacy Belagavi
KLE Academy of Higher Education and Research (KAHER), India

Vishal S. Patil
Department of Pharmacology, KLE College of Pharmacy Belagavi, KLE Academy of Higher Education and Research (KAHER), India
ICMR-National Institute of Traditional Medicine, Karnataka, India

Prakash Rajak
Department of Pharmaceutical Sciences
Dibrugarh University, Assam, India

Kala Bahadur Rawal
Department of Pharmacy Practice, NGSM Institute of Pharmaceutical Sciences (NGSMIPS)
Nitte (Deemed to be University), Karnataka, India

Bhargab Jyoti Sahariah
NETES Institute of Pharmaceutical Science
NEMCARE Group of Institutions, Assam, India

Bhanu P Sahu
Department of Pharmaceutical Science, GIPS
Girijananda Chowdhury University, Assam, India

Trideep Saikia
Department of Pharmaceutical Science, GIPS
Girijananda Chowdhury University, Assam, India

Muhammad Aqib Shabbir
Department of Biotechnology, Faculty of Science & Technology
University of Central Punjab, Lahore, Pakistan

Amit Kumar Shrivastava
Department of Oriental Pharmacy and Wonkwang-Oriental Medicine Research Institute
Wonkwang University, Korea

B. Sivakumar
Department of Pharmaceutical Chemistry, SRM College of Pharmacy
SRM Institute of Science and Technology, Tamil Nadu, India

Vrishti M Thakur
Acropolis Institute of Pharmaceutical Education and Research
Indore, India

N. Yellasubbaiah
Department of Pharmaceutical Chemistry
Santhiram College of Pharmacy, Andhra Pradesh, India

Fatima Zaidi
Department of Biotechnology, Faculty of Science & Technology
University of Central Punjab, Lahore, Pakistan

1 Alzheimer's Disease

Epidemiology, Neuropathology, and Neurochemistry

Raushan Kumar Chaudhary, Uday Venkat Mateti, Pukar Khanal, Kala Bahadur Rawal, Praneetha Jain, Vishal S Patil, Amit Kumar Shrivastava and B M Patil*

1.1 INTRODUCTION

The demographics of the world are changing to incorporate larger populations of the elderly as a result of rising life expectancy. Although people are living longer, not everyone can age healthily. Around 50 million people worldwide—mostly elderly people—suffer from dementia, which is the decline of memory and cognition. Alzheimer's disease (AD) is a neurodegenerative disease that progresses over time and results in brain atrophy, neuronal loss, and death (Cummings et al., 2021; Taudorf et al., 2021; Huang et al., 2020). AD threatens to become a global epidemic as there is no prevention or cure for the disease to date, it has cost more than one trillion USD to manage.

1.2 EPIDEMIOLOGY OF AD

1.2.1 Incidence and Prevalence

AD is the most common predisposing factor for dementia worldwide, and its incidence increases with age. Study reports revealed that the incidence of AD approximately double every 5 years after age of 65. However, early-onset AD can emerge before age 65, although it is relatively rare. Furthermore, women are more likely to get AD than men, possibly because women live longer lives than men. A study report shows that AD is the most common type of dementia, comprising up to 70% of dementia cases, and in 2023, AD poses a significant global health burden. In 2006, there were 26.6 million people living with AD worldwide, which was expected to quadruple by 2050, when AD is projected to affect 1 out of every 85 individuals on this planet (Brookmeyer et al., 2007). In addition, Nichols and colleagues forecasted that worldwide 150 million people will be living with AD and other types of dementia by 2050 (GBD 2019 Dementia Forecasting Collaborators, 2022).

1.2.2 Global Discrepancies

The prevalence of AD varies across different regions and countries. Some studies suggest that developed countries (Europe and North America) have a higher prevalence than developing countries (South America, Asia, and Africa), possibly due to differences in health care access, population aging, and lifestyle factors. However, the occurrence of AD is lower in developed countries than in developing and underdeveloped countries, which could be due to improvements in education, effective and timely management of risk factors for vascular abnormalities, and a decrease in the incidence of stroke across Europe and North America (Lv et al., 2023). In 2006, the overall prevalence of AD in Africa was 1.33 million, Asia was 12.65 million, Europe was 7.21 million, Latin America and the Caribbean was 2.03 million, North America was 3.10 million, and Oceania was

DOI: 10.1201/9781003412069-1

0.26 million. By 2050, the numbers were projected to be 6.33 million, 62.85 million, 16.51 million, 10.85 million, 8.85 million, and 0.84 million, respectively (Brookmeyer et al., 2007).

According to the Alzheimer's Association, AD is rapidly affecting American people of all ages. In 2023, around 6.7 million Americans are living with AD, which is forecasted to be 12.7 million by 2050 and 13.8 million by 2060. In America, 1 out of 9 people aged 65 and above are suffering from AD, and about two-third of American women are at risk of AD (Alzheimer's Disease Facts and Figures, 2023; Alzheimer's Associations, 2023).

According to the meta-analysis research of Niu et al., the prevalence of AD among Europeans was 5.05%, 3.31% in men and 7.13% in women. The incidence of AD was 11.08 cases per 1000 person-years, and the incidence among women was approximately double that of men (13.25 vs 7.03 cases per 1000 person-years, respectively) (Niu et al., 2017).

Chan et al. conducted a comprehensive evaluation using Chinese data from 1990 to 2010 and identified exponential growth in AD cases during that period, with an incidence rate of 6.25 cases per 1000 person-years: 1.9 million cases were reported in 1990, 3.71 million cases in 2000, and 5.69 million cases in 2010 (Chan et al., 2013). A recent study showed that 8.8 million people in India were above 60 years old, and 7.4% were living with AD and other types of dementia. Notably, the incidence rate of dementia is lower in Indian people than in other developing countries (Lee et al., 2023).

1.2.3 Survival Status of AD

In 1976, a study reported that AD was the fourth most common leading cause of death among older adults (Katzman, 1976). On average, the life expectancy of AD people is around 4 to 8 years after clinical diagnosis, although some people can survive up to 20 years or more. Early detection and treatment of the condition and other linked issues such as vascular anomalies, pneumonia, and lifestyle changes can significantly improve the survival rate (Rasmussen & Langerman, 2019). A recent study shows a declining pattern of AD incidence in some developed countries. A well-established economy, a high-quality education level, and better management of cardiovascular predisposing factors have been considered key in the declining pattern (Ponjoan et al., 2020).

1.2.4 Race

Elderly Black Americans are twice as likely to have AD and other dementia as elderly whites, and elderly Hispanics are one and a half times more likely to have AD and other dementia than elderly whites (Alzheimer's Disease Facts and Figures, 2023).

1.2.5 Economic Impact

AD has significant societal costs in the massive production losses of patients, families, and societies, and the cost of treating and managing AD is increasing exponentially. The global cost of AD and AD-related dementia was estimated at around USD 2.8 trillion in 2019, and this burden was expected to reach USD 4.7 trillion USD by 2030, USD 8.5 trillion in 2040, and USD 16.9 trillion by 2050. Moreover, 65% of the total global economic burden of the AD and AD-related dementia was accounted for in low- and middle-income nations (Nandi et al., 2022). Currently, the cost of AD in the US is USD 321 billion and is expected to reach USD 1 trillion by 2050 (Skaria, 2022). According to 2016 data, around 15.9 million unpaid caregivers (family and friends) in the United States provided 18.2 billion hours free of charge to care for AD and other forms of dementia, valued at USD 230.1 billion in national contributions (Prevalence: Alzheimer's Disease Statistics, 2023). In 2008, the cost of AD and other types of dementia in developing nations was USD 73 billion (Kalaria et al., 2008).

1.3 NEUROPATHOLOGY OF AD

AD results from abnormal conformational changes in protein due to the unusual processing and polymerization of soluble proteins (Britannica, The Editors of Encyclopaedia, 2023). AD is generally categorized into two types based on its onset of age, i.e., early-onset AD (EOAD) (<65 years of age) and late-onset AD (LOAD) (>65 years of age) (Bekris et al., 2010). EOAD is rare, accounting for about 5–10% of AD cases, and is often linked with inherited genetic mutation or the dominance of certain genes (Wingo et al., 2012). Meanwhile, LOAD constitutes about 90–95% of all AD cases and is often linked with genetic and environmental factors (Diniz et al., 2017). That is, EOAD and LOAD are familial and sporadic, respectively (Wingo et al., 2012; Diniz et al., 2017).

AD has been further classified into four different stages based on the nature of existing symptoms: preclinical, mild/early, moderate, and severe/late stage (Britannica, The Editors of Encyclopaedia, 2023; Kumar et al., 2023). The preclinical stage is also called the presymptomatic stage, characterized by initial mild changes in memory capacity with no clinical manifestation and no effect on carrying out day-to-day activity. However, early-stage pathological changes have been reported in the hippocampus and cortex during the preclinical stage (Kumar et al., 2023).

Mild-stage AD is generally characterized by loss of memory and concentration that hampers daily life activities and is associated with disorientation as to time and place, unusual mood changes, depression, etc. Moderate AD is characterized by increased memory loss including difficulty recognizing one's own family members or close ones and struggles with reading, writing, and speaking. Severe damage to the cortex region often characterizes late-stage AD due to the progressive deposition of Aβ plaque and neurofibrillary tangles, severely affecting the patient's functional and cognitive capacity, resulting in a bedridden condition (Britannica, The Editors of Encyclopaedia, 2023; Kumar et al., 2023). Several hypotheses and mechanisms have been proposed in the neuropathology of AD over the decades. However, the complex nature of the disease, its prognosis, and treatment failures have made it a mysterious and incompletely understood disease among psychiatric illnesses.

1.3.1 Amyloid β Formation

The neuronal cell membrane consists of a bitopic integral unique protein called an amyloid precursor protein (APP), and its physiological function is not well understood. APP consists of around 770 amino acids that are sequentially chopped down by 2 different membrane-associated enzymes, α-secretase or β-secretase/BACE1 (beta-site APP cleaving enzyme 1) and γ-secretase (Selkoe, 2004). Normally, the cleaved peptide formed by the sequential cleavage of APP by α-secretase and γ-secretase is soluble and non-pathogenic. This APP processing is a nonamyloidogenic pathway that is carried out by metalloproteases (AAM10 and ADAM17) called α-secretase (Selkoe, 2004; Kimberly et al., 2001).

The cleavage of APP by BACE1 and γ-secretase results in insoluble, sticky, and pathogenic peptides via the amyloidogenic pathway. The breakdown products of α-secretase and β-secretase are soluble ectodomain APPsα along with C-fragment terminal (C83) and truncated APPsβ along with C-fragment terminal (C99), respectively (Selkoe, 2004). Meanwhile, γ-secretase is an intramembrane aspartyl protease that consists of a multidomain protein such as presenilin-1 (PSEN1), nicastrin, anterior pharynx defective 1 (Aph1), and PSEN2 (Bergmans & De Strooper, 2010). The C-fragment terminal (C83 or C99) undergoes further cleavage through this γ-secretase, resulting in the production of soluble P3 peptide and insoluble Aβ, and the APPsα and P3 are responsible for neuroprotection, synaptic signalling, gene expression, etc. (Selkoe, 2004).

The proteolysis of APP results in the formation of amyloid β (Aβ) peptides which are generally 40 or 42 amino acids in length ($Aβ_{40}$ or $Aβ_{42}$) (Schilling et al., 2018). As the amyloid β (Aβ) is sticky, they build up to form Aβ plaque (Britannica, The Editors of Encyclopaedia, 2023). The N-terminal truncated and pyroglutamate Aβ peptides have been found to have augmented cytotoxicity, and these peptides have been composed of about 10–50% total Aβ peptides in the brains of AD patients (Goldblatt et al., 2017; Galante et al., 2016).

However, the mechanism behind Aβ peptides-induced cytotoxicity remains unclear to date, although disbalance in cellular ionic homeostasis and oxidative damage to neurons were proposed as mechanisms of cytotoxicity, specifically the mechanisms via which ionic homeostasis at the cellular level is altered, such as the Aβ-induced modulation of cholinergic, glutaminergic, and other ion channels located in the plasma membrane of cells (Wang & Zheng, 2019). Researchers have also proposed the Aβ-induced activation of the ryanodine receptor calcium-channel of the endoplasmic reticulum: An influx of excessive calcium ions inside the mitochondria enhances the permeability of the mitochondrial membrane, which activates the mitochondrial permeability transition pore, which leads to cell apoptosis (Wang & Zheng, 2019; Quintana et al., 2020).

Apart from this, there might be a direct formation of pores in the neuronal membranes by Aβ, which conducts ions (Peters et al., 2016). Further, neuronal hyperexcitability via the competitive inhibition of glutamate uptake by neurons via Aβ binding to glutamate receptors also contributes to AD (Zott et al., 2019). Neuronal damage due to oxidative stress occurs via several methods, such as activation of the enzyme NADPH, forming reactive oxygen species (ROS) by Aβ-coordinated redox-active metal (Cu) (Cheignon et al., 2018), or increasing the concentration of calcium in the mitochondria, leading to superoxide-led mitochondrial degeneration (Quintana et al., 2020).

1.3.2 Tau and Tangle Formation

Neuronal cells are held together by a cytoskeleton that is partly made up of microtubules. These neuronal microtubules consist of a specialized protein called tau which does not allow microtubules to break. Thus, it helps sustain neuronal cell integrity and function (Barthélemy et al., 2020). Tau often undergoes phosphorylation by the kinase enzyme activated by Aβ deposition in the neuronal environment. The phosphorylation of tau leads to changes in structure confirmation/misfolding, followed by clumping or aggregation that leads to neurofibrillary tangles (NFT). This tangle formation leads to the disruption of microtubules and contributes to neuronal dysfunction and synaptic failure; meanwhile, Aβ plaque build-up is followed by tangle formation and accumulation in the hippocampus and neocortex. Thus, Aβ deposition and tangle formation are interlinked with each other and contribute to neurotoxicity in AD (Barthélemy et al., 2020; Lee et al., 2001).

Generally, tau phosphorylation is mediated by kinases such as glycogen synthase kinase-3-β (GSK-3β), mitogen-activated protein kinase p38, certain isoforms of cyclin-dependent kinases (CDK), and the calpain–CDK5 signalling pathway (site specific) (De Strooper & Woodgett, 2003). GSK-3β is not only involved in phosphorylation-related AD pathogenesis but also participates in the cleavage of APP and the production of Aβ along with the elevation of oxidative stress in mitochondria. However, certain other kinases also contribute towards tau phosphorylation, such as protein kinase-A (PKA), protein kinase-C (PKC), MAPK, JNK, ERK-1, and ERK-2, caspase -3 or -9 and serine/threonine kinase. All these kinases are activated by Aβ plaque (Kuruva & Reddy, 2017).

AD pathogenesis also depends on the pattern of Aβ and tau spread, which commonly occurs via two distinct mechanisms, i.e., transneuronal spread (neuron-to-neuron spread via synaptic connection) and local spread (spread occurs via extracellular space or vesicles). Aβ has been reported to undergo both transneuronal and local spread, whereas tau undergoes only transneuronal spread. Tau immediately after spread transfers its characteristics to the recipient cell, i.e., misfolding and clumping (Kim et al., 2019).

Similarly, vesicles such as brain exosomes have been found to cross the blood–brain barrier (BBB) and spread Aβ and tau proteins into the peripheral blood of humans (Mustapic et al., 2017). Further, nonneuronal cells such as astrocyte cell-derived exosomes have been reported to have a role in the spread of monomeric and aggregated Aβ and tau proteins in mouse models. Other nonneuronal cells, such as microglia and oligodendrocytes, also contribute to the spreading of Aβ and tau (Narasimhan et al., 2020). This suggests that Aβ and tau from peripheral blood could be potential diagnostic biomarkers. These all ultimately result in AD progression, neurodegeneration, synaptic loss, and the disruption of neuronal signals.

Tau (both physiologic and pathogenic) undergoes cellular uptake via low-density lipoprotein receptor-related protein 1 (LRP1)-mediated endocytosis (Rauch et al., 2020). LRP1 promotes the Aβ clearance in an apolipoprotein E (ApoE) isoform-specific manner by regulating BBB via two mechanisms, i.e., cellular absorption and degradation and transcytosis (Shinohara et al., 2017).

1.3.3 Immunological Reactions

Neuroimmunomodulation is another component of pathogenesis mediated by Aβ and tau deposition. Aβ plaque and tau have been reported to mediate immunological reactions by activating immune cells in the brain such as astrocytes and microglia, which results in a neuroinflammatory cycle. The cycles include the stimulation of astrocytes, secretion of ROS and metalloproteases, and formation of more Aβ and tau, resulting in neurodegeneration and promoting neuronal and synaptic loss via reactive microgliosis and astrogliosis (Das & Chinnathambi, 2019). Microglia are responsible for maintaining plasticity and synapse via phagocytosis. They infiltrate in response to the Aβ and tau deposition and are responsible for mediating innate immunity (Heneka et al., 2015). The cell surfaces of microglia attach to the surface of Aβ via several receptors such as SCARA1, CD36, CD14, α6β1 integrin, CD47, and Toll-like receptors, which facilitates the endocytosis of the pathogenic proteins (Aβ and tau) inside microglia (Liu et al., 2005).

Inside the microglia, these pathogenic proteins are degraded by the proteases (neprilysin) and insulin-degrading enzymes in the lysosomes (Malito et al., 2008). However, in severe cases of AD, the microglial-mediated clearance of Aβ crashes due to the elevation in localized cytokine concentration, leading to a decrease in surface receptors (Hickman et al., 2008). Triggering receptor expressed on myeloid cells 2 (TREM2) is a cell surface receptor on microglia which mediates the phagocytosis of the pathogenic proteins involved in AD. However, the mutation of TREM2 leads to the inefficient clearance of Aβ by microglia, which promotes the accumulation of Aβ and tau, resulting in neurodegeneration and a severe form of AD (Neumann & Daly, 2013).

Hyperinsulinemia is another condition that has been linked with neuroinflammation, nerve cell dysfunction, apoptosis, and decreases in acetylcholine and transthyretin. Insulin receptors are responsible for glucose utilization by the neurons in the brain. However, in the case of AD, the patient's neuron lacks the insulin receptor, or there is resistance to insulin action, both of which result in hyperinsulinemia (elevated insulin in serum). This abnormal concentration of insulin is associated with AD (Britannica, The Editors of Encyclopaedia, 2023).

1.3.4 Amyloid Angiopathy and Apolipoprotein E

The deposition of Aβ plaque in the brain's microvasculature leads to amyloid-induced angiopathy, which occurs in about 90% of AD (Quintana et al., 2020). Aβ plaque in cerebral vasculature results in vascular dysfunction via stimulation of NADPH, ROS-induced increases in the calcium load in the endothelium, mitochondrial degeneration, and pericyte contraction mediated via endothelin-1, resulting in vasoconstriction and hypoperfusion. This damage to the cerebral vascular system results in AD's pathogenesis and progression, which generally occurs before the appearance of cognitive symptoms (Quintana et al., 2020; Nortley et al., 2019). Hence, vascular angiopathy is regarded as the early-onset therapeutic target in AD pathogenesis.

Cerebrovascular dysfunction in AD has been attributed to the ApoE gene, which is linked with late-onset AD in about 95% of cases. The ApoE gene is located on chromosome 19 and has 3 distinct allelic variants due to single nucleotide polymorphism (SNP), i.e., ApoEε2, ApoEε3, and ApoEε4 which occurs at likelihoods of 5–10%, 70–80%, and 10–15%, respectively, in humans (Martínez-Martínez et al., 2020). Individuals carrying the ApoEε4 variant are highly susceptible (4 times in the case of heterozygous and 15 times in the homozygous case) to AD compared with individuals with ApoEε2 and ApoEε3 (Chen et al., 2021). Unfortunately, it has been evidenced that subjects carrying ApoEε4 also have leaky BBB in the vascular system in the region brain associated with memory and cognition, i.e., hippocampus and parahippocampal gyrus, which is linked with the existence of cognitive symptoms among patients with AD (Montagne et al., 2020). ApoEε4 is also found to contribute to the elevation of Aβ in the brain of transgenic mice, leading to tau hyperphosphorylation (Tesseur et al., 2000). Further, it has been reported that ApoE and Aβ can occur together in the plaque and synaptic space in the brain. Additionally, the ApoE receptors and their corresponding proteins might be overexpressed in AD and play a crucial role in depositing the ApoE–Aβ complex in the synaptic terminals (Bilousova et al., 2019).

1.3.5 Genetics or Epigenetics

Any defect in genetic makeup or mutation, epigenetic variations, or altered gene expression has been identified to be a new and growing area in the pathogenesis of AD (Britannica, The Editors of Encyclopaedia, 2023; Nikolac et al., 2021). EOAD has been linked to autosomal dominance with APP, PSEN1, PSEN2, and PS1 gene mutation (Li et al., 2019; Dries & Yu, 2008). APP, PSEN1, and PSEN2 genes are located on chromosomes 21, 14, and 1, respectively, and about 30 APP, more than 200 PSEN1, and nearly 40 PSEN2 mutations have been reported so far (Li et al., 2019; Lanoiselée et al., 2017). Mutation of these genes is responsible for the production of Aβ, which increases the $A\beta_{42}/A\beta_{40}$ ratio, resulting in AD. Similarly, LOAD has been linked with the defect in the gene responsible for cholesterol transport, i.e., apolipoprotein E genes (ApoE gene) (Britannica, The Editors of Encyclopaedia, 2023). Apart from the ApoE, 20 other genes related to lipid metabolism, endocytosis, and innate immunity has been identified based on comprehensive genome association studies (Nikolac et al., 2021). ApoEε4 is a critical risk factor for both LOAD and EOAD compared to ApoEε2 and ApoEε3 (Kim et al., 2009). Further, the outer mitochondrial membrane 40 homologs (TOMM40) gene's translocase is linked to AD onset (Britannica, The Editors of Encyclopaedia, 2023). Further, ABCA1 mutation has been found to lower the level of HDL in the blood and promotes the accumulation of cholesterol in tissue and AD pathogenesis (Nordestgaard et al., 2015). Generally, the long form of the TOMM40 gene is associated with AD before 80 years of age, whereas the short form of TOMM40 is associated with AD after 80 years of age (Britannica, The Editors of Encyclopaedia, 2023). Further, two novel genes, CLU and BIN1, are responsible for the reduced clearance of Aβ and for Aβ and NFT formation (Foster et al., 2019). Likewise, the ApoEε4 allele and polymorphism in the oestrogen receptor have been documented to be risk factors for AD, making women 2/3rd more prone to AD than men (Zhao et al., 2015). Another gene, evolutionarily conserved signalling intermediate in the Toll pathway (ECSIT), plays a role in mitochondrial dysfunction, increases oxidative stress, and promotes inflammation by interacting with genes such as ApoE, GCDH, LONP1, PSEN1, and PSEN2, which in turn contributes to AD pathogenesis (Zhao et al., 2015). Additionally, DNA methylation, DNA hydroxymethylation, post-translation modification in histone protein, mitochondrial DNA, and regulation of noncoding RNA are epigenetically linked to AD pathogenesis (Nikolac et al., 2021).

1.3.6 Lifestyle and Prevention

Elevated homocysteine level has been associated with the pathogenesis of AD. Vitamin A controls the homocysteine level and helps in the poor prognosis of AD. Similarly, caffeine intake has been found to decrease the level of Aβ in human brains and blood. Additionally, consumption of moderate amounts of alcohol (8–14 drinks per week) among individuals of age 75 years with normal cognitive function has been associated with a 40% reduced risk of dementia (Britannica, The Editors of Encyclopaedia, 2023). Even the bacterium responsible for chronic periodontitis, i.e., *Porphyromonas gingivalis*, can induce AD. This bacterium secretes a toxin called gingipain that is responsible for the increased production of Aβ and neurotoxicity. Similarly, GM-CSF (granulocyte–macrophage colony-stimulating factor) is a protein produced in rheumatoid arthritis patients that destroys pathogenic Aβ proteins by activating immune cells (Britannica, The Editors of Encyclopaedia, 2023).

1.3.7 Risk Factors Associated with AD

Aging is a natural and irreversible process during which an individual's nervous system undergoes several changes, such as reduced brain size and weight, loss of neurons and synapses, and increased brain ventricle size. These all generally occur with aging, making diagnosing AD difficult (Hou et al., 2019).

Apart from aging, environmental factors such as metals, air pollution, diet, and infections are equally crucial for AD pathology. Toxic metals such as lead and aluminium are reported to cause AD by increasing the expression of β-secretase leading to Aβ deposition and via tau misfolding, respectively (Huat et al., 2019; Colomina & Peris-Sampedro, 2017). Air pollution has been found to increase oxidative stress and cause neuroinflammation/neurodegeneration along with the deposition of Aβ and tau in the frontal cortex region of the brain (Croze & Zimmer, 2018).

Further, malnutrition is one of the vital components which is associated with the occurrence of cognitive dysfunction. Poor cognition has been observed among individuals with deficiencies in folic acid, vitamin B12, and vitamin D (Koyama et al., 2016). Increased serum concentrations of AGE have been linked with cognitive decline and poor memory, leading to AD. AGE is a secondary toxin product of the degradation of amino groups in proteins, lipids, and nucleic acids. AGE has its receptors (RAGE) on microglia and astrocytes, which mediates its role in AD (Abate et al., 2017). Infection with Herpes simplex virus (HSV-1), *Treponema pallidum*, and *Chlamydia pneumonia* has been reported to be associated with AD (Sochocka et al., 2017; Fülöp et al., 2018). An individual with HSV-1 infection has been found to have ApoEε4 allele. Similarly, syphilitic dementia caused by *Treponema pallidum* has been documented that produces lesions similar to NFT. Further, deterioration of cognitive function and LOAD have been documented due to *Chlamydia pneumonia* infection, which occurs via the activation of astrocytes and microglia, leading to cell apoptosis (Sochocka et al., 2017; Fülöp et al., 2018). Additionally, obesity, diabetes, and cardiovascular comorbidities contribute to AD pathogenesis. Cardiovascular diseases such as atrial fibrillation, hypertension, stroke, and coronary artery disease are modifiable risk factors associated with AD (Santos et al., 2017; Alford et al., 2018).

1.4 NEUROCHEMISTRY

Neurochemical analysis of postmortem brain and cortical biopsy tissue samples has contributed to the improved comprehension of the selectivity of neuronal degeneration. While our understanding of Alzheimer's disease is still evolving, several key neurochemical features have been identified.

1.4.1 Acetylcholine

Acetylcholine (ACh) is a prominent neurotransmitter in the cholinergic system, playing a vital role in cognitive functions. In AD, there is a significant reduction in ACh levels in the brain (Contestabile, 2011). The basal forebrain nuclei send important cholinergic projections to the hippocampus, cerebral cortex, and amygdala, and the loss of cholinergic neurons in these regions contributes to memory impairment. Postmortem studies of AD patients' brains have consistently shown decreased acetylcholine levels (Ballinger et al., 2016). The diminished activity of the enzyme choline acetyltransferase in the cerebral cortex and the synthesis of ACh are believed to result from damage to the ascending cholinergic projection from the nucleus basalis to the cortex. The extent of this loss of activity correlates with the severity of dementia and the density of cortical senile plaques in AD.

Moreover, there is specific evidence indicating that a notable reduction of ACh receptors is associated with the accumulation of amyloid β plaques and neurofibrillary tangles, which are pathological hallmarks of AD (Ferreira-Vieira et al., 2016). Acetylcholine exerts its functions through muscarinic ACh and nicotinic ACh receptors (Francis et al., 1993; Perry et al., 1978; Wilcock et al., 1982; Sims et al., 1983).

1.4.2 Serotonin

Numerous studies have provided increasing evidence of serotonin's significant role in regulating mood, emotions, the sleep/wake cycle, locomotion, and various behavioural phenomena. Serotoninergic neurons in the raphe nuclei regulate these functions through different serotonin (5-HT) receptors. The brain contains seven distinct 5-HT receptors. In AD, there is a deficit in serotonergic activity, and reductions in 5-HT levels have been associated with aggression and other psychological features in various neuropsychiatric conditions. Only a small percentage (approximately 1–2%) of serotonin is in the brain. The distribution of serotonergic projections in the cortex lacks well-defined patterns, and serotonin primarily exerts an inhibitory effect. Reduced serotonergic activity has been observed in AD, but as with the previous hypothesis, these alterations in postsynaptic 5-HT receptors could be influenced by cholinergic denervation (Ellison et al., 1986; Rossor et al., 1982).

Central measurements of 5-HT activity in humans have predominantly relied on cerebrospinal fluid concentrations of 5-hydroxy indole acetic acid (5-HIAA), the primary metabolite of serotonin. The serotonergic system is complex, consisting of multiple subtypes of serotonergic receptors that exhibit region-specific differences in their expression and functional properties. For example, 5-HT1A receptors are abundant in the hippocampus, septum, neocortex, and raphe nuclei, and they are present on the terminals of nonserotonergic neurons, facilitating the release of ACh, noradrenaline (NA), and dopamine in the brain; 5-HT1B receptors show high density in the pallidum and substantia nigra, and they can modulate the release of GABA, ACh, and glutamate neurons (Olivier, 2004).

1.4.3 Dopamine

Dopamine, synthesized in the midbrain, is crucial in coordinating functions such as motor control, cognition, and reward processing. Multiple studies have highlighted the loss of dopaminergic neurons in AD. Recent research has revealed that dopaminergic neurons in the ventral tegmental area (VTA) are significantly involved in cognition, memory formation, and synaptic plasticity. Degeneration of dopaminergic neurons in the VTA occurs in the early stages of AD, while the loss of dopaminergic neurons associated with the striatum occurs in later stages. The loss of dopamine in the striatum contributes to extrapyramidal symptoms and affects 35–40% of AD patients. Substantial evidence indicates that the relative preservation of dopaminergic function in the context of diminished cholinergic activity could underlie psychotic symptoms in AD. Similarly, despite the diminished serotonergic activity, relative dopaminergic preservation in specific brain regions has been linked to aggressive or impulsive behaviours in humans. Consistent with this hypothesis, several studies have reported a negative correlation between the ratios of choline acetyltransferase (ChAT) to dopamine (DA) and ChAT to D1 receptor (D1) in the temporal cortex and aggressive behaviour (Rangel-Barajas et al., 2015; Nobili et al., 2017; Martorana & Koch, 2014).

1.4.4 Noradrenaline

The central noradrenergic system plays a role in regulating arousal and autonomic functions. Research suggests that hyperactivity of this system is associated with agitation and aggressive behaviours in various populations, including individuals with AD. The locus coeruleus (LC), which is a brain structure, is the primary source of noradrenaline in the central nervous system (CNS).

Early tau pathology in the LC can occur before the characteristic appearance of tau and amyloid pathology in the medial temporal lobe and other cortical regions in AD. As AD progresses, the LC experiences neuron loss, leading to a decrease in its volume. These findings suggest that dysfunction and degeneration of the noradrenergic system, particularly the LC, contribute to the pathogenesis of AD (Chamberlain & Robbins, 2013; Kaur et al., 2019).

1.4.5 GABA

GABA, γ-aminobutyric acid, is a crucial inhibitory neurotransmitter in the CNS. It plays a significant role in regulating neuronal excitability and maintaining a balance between excitation and inhibition in the brain. In the context of AD, evidence suggests that GABAergic dysfunction contributes to the features of the disease (Wu & Sun, 2015; Li et al., 2016).

1.4.6 The Glutamatergic System

Glutamate is the principal excitatory neurotransmitter in the brain, and it plays a crucial role in various cognitive processes, such as learning and memory. Glutamate pathways originate in the cortex and project to different brain regions, including the striatum, thalamus, substantia nigra, hippocampus, and brainstem. In AD, evidence suggests a loss of glutamatergic neurons and disruptions in glutamate activity, which have been associated with the severity of dementia and cognitive symptoms of the disorder.

In AD, there is a reduction in the number of nicotinic and muscarinic ACh receptors, mainly located on presynaptic cholinergic terminals. However, postsynaptic muscarinic receptors (M1 and M3) are relatively preserved. Changes in the coupling between muscarinic M1 receptors, their G-proteins, and second messenger systems have also been observed (Keverne & Ray, 2008).

Postmortem studies indicate that alterations in serotonergic neurotransmission in AD are more closely linked to behavioural disturbances such as depression than to cognitive dysfunction. Due to the loss of cholinergic and other pyramidal neurons, there is a profound reduction in excitatory amino acid (EAA) neurotransmission, leading to hypoactivity in pyramidal neurons. This reduction in EAA neurotransmission is compounded by the maintained levels of inhibition exerted by GABAergic neurons (Francis, 2003).

These findings highlight the complex interplay of various neurotransmitter systems in AD. These neurochemical changes interact and contribute to the complex pathophysiology of AD, leading to cognitive decline, behavioural disturbances, and other characteristic symptoms of the disease. It is important to note that AD is a multifactorial disorder, and these neurochemical alterations are part of a broader cascade of pathological changes, including the accumulation of amyloid-beta plaques and neurofibrillary tangles.

1.5 CONCLUSION

In conclusion, Alzheimer's disease is a prevalent and progressive neurodegenerative disorder characterized by the accumulation of beta-amyloid plaques and neurofibrillary tangles that leads to the degeneration of brain cells and the subsequent decline in cognitive functions. With an increasing global aging population, the prevalence of Alzheimer's disease is expected to rise significantly, necessitating continued efforts in research, early detection, and the development of effective treatments to alleviate the burden on individuals, families, and health care systems worldwide.

ENDNOTE

* Corresponding author

REFERENCES

Abate, G., Marziano, M., Rungratanawanich, W., Memo, M., & Uberti, D. (2017). Nutrition and AGE-ing: Focusing on Alzheimer's disease. *Oxidative Medicine and Cellular Longevity*, 2017, 7039816.

Alford, S., Patel, D., Perakakis, N., & Mantzoros, C. S. (2018). Obesity as a risk factor for Alzheimer's disease: Weighing the evidence. *Obesity Reviews: An Official Journal of the International Association for the Study of Obesity*, 19(2), 269–280.

Alzheimer's Disease Facts and Figures. (2023). Alzheimer's & dementia: The journal of the Alzheimer's Association. *Alzheimer's & Dementia*, 19(4), 1598–1695.

Ballinger, E. C., Ananth, M., Talmage, D. A., & Role, L. W. (2016). Basal forebrain cholinergic circuits and signaling in cognition and cognitive decline. *Neuron*, 91(6), 1199–1218.

Barthélemy, N. R., Li, Y., Joseph-Mathurin, N., Gordon, B. A., Hassenstab, J., Benzinger, T. L. S., Buckles, V., Fagan, A. M., Perrin, R. J., Goate, A. M., Morris, J. C., Karch, C. M., Xiong, C., Allegri, R., Mendez, P. C., Berman, S. B., Ikeuchi, T., Mori, H., Shimada, H., Shoji, M., & Dominantly Inherited Alzheimer Network. (2020). A soluble phosphorylated tau signature links tau, amyloid and the evolution of stages of dominantly inherited Alzheimer's disease. *Nature Medicine*, 26(3), 398–407.

Bekris, L. M., Yu, C. E., Bird, T. D., & Tsuang, D. W. (2010). Genetics of Alzheimer disease. *Journal of Geriatric Psychiatry and Neurology*, 23(4), 213–227.

Bergmans, B. A., & De Strooper, B. (2010). Gamma-secretases: From cell biology to therapeutic strategies. *The Lancet Neurology*, 9(2), 215–226.

Bilousova, T., Melnik, M., Miyoshi, E., Gonzalez, B. L., Poon, W. W., Vinters, H. V., Miller, C. A., Corrada, M. M., Kawas, C., Hatami, A., Albay, R., Glabe, C., & Gylys, K. H. (2019). Apolipoprotein E/amyloid-β complex accumulates in Alzheimer disease cortical synapses via Apolipoprotein E receptors and is enhanced by APOE4. *The American Journal of Pathology*, 189(8), 1621–1636.

Britannica, The Editors of Encyclopaedia. (2023, June 10). Alzheimer disease. *Encyclopedia Britannica*. Available from: www.britannica.com/science/Alzheimer-disease (Accessed on 18 June 2023).

Brookmeyer, R., Johnson, E., Ziegler-Graham, K., & Arrighi, H. M. (2007). Forecasting the global burden of Alzheimer's disease. *Alzheimer's & Dementia: The Journal of the Alzheimer's Association*, 3(3), 186–191.

Chamberlain, S. R., & Robbins, T. W. (2013). Noradrenergic modulation of cognition: Therapeutic implications. *Journal of Psychopharmacology* (Oxford, England), 27(8), 694–718.

Chan, K. Y., Wang, W., Wu, J. J., Liu, L., Theodoratou, E., Car, J., Middleton, L., Russ, T. C., Deary, I. J., Campbell, H., Wang, W., Rudan, I., & Global Health Epidemiology Reference Group (GHERG). (2013). Epidemiology of Alzheimer's disease and other forms of dementia in China, 1990–2010: A systematic review and analysis. *Lancet* (London, England), 381(9882), 2016–2023.

Cheignon, C., Tomas, M., Bonnefont-Rousselot, D., Faller, P., Hureau, C., & Collin, F. (2018). Oxidative stress and the amyloid beta peptide in Alzheimer's disease. *Redox Biology*, 14, 450–464.

Chen, Y., Strickland, M. R., Soranno, A., & Holtzman, D. M. (2021). Apolipoprotein E: Structural insights and links to Alzheimer disease pathogenesis. *Neuron*, 109(2), 205–221.

Colomina, M. T., & Peris-Sampedro, F. (2017). Aluminum and Alzheimer's disease. *Advances in Neurobiology*, 18, 183–197.

Contestabile, A. (2011). The history of the cholinergic hypothesis. *Behavioural Brain Research*, 221(2), 334–340.

Croze, M. L., & Zimmer, L. (2018). Ozone atmospheric pollution and Alzheimer's disease: From epidemiological facts to molecular mechanisms. *Journal of Alzheimer's Disease: JAD*, 62(2), 503–522.

Cummings, J., Lee, G., Zhong, K., Fonseca, J., & Taghva, K. (2021). Alzheimer's disease drug development pipeline: 2021. *Alzheimer's & Dementia* (New York, NY), 7(1), e12179.

Das, R., & Chinnathambi, S. (2019). Microglial priming of antigen presentation and adaptive stimulation in Alzheimer's disease. *Cellular and Molecular Life Sciences: CMLS*, 76(19), 3681–3694.

De Strooper, B., & Woodgett, J. (2003). Alzheimer's disease: Mental plaque removal. *Nature*, 423(6938), 392–393.

Diniz, L. P., Tortelli, V., Matias, I., Morgado, J., Bérgamo Araujo, A. P., Melo, H. M., Seixas da Silva, G. S., Alves-Leon, S. V., de Souza, J. M., Ferreira, S. T., De Felice, F. G., & Gomes, F. C. A. (2017). Astrocyte transforming growth factor beta 1 protects synapses against Aβ oligomers in Alzheimer's disease model. *The Journal of Neuroscience: The Official Journal of the Society for Neuroscience*, 37(28), 6797–6809.

Dries, D. R., & Yu, G. (2008). Assembly, maturation, and trafficking of the gamma-secretase complex in Alzheimer's disease. *Current Alzheimer Research*, 5(2), 132–146.

Ellison, D. W., Beal, M. F., Mazurek, M. F., Bird, E. D., & Martin, J. B. (1986). A postmortem study of amino acid neurotransmitters in Alzheimer's disease. *Annals of Neurology*, 20(5), 616–621.

Ferreira-Vieira, T. H., Guimaraes, I. M., Silva, F. R., & Ribeiro, F. M. (2016). Alzheimer's disease: Targeting the cholinergic system. *Current Neuropharmacology*, 14(1), 101–115.

Foster, E. M., Dangla-Valls, A., Lovestone, S., Ribe, E. M., & Buckley, N. J. (2019). Clusterin in Alzheimer's disease: Mechanisms, genetics, and lessons from other pathologies. *Frontiers in Neuroscience*, 13, 164.

Francis, P. T. (2003). Glutamatergic systems in Alzheimer's disease. *International Journal of Geriatric Psychiatry*, 18(Suppl 1), S15–S21.

Francis, P. T., Sims, N. R., Procter, A. W., & Bowen, D. M. (1993). Cortical pyramidal neurone loss may cause glutamatergic hypoactivity and cognitive impairment in Alzheimer's disease: Investigative and therapeutic perspectives. *Journal of Neurochemistry*, 60(5), 1589–1604.

Fülöp, T., Itzhaki, R. F., Balin, B. J., Miklossy, J., & Barron, A. E. (2018). Role of microbes in the development of Alzheimer's disease: State of the art—an international symposium presented at the 2017 IAGG congress in San Francisco. *Frontiers in Genetics*, 9, 362.

Galante, D., Ruggeri, F. S., Dietler, G., Pellistri, F., Gatta, E., Corsaro, A., Florio, T., Perico, A., & D'Arrigo, C. (2016). A critical concentration of N-terminal pyroglutamylated amyloid beta drives the misfolding of Ab1-42 into more toxic aggregates. *The International Journal of Biochemistry & Cell Biology*, 79, 261–270.

GBD 2019 Dementia Forecasting Collaborators. (2022). Estimation of the global prevalence of dementia in 2019 and forecasted prevalence in 2050: An analysis for the Global Burden of Disease Study 2019. *The Lancet Public Health*, 7(2), e105–e125.

Goldblatt, G., Cilenti, L., Matos, J. O., Lee, B., Ciaffone, N., Wang, Q. X., Tetard, L., Teter, K., & Tatulian, S. A. (2017). Unmodified and pyroglutamylated amyloid β peptides form hypertoxic hetero-oligomers of unique secondary structure. *The FEBS Journal*, 284(9), 1355–1369.

Heneka, M. T., Carson, M. J., El Khoury, J., Landreth, G. E., Brosseron, F., Feinstein, D. L., Jacobs, A. H., Wyss-Coray, T., Vitorica, J., Ransohoff, R. M., Herrup, K., Frautschy, S. A., Finsen, B., Brown, G. C., Verkhratsky, A., Yamanaka, K., Koistinaho, J., Latz, E., Halle, A., Petzold, G. C., . . . & Kummer, M. P. (2015). Neuroinflammation in Alzheimer's disease. *The Lancet Neurology*, 14(4), 388–405.

Hickman, S. E., Allison, E. K., & El Khoury, J. (2008). Microglial dysfunction and defective beta-amyloid clearance pathways in aging Alzheimer's disease mice. *The Journal of Neuroscience: The Official Journal of the Society for Neuroscience*, 28(33), 8354–8360.

Hou, Y., Dan, X., Babbar, M., Wei, Y., Hasselbalch, S. G., Croteau, D. L., & Bohr, V. A. (2019). Ageing as a risk factor for neurodegenerative disease. Nature reviews. *Neurology*, 15(10), 565–581.

Huang, L. K., Chao, S. P., & Hu, C. J. (2020). Clinical trials of new drugs for Alzheimer disease. *Journal of Biomedical Science*, 27(1), 18.

Huat, T. J., Camats-Perna, J., Newcombe, E. A., Valmas, N., Kitazawa, M., & Medeiros, R. (2019). Metal toxicity links to Alzheimer's disease and neuroinflammation. *Journal of Molecular Biology*, 431(9), 1843–1868.

Kalaria, R. N., Maestre, G. E., Arizaga, R., Friedland, R. P., Galasko, D., Hall, K., Luchsinger, J. A., Ogunniyi, A., Perry, E. K., Potocnik, F., Prince, M., Stewart, R., Wimo, A., Zhang, Z. X., Antuono, P., & World Federation of Neurology Dementia Research Group. (2008). Alzheimer's disease and vascular dementia in developing countries: Prevalence, management, and risk factors. *The Lancet Neurology*, 7(9), 812–826.

Katzman, R. (1976). Editorial: The prevalence and malignancy of Alzheimer disease. A major killer. *Archives of Neurology*, 33(4), 217–218.

Kaur, S., DasGupta, G., & Singh, S. (2019). Altered neurochemistry in Alzheimer's disease: Targeting neurotransmitter receptor mechanisms and therapeutic strategy. *Neurophysiology*, 51, 293–309.

Keverne, J., & Ray, M. (2008). Neurochemistry of alzheimer's disease. *Psychiatry*, 7(1), 6–8.

Kim, H. R., Lee, P., Seo, S. W., Roh, J. H., Oh, M., Oh, J. S., Oh, S. J., Kim, J. S., & Jeong, Y. (2019). Comparison of amyloid β and tau spread models in Alzheimer's disease. *Cerebral Cortex* (New York, NY: 1991), 29(10), 4291–4302.

Kim, J., Basak, J. M., & Holtzman, D. M. (2009). The role of apolipoprotein E in Alzheimer's disease. *Neuron*, 63(3), 287–303.

Kimberly, W. T., Zheng, J. B., Guénette, S. Y., & Selkoe, D. J. (2001). The intracellular domain of the beta-amyloid precursor protein is stabilized by Fe65 and translocates to the nucleus in a notch-like manner. *The Journal of Biological Chemistry*, 276(43), 40288–40292.

Koyama, A., Hashimoto, M., Tanaka, H., Fujise, N., Matsushita, M., Miyagawa, Y., Hatada, Y., Fukuhara, R., Hasegawa, N., Todani, S., Matsukuma, K., Kawano, M., & Ikeda, M. (2016). Malnutrition in Alzheimer's disease, dementia with Lewy bodies, and frontotemporal lobar degeneration: Comparison using serum albumin, total protein, and hemoglobin level. *PLoS ONE*, 11(6), e0157053.

Kumar, A., Sidhu, J., Goyal, A., et al. (2023, January). Alzheimer disease [Updated 2022 Jun 5]. In: *StatPearls* [Internet]. Treasure Island, FL: StatPearls Publishing. Available from: www.ncbi.nlm.nih.gov/books/NBK499922/ (Accessed on 18 June 2023).

Kuruva, C. S., & Reddy, P. H. (2017). Amyloid beta modulators and neuroprotection in Alzheimer's disease: A critical appraisal. *Drug Discovery Today*, 22(2), 223–233.

Lanoiselée, H. M., Nicolas, G., Wallon, D., Rovelet-Lecrux, A., Lacour, M., Rousseau, S., Richard, A. C., Pasquier, F., Rollin-Sillaire, A., Martinaud, O., Quillard-Muraine, M., de la Sayette, V., Boutoleau-Bretonniere, C., Etcharry-Bouyx, F., Chauviré, V., Sarazin, M., le Ber, I., Epelbaum, S., Jonveaux, T., Rouaud, O., & Collaborators of the CNR-MAJ Project (2017). APP, PSEN1, and PSEN2 mutations in early-onset Alzheimer disease: A genetic screening study of familial and sporadic cases. *PLoS Medicine*, 14(3), e1002270.

Lee, J., Meijer, E., Langa, K. M., Ganguli, M., Varghese, M., Banerjee, J., Khobragade, P., Angrisani, M., Kurup, R., Chakrabarti, S. S., Gambhir, I. S., Koul, P. A., Goswami, D., Talukdar, A., Mohanty, R. R., Yadati, R. S., Padmaja, M., Sankhe, L., Rajguru, C., Gupta, M., & Dey, A. B. (2023). Prevalence of dementia in India: National and state estimates from a nationwide study. *Alzheimer's & Dementia: The Journal of the Alzheimer's Association*. Advance online publication. http://doi.org/10.1002/alz.12928.

Lee, V. M., Goedert, M., & Trojanowski, J. Q. (2001). Neurodegenerative tauopathies. *Annual Review of Neuroscience*, 24, 1121–1159.

Li, N. M., Liu, K. F., Qiu, Y. J., Zhang, H. H., Nakanishi, H., & Qing, H. (2019). Mutations of beta-amyloid precursor protein alter the consequence of Alzheimer's disease pathogenesis. *Neural Regeneration Research*, 14(4), 658–665.

Li, Y., Sun, H., Chen, Z., Xu, H., Bu, G., & Zheng, H. (2016). Implications of GABAergic neurotransmission in Alzheimer's disease. *Frontiers in Aging Neuroscience*, 8, 31.

Liu, Y., Walter, S., Stagi, M., Cherny, D., Letiembre, M., Schulz-Schaeffer, W., Heine, H., Penke, B., Neumann, H., & Fassbender, K. (2005). LPS receptor (CD14): A receptor for phagocytosis of Alzheimer's amyloid peptide. *Brain: A Journal of Neurology*, 128(Pt 8), 1778–1789.

Lv, B., Liang, L., Chen, A., Yang, H., Zhang, X., Guo, F., & Qian, H. (2023). Mortality of Alzheimer's disease and other dementias in China: Past and future decades. *International Journal of Public Health*, 68, 1605129.

Malito, E., Hulse, R. E., & Tang, W. J. (2008). Amyloid beta-degrading cryptidases: Insulin degrading enzyme, presequence peptidase, and neprilysin. *Cellular and Molecular Life Sciences: CMLS*, 65(16), 2574–2585.

Martínez-Martínez, A. B., Torres-Perez, E., Devanney, N., Del Moral, R., Johnson, L. A., & Arbones-Mainar, J. M. (2020). Beyond the CNS: The many peripheral roles of APOE. *Neurobiology of Disease*, 138, 104809.

Martorana, A., & Koch, G. (2014). Is dopamine involved in Alzheimer's disease? *Frontiers in Aging Neuroscience*, 6, 252.

Montagne, A., Nation, D. A., Sagare, A. P., Barisano, G., Sweeney, M. D., Chakhoyan, A., Pachicano, M., Joe, E., Nelson, A. R., D'Orazio, L. M., Buennagel, D. P., Harrington, M. G., Benzinger, T. L. S., Fagan, A. M., Ringman, J. M., Schneider, L. S., Morris, J. C., Reiman, E. M., Caselli, R. J., Chui, H. C., . . . Zlokovic, B. V. (2020). APOE4 leads to blood-brain barrier dysfunction predicting cognitive decline. *Nature*, 581(7806), 71–76.

Mustapic, M., Eitan, E., Werner, J. K., Jr, Berkowitz, S. T., Lazaropoulos, M. P., Tran, J., Goetzl, E. J., & Kapogiannis, D. (2017). Plasma extracellular vesicles enriched for neuronal origin: A potential window into brain pathologic processes. *Frontiers in Neuroscience*, 11, 278.

Nandi, A., Counts, N., Chen, S., Seligman, B., Tortorice, D., Vigo, D., & Bloom, D. E. (2022). Global and regional projections of the economic burden of Alzheimer's disease and related dementias from 2019 to 2050: A value of statistical life approach. *EClinicalMedicine*, 51, 101580.

Narasimhan, S., Changolkar, L., Riddle, D. M., Kats, A., Stieber, A., Weitzman, S. A., Zhang, B., Li, Z., Roberson, E. D., Trojanowski, J. Q., & Lee, V. M. Y. (2020). Human tau pathology transmits glial tau aggregates in the absence of neuronal tau. *The Journal of Experimental Medicine*, 217(2), e20190783.

Neumann, H., & Daly, M. J. (2013). Variant TREM2 as risk factor for Alzheimer's disease. *The New England Journal of Medicine*, 368(2), 182–184.

NikolacPerkovic, M., VideticPaska, A., Konjevod, M., Kouter, K., SvobStrac, D., Nedic Erjavec, G., & Pivac, N. (2021). Epigenetics of Alzheimer's disease. *Biomolecules*, 11(2), 195. https://doi.org/10.3390/biom11020195.

Niu, H., Álvarez-Álvarez, I., Guillén-Grima, F., & Aguinaga-Ontoso, I. (2017). Prevalence and incidence of Alzheimer's disease in Europe: A meta-analysis. Prevalencia e incidencia de la enfermedad de Alzheimer en Europa: Metaanálisis. *Neurologia* (Barcelona, Spain), 32(8), 523–532.

Nobili, A., Latagliata, E. C., Viscomi, M. T., Cavallucci, V., Cutuli, D., Giacovazzo, G., Krashia, P., Rizzo, F. R., Marino, R., Federici, M., De Bartolo, P., Aversa, D., Dell'Acqua, M. C., Cordella, A., Sancandi, M., Keller, F., Petrosini, L., Puglisi-Allegra, S., Mercuri, N. B., Coccurello, R., . . . D'Amelio, M. (2017). Dopamine neuronal loss contributes to memory and reward dysfunction in a model of Alzheimer's disease. *Nature Communications*, 8, 14727.

Nordestgaard, L. T., Tybjærg-Hansen, A., Nordestgaard, B. G., & Frikke-Schmidt, R. (2015). Loss-of-function mutation in ABCA1 and risk of Alzheimer's disease and cerebrovascular disease. *Alzheimer's & Dementia: The Journal of the Alzheimer's Association*, 11(12), 1430–1438.

Nortley, R., Korte, N., Izquierdo, P., Hirunpattarasilp, C., Mishra, A., Jaunmuktane, Z., Kyrargyri, V., Pfeiffer, T., Khennouf, L., Madry, C., Gong, H., Richard-Loendt, A., Huang, W., Saito, T., Saido, T. C., Brandner, S., Sethi, H., & Attwell, D. (2019). Amyloid β oligomers constrict human capillaries in Alzheimer's disease via signaling to pericytes. *Science* (New York, NY), 365(6450), eaav9518.

Olivier, B. (2004). Serotonin and aggression. *Annals of the New York Academy of Sciences*, 1036, 382–392.

Perry, E. K., Tomlinson, B. E., Blessed, G., Bergmann, K., Gibson, P. H., & Perry, R. H. (1978). Correlation of cholinergic abnormalities with senile plaques and mental test scores in senile dementia. *British Medical Journal*, 2(6150), 1457–1459.

Peters, C., Bascuñán, D., Opazo, C., & Aguayo, L. G. (2016). Differential membrane toxicity of amyloid-β fragments by pore forming mechanisms. *Journal of Alzheimer's Disease: JAD*, 51(3), 689–699.

Ponjoan, A., Garre-Olmo, J., Blanch, J., Fages, E., Alves-Cabratosa, L., Martí-Lluch, R., Comas-Cufí, M., Parramon, D., Garcia-Gil, M., & Ramos, R. (2020). Is it time to use real-world data from primary care in Alzheimer's disease? *Alzheimer's Research & Therapy*, 12(1), 60.

Prevalence: Alzheimer's Associations. Available from: www.alz.org/alzheimers-dementia/facts-figures#:~:text=An%20estimated%206.7%20million%20Americans,Americans%20with%20Alzheimer's%20are%20women (Accessed on 10 June 2023).

Prevalence: Alzheimer's Disease Statistics. Available from: https://alzheimersnewstoday.com/alzheimers-disease-statistics/#:~:text=There%20is%20a%20rapid%20growth,a%20related%20form%20of%20dementia (Accessed on 10 June 2023).

Quintana, D. D., Garcia, J. A., Anantula, Y., Rellick, S. L., Engler-Chiurazzi, E. B., Sarkar, S. N., Brown, C. M., & Simpkins, J. W. (2020). Amyloid-β causes mitochondrial dysfunction via a Ca2+-driven upregulation of oxidative phosphorylation and superoxide production in cerebrovascular endothelial cells. *Journal of Alzheimer's Disease: JAD*, 75(1), 119–138.

Rangel-Barajas, C., Coronel, I., & Florán, B. (2015). Dopamine receptors and neurodegeneration. *Aging and Disease*, 6(5), 349–368.

Rasmussen, J., & Langerman, H. (2019). Alzheimer's disease—why we need early diagnosis. *Degenerative Neurological and Neuromuscular Disease*, 9, 123–130.

Rauch, J. N., Luna, G., Guzman, E., Audouard, M., Challis, C., Sibih, Y. E., Leshuk, C., Hernandez, I., Wegmann, S., Hyman, B. T., Gradinaru, V., Kampmann, M., & Kosik, K. S. (2020). LRP1 is a master regulator of tau uptake and spread. *Nature*, 580(7803), 381–385.

Rossor, M. N., Garrett, N. J., Johnson, A. L., Mountjoy, C. Q., Roth, M., & Iversen, L. L. (1982). A postmortem study of the cholinergic and GABA systems in senile dementia. *Brain: A Journal of Neurology*, 105(Pt 2), 313–330.

Santos, C. Y., Snyder, P. J., Wu, W. C., Zhang, M., Echeverria, A., & Alber, J. (2017). Pathophysiologic relationship between Alzheimer's disease, cerebrovascular disease, and cardiovascular risk: A review and synthesis. *Alzheimer's & Dementia* (Amsterdam, Netherlands), 7, 69–87.

Schilling, S., Rahfeld, J. U., Lues, I., & Lemere, C. A. (2018). Passive Aβ immunotherapy: Current achievements and future perspectives. *Molecules (Basel, Switzerland)*, 23(5), 1068.

Selkoe, D. J. (2004). Cell biology of protein misfolding: The examples of Alzheimer's and Parkinson's diseases. *Nature Cell Biology*, 6(11), 1054–1061.

Shinohara, M., Tachibana, M., Kanekiyo, T., & Bu, G. (2017). Role of LRP1 in the pathogenesis of Alzheimer's disease: Evidence from clinical and preclinical studies. *Journal of Lipid Research*, 58(7), 1267–1281.

Sims, N. R., Bowen, D. M., Allen, S. J., Smith, C. C., Neary, D., Thomas, D. J., & Davison, A. N. (1983). Presynaptic cholinergic dysfunction in patients with dementia. *Journal of Neurochemistry*, 40(2), 503–509.

Skaria, A. P. (2022). The economic and societal burden of Alzheimer disease: Managed care considerations. *The American Journal of Managed Care*, 28(10 Suppl), S188–S196.

Sochocka, M., Zwolińska, K., & Leszek, J. (2017). The infectious etiology of Alzheimer's disease. *Current Neuropharmacology*, 15(7), 996–1009.

Taudorf, L., Nørgaard, A., Waldemar, G., & Laursen, T. M. (2021). Mortality in dementia from 1996 to 2015: A national registry-based cohort study. *Journal of Alzheimer's Disease: JAD*, 79(1), 289–300.

Tesseur, I., Van Dorpe, J., Spittaels, K., Van den Haute, C., Moechars, D., & Van Leuven, F. (2000). Expression of human apolipoprotein E4 in neurons causes hyperphosphorylation of protein tau in the brains of transgenic mice. *The American Journal of Pathology*, 156(3), 951–964.

Wang, X., & Zheng, W. (2019). Ca^{2+} homeostasis dysregulation in Alzheimer's disease: A focus on plasma membrane and cell organelles. *FASEB Journal: Official Publication of the Federation of American Societies for Experimental Biology*, 33(6), 6697–6712.

Wilcock, G. K., Esiri, M. M., Bowen, D. M., & Smith, C. C. (1982). Alzheimer's disease. Correlation of cortical choline acetyltransferase activity with the severity of dementia and histological abnormalities. *Journal of the Neurological Sciences*, 57(2–3), 407–417.

Wingo, T. S., Lah, J. J., Levey, A. I., & Cutler, D. J. (2012). Autosomal recessive causes likely in early-onset Alzheimer disease. *Archives of Neurology*, 69(1), 59–64.

Wu, C., & Sun, D. (2015). GABA receptors in brain development, function, and injury. *Metabolic Brain Disease*, 30(2), 367–379.

Zhao, L., Woody, S. K., & Chhibber, A. (2015). Estrogen receptor β in Alzheimer's disease: From mechanisms to therapeutics. *Ageing Research Reviews*, 24(Pt B), 178–190.

Zott, B., Simon, M. M., Hong, W., Unger, F., Chen-Engerer, H. J., Frosch, M. P., Sakmann, B., Walsh, D. M., & Konnerth, A. (2019). A vicious cycle of β amyloid-dependent neuronal hyperactivation. *Science* (New York, NY), 365(6453), 559–565.

2 Preclinical Approach
In vivo and In vitro Methods for Evaluating Anti-Alzheimer's Drugs

Nilutpal Sharma Bora, Bhargab Jyoti Sahariah, Chanam Melody Devi and Banshongdor H. Mawlieh*

2.1 INTRODUCTION

Alzheimer's disease (AD) is a slow-progressing neurological disorder that predominantly affects the elderly population, with the majority of cases occurring in individuals over the age of 65 years (Ferri et al., 2005). AD is the most frequent cause of dementia, accounting for approximately 60–80% of the total number of cases of dementia globally (Alzheimer's Association, 2022). The hallmark pathological features of AD include the accumulation of extracellular amyloid beta (Aβ) plaques and intracellular neurofibrillary tangles (NFTs) in the brain, which lead to the progressive loss of neurons and cognitive decline (Hardy & Selkoe, 2002).

The precise aetiology of AD is yet to be fully ascertained, but it is supposed to be aetiologically diverse, involving genetic, environmental, and lifestyle factors (Alzheimer's Association, 2022). Genetic mutations within the amyloid precursor protein (APP), presenilin-1 (PSEN1), and presenilin-2 (PSEN2) genes have been correlated with the early onset of genetic AD, which makes up fewer than 5% of all AD cases (Gatz et al., 2006). However, the majority of AD cases are erratic, and risk factors such as age, gender, cardiovascular disease, diabetes, and head injury have been associated with an increased risk of developing AD (Ferri et al., 2005).

The clinical signs of AD include a steady decline in cognitive function, diminished memory, language struggle, confusion, and changes in personality and character (Alzheimer's Association, 2022). These symptoms can profoundly deteriorate the quality of life of those suffering and of caretakers, leading to increased health care costs and societal burdens.

To date, no cure for AD has been developed,; the present treatments aim to mitigate the symptoms and slow the progression of the disease. Cholinesterase inhibitors such as donepezil, galantamine, and rivastigmine as well as memantine, an N-methyl-D-aspartate (NMDA) receptor antagonist, are currently licensed by the US Food and Drug Administration for the treatment of AD (Alzheimer's Association, 2022). These drugs can recover cognitive functions, but their effects are modest and temporary.

Recent research has focused on improving disease-altering therapies that target the underlying pathophysiology of AD, such as Aβ and tau. Several approaches are being investigated, including immunotherapies, small-molecule inhibitors, and gene therapies (Golde et al., 2018). However, these therapies are still in the experimental stage, and their efficacy and safety need to be further evaluated in clinical trials.

Alzheimer's disease leads to significant cognitive mutilation and functional decline in the elderly. The pathophysiology of AD is multifaceted and involves a magnitude of genetic, environmental, and lifestyle aspects. Currently, available treatments can only assuage the symptoms and delay the development of the disease. Novel disease-modifying therapies are under investigation, which holds promise for future management of AD.

DOI: 10.1201/9781003412069-2

2.1.1 Pathogenesis of Alzheimer's Disease

The pathogenesis of AD is multifactorial and involves various molecular, cellular, and physiological processes. The accumulation of Aβ peptides in the form of extracellular plaques is one of the key features of AD pathology. Aβ is derived from the processing of APP by enzymes called beta-secretase and gamma-secretase (Golde et al., 2018). Aβ peptides are toxic to neurons and disrupt normal neuronal function, which leads to the progressive deterioration of synapses and neurons (LaFerla et al., 2007).

Tau is another vital participant in the pathogenesis of AD. Tau is a microtubule-associated protein that serves a crucial role in maintaining the microtubules in neuronal axons. In AD, tau is hyperphosphorylated, resulting in NFTs in the neurons, leading to their dysfunction and death (Gómez-Isla et al., 1997).

The pathogenesis of AD also involves multiple inflammation and oxidative stress markers. Microglial cells, which are also known as resident immune cells of the brain, are triggered in response to Aβ accumulation and liberate inflammatory cytokines and reactive oxygen species (ROS), leading to neuronal damage and death (Heneka et al., 2015).

The involvement of genetics in the pathophysiology of AD is also firmly established. Genetic variations in the genes that encode APP, PS1, and PS2, which are constituents of the gamma-secretase enzymatic complex, are related to early-onset hereditary AD (Tanzi & Bertram, 2005). Variants of the apolipoprotein E (*APOE*) gene have been associated with late-onset AD (C.-C. Liu et al., 2013).

Recent studies have highlighted the role of epigenetics in the pathogenesis of AD. Epigenetic modifications, such as DNA methylation and histone modifications, can regulate the expression of genes involved in AD pathology (Sanchez-Mut & Gräff, 2015). Changes in DNA methylation patterns have been reported in the brains of AD patients, suggesting a role for epigenetics in the regulation of AD-associated genes (Lunnon et al., 2014). The accumulation of Aβ and tau, inflammation, oxidative stress, and genetic and epigenetic factors all contribute to the progressive loss of neurons and synapses in the brain. Understanding the key players in the pathogenesis of AD is critical for the development of novel therapeutic strategies to treat this devastating disorder.

2.1.2 Diagnosis of Alzheimer's Disease

The diagnosis of AD is generally based on clinical examinations, including an exhaustive medical history, a physical checkup, and a neuropsychological evaluation (Dubois et al., 2016). The National Institute on Aging-Alzheimer's Association diagnostic criteria for AD recommends a stepwise approach, starting with the evaluation of cognitive impairment, followed by the assessment of biomarkers of AD pathology, including cerebrospinal fluid (CSF) biomarkers and neuroimaging (Jack et al., 2018).

Neuroimaging methods such as magnetic resonance imaging (MRI) and positron emission tomography (PET) can identify functional and structural alterations in the brain, including the existence of Aβ lesions and NFTs, which are characteristic markers of AD (Gordon et al., 2019). PET imaging using radiotracers that bind to Aβ and tau proteins, such as florbetapir and flortaucipir, respectively, has been shown to improve the accuracy of AD diagnosis (Johnson et al., 2013). CSF biomarkers such as Aβ, total tau (T-tau), and phosphorylated tau (P-tau) reflect the underlying pathological changes in the brain and can aid in the diagnosis of AD (Jack et al., 2018). Reduced levels of Aβ and elevated levels of T-tau and P-tau have been linked to AD pathology (Blennow et al., 2010).

Recent advances in blood-based biomarkers have shown promise for the early identification of AD. Several studies have reported that blood levels of Aβ42 and tau proteins are altered in patients with AD and can predict the risk of cognitive degeneration (Janelidze et al., 2020). The utilization of blood-based biomarkers for the detection of AD is still in the experimental stage and requires further validation in larger cohorts.

The diagnosis of Alzheimer's disease is primarily collectively dependent upon clinical assessment, neuropsychological testing, and the evaluation of biomarkers of AD pathology, such as CSF biomarkers and neuroimaging. PET imaging using radiotracers that bind to Aβ and tau proteins and biomarkers present in the blood have shown promise for the timely diagnosis of AD. Early diagnosis of AD is essential for initiating appropriate treatment and support and planning for future care needs.

2.1.3 Worldwide Prevalence and Causes of AD

It is predicted that roughly 50 million individuals worldwide are living with AD or associated dementias, with this figure likely to quadruple by 2050 (World Health Organization, 2023). Although the exact cause of AD is still unknown, genetics has been identified as a major risk factor for the disease. The worldwide prevalence of Alzheimer's disease varies significantly across different regions and populations. According to a systematic meta-analytical review conducted by Alzheimer's Disease International in 2015, the incidence of dementia in people aged 60 and over was estimated to be 5–7% in high-income countries, 4–5% in middle-income countries, and 2–3% in low-income countries (Alzheimer's Disease International, 2015). In 2019, it was anticipated that 5.8 million individuals in the United States would be diagnosed with Alzheimer's disease, with this number being projected to climb to 13.8 million by 2050 (Alzheimer's Association, 2019). In Europe, it is estimated that approximately 10 million people are living with dementia, with this number expected to increase twofold by 2050 (World Health Organization, 2023).

Mutations in APP, PSEN1, and PSEN2 are associated with early-onset familial AD, while the APOE ε4 variant is associated with late-onset AD. In addition to APP, PSEN1, and PSEN2, variants of *APOE* have also been highly related to the likelihood of acquiring late-onset AD. *APOE* is involved in the metabolism and clearance of beta-amyloid protein, and the APOE ε4 variant has been shown to boost the chances of developing AD by up to 4 times compared to individuals without the variant (Corder et al., 1998).

Recent developments in DNA studies have also uncovered new genetic indicators for AD. Genome-wide association studies have found over 20 genetic loci related to the risk of acquiring AD, including genes involved in immunological function, lipid utilization, and synaptic activity (Karch & Goate, 2015). However, the effect sizes of these genetic variants are generally small, and their exact contributions to AD pathology are still being investigated. Meanwhile, although genetics has been identified as a major risk factor for AD, environmental and lifestyle factors, have also been implicated in the development and progression of the disease.

2.1.4 Environmental Factors

Several environmental variables have been connected to the development of AD, which includes exposure to fine particulate matter (PM2.5) (Cacciottolo et al., 2017). Heavy metal exposure, particularly lead and mercury, has also been implicated in AD pathology (Sensi et al., 2018). Additionally, chronic stress has been shown to contribute to AD pathology through dysregulation of the hypothalamic–pituitary–adrenal axis and increased inflammation (Morozova et al., 2022).

2.1.5 Lifestyle Factors

An array of lifestyle factors has also been linked to the development and progression of AD such as chronic alcohol consumption (Beydoun et al., 2014), a sedentary lifestyle and physical inactivity (Livingston et al., 2020), and diets high in saturated and trans fats (Morris et al., 2014).

2.1.6 Infectious Agents

Several infectious agents have been linked to the development of AD. Herpes simplex virus type 1) has been shown to contribute to AD pathology by stimulating beta-amyloid protein production and neuroinflammation (Khokale et al., 2020). Other viruses, including cytomegalovirus and hepatitis C virus, have also been linked with AD progression (Khokale et al., 2020).

2.1.7 Treatment and Management of Alzheimer's Disease

Although presently there is no cure for AD, treatment options are available to help manage symptoms and delay the progression of the disease. The treatment options for AD include drug therapy, supportive care, diet, and lifestyle changes.

2.1.8 Available Treatment Options

2.1.8.1 Acetylcholinesterase Inhibitors

Acetylcholinesterase inhibitors (AChEIs) are a family of medications that are used for managing the cognitive symptoms of AD. These medications function by boosting the availability of acetylcholine, a neurotransmitter that is reduced in AD. The three AChEIs that have been authorized for the treatment of AD are donepezil, rivastigmine, and galantamine. AChEIs have been shown to increase cognitive functionality, global clinical status, and daily living patterns of patients with mild to moderate AD (Khokale et al., 2020).

2.1.8.2 N-methyl-D-aspartate Receptor Antagonists

N-methyl-D-aspartate (NMDA) receptor antagonists are a family of medications that are used to treat the cognitive symptoms of AD. These medications function by suppressing the action of NMDA receptors, which are important in learning and memory processes. The NMDA receptor antagonist memantine is now licensed for the treatment of mild to severe AD (Olivares et al., 2012). Memantine has been demonstrated to enhance cognitive performance and general clinical status in people with moderate to severe AD (McShane et al., 2019).

2.1.8.3 Combination Therapy

Combination therapy, which involves the use of both AChEIs and memantine, a combination of donepezil and memantine, has been observed to be of great help in the management of AD symptoms. Combination therapy has been shown to provide greater cognitive benefits than either drug alone and can also delay institutionalization (Guo et al., 2020).

2.1.8.4 Other Therapies

Other drug therapies that have been investigated for the treatment of AD include anti-inflammatory drugs, antioxidants, and hormone replacement therapy. While these therapies have shown some effects in animal studies, clinical trials have not demonstrated significant benefits (Mallah et al., 2020).

2.1.8.5 Diet and Lifestyle Changes

Recent research has shown that diet and lifestyle changes can help reduce the risk of developing AD and improve symptoms in those who have already been diagnosed. Studies have shown that a healthy diet can reduce the risk of developing AD and can also help slow the progression of the disease. The Mediterranean diet, which is rich in whole grains, fish, fruits, vegetables, and healthy fats, has been associated with a reduced risk of AD (Trichopoulou et al., 2014). This diet has been demonstrated to improve cognitive function and reduce risks of cognitive decline in older adults (Féart et al., 2010).

Another dietary approach that has been studied for its potential benefits in AD is the ketogenic diet, which is high in fat and low in carbohydrates and has been scientifically proven to reduce cognitive decline and improve memory function in animal models of AD (Rusek et al., 2019).

However, more research is required to determine the efficacy of this diet in humans. In addition to dietary changes, lifestyle changes such as physical activity and cognitive stimulation can also help reduce the risk of developing AD and improve symptoms in those who have already been diagnosed. Regular physical activity improves cognitive function and reduces the risk of cognitive degeneration in the geriatric population (Meng et al., 2020).

Additionally, cognitive stimulation activities such as reading, playing games, and learning new skills delay the onset of AD (Schultz et al., 2015). Stress reduction techniques such as meditation and mindfulness can also be beneficial for those with AD. Studies have shown that these practices can improve cognitive function and reduce the risk of cognitive decline (Khalsa, 2015).

2.1.9 Ongoing Clinical Trials for the Treatment of Alzheimer's Disease

Although there are medications available that can improve symptoms, there is a pressing need for effective disease-modifying therapies. Numerous clinical trials are currently underway to develop new treatments for AD, targeting various aspects of the disease's pathogenesis.

2.1.9.1 Beta-Amyloid Targeted Therapies

One of the characteristics of AD is the buildup of Aβ protein in the brain, and numerous clinical trials are currently underway to develop therapies that target beta-amyloid. One such therapy is aducanumab, an antibody that targets beta-amyloid and has shown promising results in phase 1b and phase 3 clinical trials (Beshir et al., 2022). Other Aβ-targeted therapies in clinical trials include BAN2401, a monoclonal antibody, and donanemab, a humanized monoclonal antibody (Vitek et al., 2023).

2.1.9.2 Tau-Targeted Therapies

Another hallmark of AD is the accumulation of tau in the brain, and numerous clinical trials are currently underway to develop therapies that target tau. LMTX (TRx0237) is one such therapy which targets tau aggregation and has shown promising results in phase 2 clinical trials (Lee et al., 2019). Other tau-targeted therapies in clinical trials include anti-tau antibodies, small-molecule tau inhibitors, and antisense oligonucleotides.

2.1.9.3 Inflammation-Targeted Therapies

Inflammation plays a role in the pathogenesis of AD, and numerous clinical trials are currently underway to develop therapies that target inflammation. Inflammation-targeted therapies in clinical trials include ponesimod, a sphingosine-1-phosphate receptor modulator, and VX-765, an interleukin-1 beta inhibitor (Asle-Rousta et al., 2013; Bierhansl et al., 2022; Flores et al., 2020).

2.1.9.4 Other Targeted Therapies

Numerous other therapies are in clinical trials for the treatment of AD. A marine-derived oligosaccharide, GV-971, that targets gut bacteria has shown promising results in phase 3 clinical trials (Cao et al., 2022). Another therapy is TRV027, a biased ligand of the angiotensin II type 1 receptor that has shown promising results in phase 2a clinical trial (Pang et al., 2017).

Numerous clinical trials are currently underway to develop new treatments for Alzheimer's disease, targeting various aspects of the disease's pathogenesis. Beta-amyloid targeted therapies such as aducanumab, tau-targeted therapies such as LMTX, and inflammation-targeted therapies such as CNP520 are all in clinical trials. Other targeted therapies in clinical trials include GV-971 and TRV027. Although the results of these trials are still pending, they offer hope for the development of effective disease-modifying therapies for AD. However, it is noteworthy that 99.9% of the clinical trials for AD drugs have failed (Asher & Priefer, 2022)

Because AD is a condition with no specific treatment options, there is a huge avenue for the development of newer and more effective drugs. Because drugs reaching the clinical trial phase must undergo multiple safety and efficacy tests, in vivo and in vitro preclinical approaches for the evaluation of drugs for the treatment of Alzheimer's disease is of utmost importance. Therefore, this chapter will deal with the different in vivo and in vitro preclinical models available for the testing of Alzheimer's disease. Moreover, it will also shed light on a few of the latest developments, which have led to the establishment of novel models in this field.

2.2 IN VITRO EVALUATION METHODS OF ANTI-ALZHEIMER'S DRUGS

2.2.1 Determining Cholinesterase Inhibitory Activity

2.2.1.1 Microplate Assay Based on Ellman's Method

By employing either colorimetry or enhanced spectrophotometry on a microplate, it is possible to quantify the active ingredient's inhibitory effect on acetylcholinesterase (AChE) and butyrylcholinesterase (BuChE) enzymes. AChE and BuChE activity are typically measured using a microplate reader with 96 well plates. Acetylthiocholine iodide (ATCI), butyrylthiocholine iodide (BTCI), bovine serum albumin (BSA), Tris-hydrochloric acid buffer, and Ellman's reagent (5,5-dithiobis-(2-nitrobenzoic acid, DTNB) are the reagents used for the experiment.

After adding DTNB, AChE, buffer, and the active substance to the 96-well plates, the absorbance needs to be determined after 15 minutes of room temperature incubation at various intervals. In the enzymatic process, acetylthiocholine or butyrylthiocholine is hydrolysed to produce thiocholine, further reacting with Ellman's reagent (DTNB) to produce 2-nitrobenzoate-5-mercaptothiocholine and 5-thio-2-nitrobenzoate, which could be determined at the compound's designated wavelength (Mathew & Subramanian, 2014; Saber et al., 2021). Triplicates of each reaction must be carried out using the following formula to obtain the percentage inhibition:

$$\%\ inhibition = \frac{(E-S)}{E} \times 100$$

where E is the enzyme's activity in the absence of an active compound and S is the enzyme's activity in the presence of an active compound. Excel can be used to examine data and determine the mean as well as standard deviation. With the aid of statistical programs such as Prism and SPSS, the IC_{50} can be computed.

2.2.1.2 Bio-Autography Assay

To measure the AChE inhibitory activity, a TLC or HPTLC bioautography assay can be used. The compounds can be analysed using TLC on silica plates with the proper mobile phase for separation. The stock solution is prepared by dissolving acetylcholinesterase (1000 U) in a Tris-hydrochloric acid buffer at a pH of 7.8, incorporating BSA to keep the enzyme stable throughout the bioassay and maintaining the resultant mixture at 4 °C (Sibanyoni et al., 2020; Marston et al., 2002).

Prior to use, TLC plates must be completely dried and eluted with the appropriate solvent. The plate should be dried after the sample has migrated in a suitable solvent to ensure that all solvent has been removed. The enzyme stock solution is sprayed over the plate once it has dried completely. For the enzyme incubation, a flat plate is placed in a tank with a small amount of water so that the water will not directly contact the plate, but the environment still needs to be kept humid. The tank is covered, and the incubation process is carried out for 20 minutes at 37 °C. Both 1-naphthyl acetate in ethanol and Fast Blue B salt in distilled water can be mixed just prior to use (to prevent decomposition) and used for determining the enzyme after the TLC plate has incubated. After receiving a spray with a mixture of naphthyl acetate solution and Fast Blue B Salt Solution, the plate turns purple after 1 to 2 minutes (Adhami et al., 2013).

2.2.2 In vitro Aβ Disaggregation Study

2.2.2.1 Scanning Electron Microscopy (SEM)

The first step is to dilute $A\beta_{42}$ in distilled water and incubate it for 72 hours at 37 °C to produce aggregation. A varied concentration of $A\beta_{42}$ must then be incubated for 48 hours at 37 °C in a serum-free medium containing the bioactive component diluted in a suitable vehicle in 12-well plates with poly-L-lysine-coated cover slips. Before being post-fixed for an hour at 4 °C in a solution of 3% glutaraldehyde with phosphate buffer 0.1 M (pH 7.4), all samples must be rinsed 3 times for 10 minutes with phosphate buffer. Then the samples must be dehydrated in a series of acetone, dried, and post-fixed. Under a scanning electron microscope, samples that have been coated in gold are examined. Reproducibility demands that the experiment be run 3 times (Vion et al., 2018).

2.2.2.2 In vitro Cell-based Method

Using the pET28a vector, *Escherichia coli*-competent BL21 (DE3) cells with the DNA sequence $A\beta_{42}$ can be transformed. The initiation codon ATG induces the peptide to be over-expressed, which results in the insertion of a second methionine residue to its N-terminus (MDAEFRHDSGYEVHHQKLVFFAEDVGSNKGAIIGLMVGGVVIA). In order to produce overnight culture, just one BL21 (DE3) colony carrying the plasmid to be expressed at 37 °C is inoculated with a known dose of kanamycin in M9 minimal media. Furthermore, a known volume of overnight culture should be placed into Eppendorf tubes with new M9 minimum media containing kanamycin and thioflavin-S (Th-S) in order to produce the $A\beta_{42}$ peptide.

After that, the compound to be tested must be produced in an appropriate solvent, such as DMSO, 1-thio-β-D-galactopyranoside (IPTG), or a mixture of the two at varying concentrations. The samples must be grown using a Thermomixer overnight at 37 °C and 1400 rpm. In order to make a sample that will serve as negative control (maximal amyloid production), a similar quantity of DMSO must be used excluding the compound. Additionally, in order to act as a positive control, noninduced samples must be treated in distilled water without the presence of amyloid. The absorbances of all samples must be assessed as well at a certain wavelength of the compound in order to determine ideal bacterial growth. The concentrations at the IC_{50} level of the compounds with bioactive properties should be established via the same procedure, the only difference being a variation in the initial concentration for each compound.

2.2.2.3 Thioflavin T (ThT)-induced Fluorescence

A final known dose of lyophilized $A\beta_{1-42}$, either containing or without the bioactive substance, should be treated after being incubated for 24 hours at 37 °C and thoroughly dissolved in ultrapure water. Fluorescence must be measured at specified wavelengths (λ excitation and λ emission) of the particular compound in a study at various time intervals after incubation. An LS-55 spectrofluorometer that has a water-jacketed cell holder enabling control of temperature can be used to record the fluorescence emission. Cells having 1.0 cm path length at 37 °C must be used for every experiment. By analysing the fluorescence signal at a given wavelength for the stated incubation durations, the duration curve of ThT fluorescence emission can be produced. At the completion of incubation, the percent $A\beta_{1-42}$ oligomerization inhibition was determined (Hernandez-Rodriguez et al., 2014).

2.2.2.4 Polyacrylamide Gel Electrophoresis

The goal of the investigation is to ascertain whether incubating $A\beta_{1-42}$, either containing or without the compound, favours the production of $A\beta_{1-42}$ oligomers. In order to prepare aliquots of $A\beta_{1-42}$, ultrapure water must be used for dilution of the compound at concentrations of 50 μM and 100 μM. Following this, the aliquots must be incubated for 24 hours at 37 °C. Samples of $A\beta_{1-42}$ (6 μL) should be mixed with 1 μL of loading buffer led by heating on a boiling water bath for a duration of 5 mins.

Furthermore, the samples must undergo electrophoresis on 12% polyacrylamide gel, followed by staining with Coomassie blue dye. Comparison against the reference molecular weight marker must be done to determine the bands' molecular weight (Castro-Silva et al., 2021).

2.2.2.5 Atomic Force Microscopy (AFM)

The samples of $A\beta_{1-42}$ peptide (5 µL) at 50µM concentration with or without the tested compounds (100 µM) after 24 hr of incubation should be deposited onto glass and dried. The tapping mode of AFM can be used to scan the images in the air (Hernandez-Rodriguez et al., 2014).

2.2.3 In vitro Tau Phosphorylation Inhibitors Study

2.2.3.1 Cell Culture

In the experiment, 293T cells can be grown at a temperature of 37 °C in a medium that contains 5% CO2 with Dulbecco's modification of Eagle's medium (DMEM), 10% fetal bovine serum (FBS), 100 units/ml penicillin, and 100 g/ml streptomycin. For transfection, the cells must be arranged in 12-well poly-D-lysine coated plates (BD) with a count of 2.2 105 cells per well. After a 40-hour incubation period, each well plate must be transfected with 1 µg of the plasmid DNA expressing human 4R2N tau and/or HA-tagged human MARK4 employing lipofectamine 2000. Following transfection for 24 hr, the cells must be treated with the tested compounds forup to 3 hr at 37 °C. Further, the cells must be pretreated in 500 nM MG132 for up to 1 h, maintaining a temperature of 37 °C for treatment with a test compound in the presence of MG132, if MG132 is used to inhibit proteasomes (Sun et al., 2016).

2.2.3.2 Whole-Cell Lysate Kinase Assay

The aberrant tau phosphorylation by GSK3 reduces tau's ability to interact normally with microtubules. Therefore, a possible therapeutic approach for the management of AD is the development of GSK3 inhibitors that lower tau hyperphosphorylation. The neuroblastoma SH- SY5Y cells' whole-cell lysate can be used to develop a simple in vitro direct GSK3 inhibition that prompts the GSK3 assay to detect the therapeutic effect on tau phosphorylation. Phosphate-buffered saline should be used for washing SH-SY5Y cells, and phosphate protein extraction buffer consisting of a protease inhibitor and 1 mM phenylmethanesulfonyl fluoride should be used to lyse the cells. To achieve 4 µg/µL concentration of total protein, lysate must be diluted with kinase buffer and divided into aliquots. At a final 0.25% (w/w) concentration of total protein, lysate aliquots should be fortified with recombinant human GSK3. As a negative control, a sample aliquot spiked with heat-inactivated GSK3 must be used. The fortified lysate aliquots should be incubated with test sample or 5% DMSO vehicle at 37 °C for 2 h, followed by western blot examination. As a reference control, the GSK3 inhibitor must be employed at 10 mM (Liang et al., 2016).

2.2.4 In vitro Antioxidant Study

A radical scavenging activity (RSA) test on the DPPH and ABTS radicals, as well as a Fe3+/Fe2+ reducing power assay, can be performed using the compounds at a concentration of 1 mg/ml. Reducing power must be represented as absorbances, with higher values denoting higher reducing power, whereas RSA must be expressed as percent inhibition relative to a blank containing a suitable solvent. As the positive control, butylated hydroxytoluene (BHT, E321) at a concentration of 1 mg/mL must be used (Custodio et al., 2015).

2.3 EX VIVO METHODS FOR EVALUATION OF ANTI-ALZHEIMER'S DRUGS

Ex vivo models are useful for evaluating medications used for the treatment of AD. Ex vivo models entail the use of isolated tissues or cells that are maintained in vitro, allowing for the investigation of medication effects on specific cellular pathways involved in AD pathogenesis. This technique provides a simplified paradigm for examining the underlying mechanisms of AD and finding new treatment targets (Gameiro et al., 2017). One example of an ex vivo model used for testing medications in AD research is the Aβ aggregation assay. This assay analyses the capacity of putative medicines to suppress the production of beta-amyloid plaques, a hallmark of AD pathogenesis (O'Brien & Wong, 2011). Another ex vivo model utilized in AD research is the tau hyperphosphorylation test, which assesses the capacity of medicines to inhibit the aberrant phosphorylation of tau, a critical hallmark of AD pathology (Miao et al., 2019). Ex vivo models have various advantages over other in vitro models, such as primary neuronal cultures or cell lines. For example, ex vivo models can provide a more physiologically appropriate environment for drug testing, as they comprise more diverse cell types and maintain tissue architecture more faithfully than simpler models. In addition, ex vivo models can be generated from human tissue, allowing for a more direct translation of findings to clinical patients (Mobini et al., 2019). Ex vivo models allow for the investigation of medication effects on specific pathways implicated in AD pathogenesis and can provide insight into the underlying mechanisms of the illness. While there are limitations to this strategy, current advancements in culture techniques and experimental design continue to strengthen the utility of ex vivo models for drug discovery in AD.

2.3.1 Brain Slice Cultures

Brain slice cultures have become an important method for testing medications used for the treatment of AD. The cultures entail creating tiny pieces of brain tissue that are maintained in vitro, enabling the investigation of pharmacological effects on neural cells and their related networks. This method provides a unique opportunity to examine the underlying mechanisms of AD pathogenesis and to discover possible treatment targets (Cho et al., 2007).

Brain slice cultures provide significant advantages over other in vitro models, such as primary neuronal cultures or cell lines. For example, brain slice cultures contain a range of cell types, including neurons, astrocytes, and microglia, which are known to play essential roles in AD pathogenesis. In addition, brain slice cultures preserve a more physiologically appropriate three-dimensional architecture, which can alter medication effects (Croft et al., 2019; Li et al., 2014). Numerous studies have established the efficacy of brain slice cultures in evaluating medicines used for the treatment of AD.

For example, one study demonstrated that the drug memantine, which is now used for the treatment of moderate to severe AD, protected against glutamate-induced toxicity in brain slice cultures from the hippocampus (Anderson et al., 2004; Danysz & Parsons, 2012; J. Liu et al., 2019). Another study indicated that the drug donepezil, which is used to treat mild to moderate AD, enhanced the release of the neurotransmitter acetylcholine in brain slice cultures from the neocortex (Cacabelos, 2007). Brain slice cultures have also been utilized to examine the mechanisms of action of prospective novel medicines for the treatment of AD.

For example, one study evaluated the effects of a new chemical called PBT434 that targets alpha-synuclein aggregation, a characteristic of both AD and Parkinson's disease. The study demonstrated that PBT434 protected against amyloid-beta-induced toxicity in brain slice cultures from the cortex and hippocampus, suggesting that this drug may have potential as a new treatment for AD (Finkelstein et al., 2022; Richter et al., 2020). Notwithstanding the advantages of brain slice cultures, there are significant limits to this approach.

One key drawback is the difficulty in sustaining slice viability over an extended period of time, as slices might become damaged or undergo cellular alterations during the culture phase. This can alter medication responses and potentially limit the utility of brain slice cultures for long-term research. Another disadvantage of brain slice cultures is the potential for regional heterogeneity in medication responses. Various brain areas can demonstrate variances in drug sensitivity, which can be a barrier when investigating medications that target certain pathways or cell types (Cho et al., 2007). However, these limitations can be overcome with proper experimental design and suitable controls. The brain slice culture model enables the investigation of medication effects on complex brain networks and can provide insight into the underlying mechanisms of AD. Although there are limits to this approach, current advancements in culture techniques and experimental design continue to enhance the utility of brain slice cultures for drug discovery in AD.

2.3.2 Primary Neuronal Cell Cultures

Primary neuronal cultures have been widely employed as an ex vivo model for testing medications used for the treatment of AD. These models involve the separation and cultivation of neurons from embryonic or postnatal brain tissue, enabling the study of pharmacological effects on neuronal function and survival (Slanzi et al., 2020). Pure neuronal cultures offer various advantages over in vivo models, including greater experimental control and the capacity to explore specific cell types and pathways involved in AD pathogenesis. One strategy for researching AD pathology using primary neuronal cells is producing Aβ toxicity. Aβ peptides are known to form toxic aggregates in the brains of AD patients, leading to neuronal damage and death (Fontana et al., 2020; Murphy & LeVine, 2010); primary neuronal cells can be subjected to Aβ peptides in vitro, allowing for the examination of potential neuroprotective effects of medicines on Aβ-induced damage (Mansor et al., 2020).

Another strategy for researching AD pathogenesis using primary neuronal cells is inducing tau hyperphosphorylation. Tau is a significant component of neurofibrillary tangles, another hallmark of AD pathogenesis (Mandelkow & Mandelkow, 2012). Primary neuronal cultures can be subjected to substances that generate tau hyperphosphorylation in vitro, allowing for the exploration of possible medicines that can prevent or reverse this process (Yadikar et al., 2020).

Primary neuronal cultures have also been employed to explore the function of oxidative stress in AD pathogenesis. Oxidative stress is recognized as playing a significant role in neuronal damage and death in AD (Ionescu-Tucker & Cotman, 2021; Singh et al., 2019). Primary neuronal cultures can be exposed to substances that generate oxidative stress in vitro, allowing for the exploration of prospective medicines that can prevent or ameliorate oxidative damage. Primary neuronal cultures constitute a promising ex vivo model for researching AD pathophysiology and testing medicines for the treatment of AD. These models allow for the analysis of specific cell types and pathways involved in AD pathogenesis and provide greater experimental control than in vivo models. While there are drawbacks to this approach, including the lack of nonneuronal cell types and the simplified in vitro environment, current advancements in culture techniques and experimental design continue to enhance the utility of primary neuronal cultures for drug discovery in AD.

2.4 IN VIVO METHODS TO ASSESS ANTI-ALZHEIMER'S DRUGS

Animal models that are transgenic and nontransgenic are both employed to study the pathological alterations connected to AD. Mammalian models such as rats and mice are mostly employed. Animal models allow an in-depth understanding of AD pathogenesis, and their use is a crucial part of assessing the safety of various novel drugs. The complex body systems of the animals give insights into human pharmacotherapeutics as well as improved toxicity (Blaikie et al., 2022).

2.4.1 Nontransgenic Animal Models

2.4.1.1 Spatial Discrimination Test in Rodents

This test evaluates the memory and learning abilities of mice and rats. The Y-maze is a tool that tests visual cognitive processing by enabling animals to discover each of the three arms within the maze. It relies upon rats' natural earnestness to search the unexplored territory. A mouse with a sound memory and consequently unimpaired prefrontal cortex will recall the arms that have already been visited and will have a propensity to move into a less frequently used arm. The test is also useful for assessing reasonable behaviour, which is supported by the hippocampus, by confining one arm of the test sample mice to the Y-maze throughout training. The animal needs to recollect whatever section it has not previously examined and should look up for this division more frequently after a trial interval of, say one hour (Kraeuter et al., 2019). The test is scored by counting the instances in which the mouse selects the appropriate arm for exploration. In comparison with old, stressed, or even learning or memory-impaired animals, this number is higher in healthy mice (Kwon et al., 2010; Yoshizaki et al., 2020).

2.4.1.2 Radial Arm Maze (RAM) Spatial Learning Test

The neurobiological mechanism behind spatial learning in rodents that enables rats to explore, learn, and recall various spatial positions of a food reinforcement has been extensively studied using the radial arm-maze test (Parmar & Prakash, 2006). At the end of each arm is a little piece of bait. Only one "win" is reinforced for the animal to enter each maze arm before it is compelled to "shift" to another baited arm. In the RAM, tests can be performed on both working and reference memory. A reference (long-term) memory error results from entering a nonbaited arm, an arm that does not contain food. Reentering a baited arm after the food has already been consumed and the status has changed from baited to nonbaited is considered to be a working (short-term) memory error. Throughout the session, it should be counted how many nonbaited entries were made (Vorhees & Williams, 2006).

2.4.1.3 Morris Water Maze (MWM)

The MWM involves assessing the effects of drugs on learning and memory; for this test, Wistar or Sprague-Dawley rats are employed. For the test, the rat attempts to swim in a tank of water to locate a hiding platform buried beneath the surface. The equipment is a circular water tank that is 20 cm deep and filled with 25 °C water. The initial locations are 4 spots that are evenly distributed along the tank's edge (Parmar & Prakash, 2006). The tank is split into 4 equal parts, and one of the portions has a small platform measuring 19 cm with 1 cm over the water's surface. On all training days, the platform remains in the same location (Kraeuter et al., 2019). Most trials have a time limit of 2 minutes for rats and 1 minute for mice; animals that are unable to locate the platform within that time are either put there or guided there (Vorhees & Williams, 2006). The trial phase is conducted for 2–4 days until the rats are able to climb onto the platform and escape, and the amount of time the rat spends on the escape path is recorded. The MWM also makes it clear that older rats have less memory than younger and middle-aged ones (Parmar & Prakash, 2006).

2.4.1.4 Olfactory Learning in Rats

The development of olfactory discrimination in rats can be used as another paradigm to compare the sense of smell to that of higher primates. Animals are trained to discriminate between a specific positive odour (rose) and a defined negative odour (garlic) in order to receive a reward. Sprague-Dawley rats may be used. If there is a good smell, water will be given; if there is a bad smell, a flashlight will be given. Usually, there is a 15-second inter-trial interval before a fresh odour is detected. The session ends after 40 trials or when the rat correctly chooses 90% of the time. Results are shown as the percentage of correct responses (Parmar & Prakash, 2006).

2.4.2 Transgenic Mouse Models Expressing APP Mutations

Transgenic mice are employed to replicate a total ailment connected with Alzheimer's disease. Despite the models' limitations, they have provided insight into understanding the mechanism of noxious beta-amyloid, especially in regards to the result of individual beta-amyloid type and the probable infective role of beta-amyloid peptide (Elder et al., 2010).

2.4.2.1 PDAPP Mice

The primary AD mouse replica developed to display Aβ deposition was PDAPP. This is because it results in the platelet-derived growth factor supporter-driven human APP gene with the Indiana mutation (V717F). Between the ages of 6 and 9 months, human Aβ deposits start to accumulate in the cerebral cortex of PDAPP mice, coinciding with an aging-related increase in gliosis. Dendritic and synaptic levels in the molecular section of the hippocampus dentate lobe were declining at around 35 weeks of age. At about 13 weeks of age, PDAPP mice develop memory problems influenced by age (Yokohama et al., 2022).

2.4.2.2 Tg2576 Mice

The Tg2576 rodent, which has an enhanced APP genome through the Swedish variant (KM670/671NL) via the hamster prion promoter, serves as one of the most widely used models for ADs. Amyloid plaques form in Tg2576 animals between the ages of 11 and 13 months, and microglia are prominent within or adjacent to lesions in all areas that comprise the neural cortex and the hippocampus between 44 and 69 weeks of age. Although evidence that there is no brain damage, several cognitive skills, including contextual switching and spatial cue learning, are observed to be compromised by an average age of 10 months (Yokohama et al., 2022).

2.4.2.3 APP23 Mice

The Swedish variation (KM670/671NL) in the human APP genome is translated in APP23 mice by means of the mammalian Thyl transcription. APP23 mice start forming Aβ deposits in their brains around the age of 6 months, which increases with age along with quantity and size. Around 24 months old, these deposits have grown older and more numerous, taking up more space in the neocortex and hippocampus. Brain areas with Aβ plaques exhibit a significant glial response. Near the Aβ deposits, there is apparent neuronal loss in 52 weeks old mice, which predicts that cognitive damage starts three months before amyloid accumulation (Yokohama et al., 2022).

2.4.2.4 J20 Mice

When the platelet-derived growth factor promoter is present, 2 FAD-related variations of the human APP gene, the Swedish variant (KM670/671NL) and the Indiana mutant, have elevated expression in J20 mice. These animals start exhibiting disseminated plaques at 5–7 months of age, and around 8–10 months old, they have developed heavy hippocampus lesion burdens. In the limbic brain of newborns between the ages of 6 and 9 months, there are more active astrocytes and microglia. This model illustrates the accumulation of amyloid, neuronal degeneration, and some cognitive problems (Yokohama et al., 2022).

2.4.2.5 TgCRND8 Mice

The individual APP gene, which carries the Swedish mutant (KM670/671NL) and the Indiana mutation (V717F), both of which have been linked to FAD, is upregulated in TgCRND8 mice covered by hamster prion promoter. Aβ is deposited in this mouse stereotype starting at a period of 13 weeks, and the volume of Aβ accumulation rises with time. Amyloid plaque development and activated microglia develop together at 3 months of age. Then, very shortly thereafter, there is a significant glial cell reaction. Acquisition of spatial knowledge and learning reversal are weakened by 13 weeks of age (Yokohama et al., 2022).

2.4.2.6 PS2APP Mice

For the development of PS2APP mice, a pair of isolated modified species—APPswe mice and PS2 (N141I) mice—were intertwined. According to the direction of the animal prion promoter as well as the Thy1.2 booster, this system upregulates the human genome's APP gene with the Swedish mutation (KM670/671NL) and the human PSEN2 gene with the N141I mutation. Amyloid lesions begin to form in the subiculum and front-lateral layer of grey matter at the age of 9 months. By 13–17 months, proteins extend across the brain and the nuclei of pontine neurons as well as the vast majority of the neocortex that expand to the cerebral cortex and amygdala. Gliosis is visible around Aβ plaques at 9 months. Mice older than 8 months have reduced cognitive function (Yokohama et al., 2022).

2.4.2.7 AppNL-G-F Knock-In Mice

In developing AppNL-G-F species, A42 levels could be increased without the human APP gene being overexpressed. In addition to the humanized Aβ sequence, the Swedish (KM670/671NL), Arctic (E693G), and Iberian (I716F) mutations were added to the App mouse gene. By 2 months, cerebral cortex Aβ buildup had begun, and by 7 months, it approached saturation coupled with significant gliosis. The potential to learn, recall information, and perform effectively in a spatial alternation task are all affected in a period of 6 months. In AppNL-G-F knock-in mice, there is no apparent neuronal loss (Yokohama et al., 2022).

2.4.2.8 Mutations in MAPT or Seed Injection-Based Tauopathy Mouse Models

Tau seed injection design and genetically engineered representations are the two main categories of tauopathy model mice. The genetic engineering methods used in these models (transgenic or knock-in), the pronounced tau isoforms (0N4R, 1N4R, etc.), the amount of tau spreading, and the sections of the brain where the inserted tau is expressed prominently all differ (Yokohama et al., 2022)

2.4.2.9 JNPL3 Mice

The initially developed mouse strain with a MAPT defect was JNPL3 mice. They result in the expression of human 0N4R tau with the P301L mutation operated by the mouse prion protein promoter. Severe motor and behavioural deficits, including as the result of spinal cord neuropathological changes, characterize this model of behaviour. The deficits present as neurological disorders and tangle degeneration in the diencephalon, cerebral cortex, and cerebellum nuclei including the cervical spine by the time the mice are 19.5 weeks. Starting at about age 10 months, neurons begin to die off, especially throughout the spinal cord, and astrogliosis emerges in the brainstem, diencephalon, and basal telencephalon (Yokohama et al., 2022).

2.4.2.10 PS19 Mice

The mouse prion promoter controls the expression of human 1N4R tau in PS19 mice that have the P301S mutation. PS19 rats exhibit premature synaptic loss and increased microglial activity prior to neurofibrillary tangle disease and neuronal degeneration. As young as 3 months old, PS19 mice exhibit the initial symptoms of microgliosis; within 6 months, widespread inclusions that mimic neurofibrillary tangles are detected in the neocortex, amygdala, hippocampus, brain stem, and spinal cord. Brain atrophy and neuronal death are visible by the age of 8 months, predominantly in the hippocampus but also in the neocortex and entorhinal cortex. When compared with 3 months old, astrogliosis is far more common at 6 months old. Limb weakness is followed by motor impairments and areas such as grasping and extremity withdrawal when flung by the tail. In 7 to 10 months, such deficiencies proceed to paralysis, and by 12 months, 80% of PS19 mice have passed away (Yokohama et al., 2022).

2.4.2.11 3xTg Mice

To create these mice, homogeneous Psen1 M146V knock-in animals were co-administering with 2 FAD alleles from human genomes associated with AD:APP with the Swedish mutation (KM670/671NL) and MAPT with the P301L mutation. The 2-gene expression is controlled by the mouse Thyl.2 promoter. These animals exhibit increased pathology, comprising both internal and external Aβ sediments and phosphorylated tau clumps with altered conformation (Yokohama et al., 2022).

2.4.2.12 Human MAPT Knock-In Mice

Contrary to the naturally existing murine MAPT gene, these rodents have a redefined WT MAPT gene. There is production of all 6 human tau isoforms, with the amount of 4R mRNA being around 70% that of 3R mRNA. AppNL-G-F/MAPT double knock-in mice develop amyloid plaques and Aβ-related conditions, but neither NFTs nor neurodegeneration are observed in these mice. In particular, tau transformation drives up the spread of tau from one cell to another in AD patients' brain tissue, whether or not there is Aβ deposition (Yokohama et al., 2022).

2.4.2.13 Mouse Models Modified TREM2 Signalling

Myeloid cells, especially microglia in the brain, exhibit TREM2, a protein in their genome that increases the chance of contracting Alzheimer's disease. The R47H TREM2 mutation corresponds with the initial phases of late-onset AD and has an elevated risk proportion. Emergence of an individualized genetic and functional fingerprint in astrocytes as a reaction to plaque buildup revealed a neurological lesion. These microglia reduce the rate of development of neurodegeneration in some animal models, while in other mouse species, incorrect microglial activation hastens the disease's progression. The absence of TREM2 in APP model mice was found to enhance tau aggregation and variation, demonstrating the crucial role of TREM2 in the catastrophic interaction between Aβ and tau (Yokohama et al., 2022).

2.4.3 Experimental Transgenic Rats

Compared with transgenic mice, transgenic rats have a variety of probable benefits, including greater physiological, morphological, and genetic similarity to humans, an easier time collecting CSF, easier electrophysiology, and the ability to perform more complex behavioural testing due to their richer behavioural phenotype (Drummond & Wisniewski, 2017).

2.4.4 Rats That Are Transgenic for R-Thy1-APP (McGill)

McGill-R-Thyl-APP Transgenic rats (McGill) have 1 transgene (1 copy per haploid genome) that encodes human APP751 along with Swedish and Indiana mutations governed by murine Thyl.2 promoter on the inherited framework of HsdBrl:WH Wistar rats. The transgenic product cannot be found in the cerebellum or peripheral organs and is primarily expressed by telencephalic neurons. The trait in the McGill rat model includes gait challenges, emotional abnormalities, interpersonal behaviour changes, and circadian profiles in addition to the cognitive realm (Petrasek, 2018).

2.4.5 Nonhuman Primates

These are species with the highest level used in analysing the neuropathological signs of AD. Nonhuman primates are a good choice for modelling AD because their physiological resemblance to humans, complicated behaviour, and extensive brains size are advantageous for visualizing the study of CSF pooling and naturally occurring Aβ-protein accumulation with 100% sequence homology to human Aβ protein. Due to the long lifetime of the great primates (chimpanzees, gorillas, and orangutans), 1 or 2 AD investigations of marked AD pathology in the great apes have been studied along with the ethical concerns in the use of great apes in research.

Great apes develop amyloid plaques and CAA as they age because of the buildup of amyloid β in the brain. Plaques tend to be widespread and less common than that found in human AD. In general, CAA is more frequent in great apes, and fibrillar Aβ has a greater probability than plaques. Tauopathy is a rare condition despite the significant chronological homogeneity linking man and great ape tau (chimpanzee and gorilla tau share 100% and 99.5% of their genome sequences, respectively, with human tau). Gorillas have phosphorylated tau-containing focal neurons and glia, but they lack tau-positive dystrophic neurites and NFTs. Although this is a rare occurrence that has only been noted in 1 chimpanzee investigated, great apes are capable of producing NFTs. The development of NFT, in this case, is most likely a result of added AD threats, including obesity, hyperlipidaemia, and stroke. In addition, contrary to the severe cognitive decline seen in AD, memory problems seem to be relatively small and comparable with ordinary age-related memory decline (Yeo et al., 1985).

Research has also been conducted on ancient monkeys (e.g., rhesus monkeys, cynomolgus monkeys, baboons, and vervets), mostly rhesus monkeys; again, the human and rhesus monkey Aβ sequences are 100% homologous. Age-related increases in Aβ levels lead to levels in the brain that are comparable with those observed in human AD, and Aβ42 is frequently more prevalent than Aβ40. More lesions are present in the cortex than the hippocampus in rhesus monkeys above the age of 25 years, and their distribution is comparable with that in humans. Compared with great apes, rhesus monkeys have a higher prevalence of parenchymal plaques than CAA.

Even in parts of the brain where there is a lot of plaque, there is no indication of extensive neuronal death; around compact plaques, only slight neuronal death is readily visible. According to the two major investigations that explored whether plaques existed in rhesus monkeys, 40% of older animals (25–31 years) died naturally without showing any signs of plaques or CAA However, given that 100% of the less extremely old animals (33–39 years) studied did have plaques, this could have been because death occurs before plaque formation. Despite the fact that human and rhesus monkey tau have a lot in common in terms of sequence, rhesus monkeys do not suffer tauopathy (Yoshizaki et al., 2020).

It is significant to observe that 90% of baboons over the age of 26 have a substantial yet highly localized tauopathy in the hippocampus, and this condition worsens over time. Additional regions of the encephalon such as the cortex, which contains plaques and CAA, are free of NFTs. Baboons have mild to moderate Aβ deposition, and plaques and tangles do not seem to be related to one another.

A few researchers have also looked into the neuropathology linked to AD in vervets. In captivity, vervets can live up to 30 years, and as they age, they show signs of Aβ deposition, gliosis, and neural degeneration. The onset of amyloid deposition is seen around the age of 15, and it first manifests in the vasculature before parenchymal plaques. AN3pE is found in recently formed plaques and CAA in a proportion of about 1:1 with common Aβ, and plaques that are both scattered and dense are found. Neuritic plaques can be seen, some of which have dystrophic neurites that are immunoreactive to phosphorylated tau. NFTs are not found. There is much animal-to-animal heterogeneity in the prevalence of physiology in vervets, just like in other nonhuman primates.

Squirrel monkeys, which have been the subject of the most research, spontaneously acquire neuropathologies that are comparable with those in AD. After the age of 12 years old, squirrel monkeys exhibit significant Aβ buildup, particularly arterioles and capillaries in the shape of CAA. This model is especially suitable for determining if a treatment plan may be related to ARIA as a patient consequence due to the prevalence of CAA in squirrel monkeys.

Plaques are also present that are smaller than plaques in human AD and that can be either extensive or compact. A40 and A42 are both present in plaques and capillary CAA, whereas Aβ 40 predominates in arteriolar CAA. Most of the accumulation of Aβ appears in the cortex and basal ganglia, with minimal Aβ precipitation in the limbic system. According to a current mass spectroscopy investigation, squirrel monkeys carry oxidized Aβ, AβN3pE, Aβ1–40, Aβ1–42, Aβ1–34, Aβ4–40, and Aβ4–42, as well as any additional essential Aβ species present in the human brain.

Grey mouse lemurs, animals incorporated in preclinical studies on AD, have also been identified as having neuropathology linked with AD. These prosimians have an 18-year maximum lifetime when kept in captivity. Immature (8 years old) grey mouse lemurs have been found to have plaques, primarily in the cortex, and both diffuse and compact plaques have been found. Plaques are more frequently seen than CAA, and whereas CAA also contains Aβ40 and Aβ42, plaques mostly consist of Aβ42.

Intraneuronal phosphorylated tau accumulates in aged grey mouse lemurs despite the fact that this is mostly displayed by the cortex rather than the hippocampus. In a subset of animals older than 3 years, cortical shrinkage is seen, and this coincides with age-related cognitive deterioration. In conclusion, age-related Aβ pathology is common in nonhuman primates, whereas tauopathy is extremely rare. The squirrel monkey is currently the most suitable nonhuman mammal for exploring CAA, whereas the rhesus monkey is the most efficient nonhuman mammal stereotype for investigating AD (Drummond & Wisniewski, 2017).

ENDNOTE

* Corresponding author

REFERENCES

Adhami, H. R., Lutz, J., Kählig, H., Zehl, M., & Krenn, L. (2013). Compounds from gum ammoniacum with acetylcholinesterase inhibitory activity. *Scientia Pharmaceutica*, *81*(3), 793–806.

Alzheimer's Association. (2019). *Alzheimer's Disease Facts and Figures*. Alzheimer's Disease Report. www.alz.org/alzheimers-dementia/facts-figures

Alzheimer's Association. (2022). *What Is Alzheimer's Disease? Symptoms & Causes*. Alzheimer's & Dementia. www.alz.org/alzheimers-dementia/what-is-alzheimer

Alzheimer's Disease International. (2015). *World Alzheimer Report 2015*. www.alzint.org/resource/world-alzheimer-report-2015/

Anderson, E. R., Gendelman, H. E., & Xiong, H. (2004). Memantine protects hippocampal neuronal function in murine human immunodeficiency virus type 1 encephalitis. *The Journal of Neuroscience*, *24*(32), 7194–7198.

Asher, S., & Priefer, R. (2022). Alzheimer's disease failed clinical trials. *Life Sciences*, *306*, 120861.

Asle-Rousta, M., Oryan, S., Ahmadiani, A., & Rahnema, M. (2013). Activation of sphingosine 1-phosphate receptor-1 by SEW2871 improves cognitive function in Alzheimer's disease model rats. *EXCLI Journal*, *12*, 449–461.

Bertram, L., Lill, C. M., & Tanzi, R. E. (2010). The genetics of Alzheimer disease: Back to the future. *Neuron*, *68*(2), 270–281.

Beshir, S. A., Aadithsoorya, A. M., Parveen, A., Goh, S. S. L., Hussain, N., & Menon, V. B. (2022). Aducanumab therapy to treat Alzheimer's disease: A narrative review. *International Journal of Alzheimer's Disease*, *2022*, 1–10.

Beydoun, M. A., Beydoun, H. A., Gamaldo, A. A., Teel, A., Zonderman, A. B., & Wang, Y. (2014). Epidemiologic studies of modifiable factors associated with cognition and dementia: Systematic review and meta-analysis. *BMC Public Health*, *14*(1), 643.

Bierhansl, L., Hartung, H.-P., Aktas, O., Ruck, T., Roden, M., & Meuth, S. G. (2022). Thinking outside the box: Non-canonical targets in multiple sclerosis. *Nature Reviews Drug Discovery*, *21*(8), 578–600.

Blaikie, L., Kay, G., Maciel, P., & Lin, P. K. T. (2022). Experimental modelling of Alzheimer's disease for therapeutic screening. *European Journal of Medicinal Chemistry Reports*, 100044.

Blennow, K., Hampel, H., Weiner, M., & Zetterberg, H. (2010). Cerebrospinal fluid and plasma biomarkers in Alzheimer disease. *Nature Reviews Neurology*, *6*(3), 131–144. https://doi.org/10.1038/nrneurol.2010.4

Cacabelos, R. (2007). Donepezil in Alzheimer's disease: From conventional trials to pharmacogenetics. *Neuropsychiatric Disease and Treatment*, *3*(3), 303–333.

Cacciottolo, M., Wang, X., Driscoll, I., Woodward, N., Saffari, A., Reyes, J., Serre, M. L., Vizuete, W., Sioutas, C., Morgan, T. E., Gatz, M., Chui, H. C., Shumaker, S. A., Resnick, S. M., Espeland, M. A., Finch, C. E., & Chen, J. C. (2017). Particulate air pollutants, APOE alleles and their contributions to cognitive impairment in older women and to amyloidogenesis in experimental models. *Translational Psychiatry*, *7*(1), e1022.

Cao, Y., Yu, F., Lyu, Y., & Lu, X. (2022). Promising candidates from drug clinical trials: Implications for clinical treatment of Alzheimer's disease in China. *Frontiers in Neurology*, *13*.

Castro-Silva, E. S., Bello, M., Rosales-Hernández, M. C., Correa-Basurto, J., Hernández-Rodríguez, M., Villalobos-Acosta, D., Méndez-Méndez, J. V., Estrada-Pérez, A., Murillo-Álvarez, J., & Muñoz-Ochoa, M. (2021). Fucosterol from Sargassum horridum as an amyloid-beta (Aβ1–42) aggregation inhibitor: In vitro and in silico studies. *Journal of Biomolecular Structure and Dynamics*, *39*(4), 1271–1283.

Cho, S., Wood, A., & Bowlby, M. R. (2007). Brain slices as models for neurodegenerative disease and screening platforms to identify novel therapeutics. *Current Neuropharmacology*, *5*(1), 19–33.

Corder, E. H., Lannfelt, L., Bogdanovic, N., Fratiglioni, L., & Mori, H. (1998). The role of APOE polymorphisms in late-onset dementias. *Cellular and Molecular Life Sciences CMLS*, *54*(9), 928–934.

Croft, C. L., Cruz, P. E., Ryu, D. H., Ceballos-Diaz, C., Strang, K. H., Woody, B. M., Lin, W.-L., Deture, M., Rodríguez-Lebrón, E., Dickson, D. W., Chakrabarty, P., Levites, Y., Giasson, B. I., & Golde, T. E. (2019). rAAV-based brain slice culture models of Alzheimer's and Parkinson's disease inclusion pathologies. *Journal of Experimental Medicine*, *216*(3), 539–555.

Custodio, L., Patarra, J., Alberício, F., da Rosa Neng, N., Nogueira, J. M. F., & Romano, A. (2015). Phenolic composition, antioxidant potential and in vitro inhibitory activity of leaves and acorns of Quercus suber on key enzymes relevant for hyperglycemia and Alzheimer's disease. *Industrial Crops and Products*, *64*, 45–51.

Danysz, W., & Parsons, C. G. (2012). Alzheimer's disease, β-amyloid, glutamate, NMDA receptors and memantine—searching for the connections. *British Journal of Pharmacology*, *167*(2), 324–352.

Drummond, E., & Wisniewski, T. (2017). Alzheimer's disease: Experimental models and reality. *Acta Neuropathologica*, *133*, 155–175.

Dubois, B., Hampel, H., Feldman, H. H., Scheltens, P., Aisen, P., Andrieu, S., Bakardjian, H., Benali, H., Bertram, L., Blennow, K., Broich, K., Cavedo, E., Crutch, S., Dartigues, J., Duyckaerts, C., Epelbaum, S., Frisoni, G. B., Gauthier, S., Genthon, R., . . . & Jack, C. R. (2016). Preclinical Alzheimer's disease: Definition, natural history, and diagnostic criteria. *Alzheimer's & Dementia*, *12*(3), 292–323.

Elder, G. A., Gama Sosa, M. A., & De Gasperi, R. (2010). Transgenic mouse models of Alzheimer's disease. *Mount Sinai Journal of Medicine: A Journal of Translational and Personalized Medicine: A Journal of Translational and Personalized Medicine*, *77*(1), 69–81.

Féart, C., Samieri, C., & Barberger-Gateau, P. (2010). Mediterranean diet and cognitive function in older adults. *Current Opinion in Clinical Nutrition and Metabolic Care*, *13*(1), 14–18.

Ferri, C. P., Prince, M., Brayne, C., Brodaty, H., Fratiglioni, L., Ganguli, M., Hall, K., Hasegawa, K., Hendrie, H., Huang, Y., Jorm, A., Mathers, C., Menezes, P. R., Rimmer, E., & Scazufca, M. (2005). Global prevalence of dementia: A Delphi consensus study. *The Lancet*, *366*(9503), 2112–2117.

Finkelstein, D. I., Shukla, J. J., Cherny, R. A., Billings, J. L., Saleh, E., Stefanova, N., Barnham, K. J., & Adlard, P. A. (2022). The compound ATH434 prevents alpha-synuclein toxicity in a murine model of multiple system atrophy. *Journal of Parkinson's Disease*, *12*(1), 105–115.

Flores, J., Noël, A., Foveau, B., Beauchet, O., & LeBlanc, A. C. (2020). Pre-symptomatic Caspase-1 inhibitor delays cognitive decline in a mouse model of Alzheimer disease and aging. *Nature Communications*, *11*(1), 4571.

Fontana, I. C., Zimmer, A. R., Rocha, A. S., Gosmann, G., Souza, D. O., Lourenco, M. V., Ferreira, S. T., & Zimmer, E. R. (2020). Amyloid-β oligomers in cellular models of Alzheimer's disease. *Journal of Neurochemistry*, *155*(4), 348–369.

Gameiro, M., Silva, R., Rocha-Pereira, C., Carmo, H., Carvalho, F., Bastos, M., & Remião, F. (2017). Cellular models and in vitro assays for the screening of modulators of P-gp, MRP1 and BCRP. *Molecules*, *22*(4), 600.

Gatz, M., Reynolds, C. A., Fratiglioni, L., Johansson, B., Mortimer, J. A., Berg, S., Fiske, A., & Pedersen, N. L. (2006). Role of genes and environments for explaining Alzheimer disease. *Archives of General Psychiatry*, *63*(2), 168.

Golde, T. E., DeKosky, S. T., & Galasko, D. (2018). Alzheimer's disease: The right drug, the right time. *Science*, *362*(6420), 1250–1251.

Gómez-Isla, T., Hollister, R., West, H., Mui, S., Growdon, J. H., Petersen, R. C., Parisi, J. E., & Hyman, B. T. (1997). Neuronal loss correlates with but exceeds neurofibrillary tangles in Alzheimer's disease. *Annals of Neurology*, *41*(1), 17–24.

Gordon, B. A., Blazey, T. M., Christensen, J., Dincer, A., Flores, S., Keefe, S., Chen, C., Su, Y., McDade, E. M., Wang, G., Li, Y., Hassenstab, J., Aschenbrenner, A., Hornbeck, R., Jack, C. R., Ances, B. M., Berman, S. B., Brosch, J. R., Galasko, D., . . . & Benzinger, T. L. S. (2019). Tau PET in autosomal dominant Alzheimer's disease: Relationship with cognition, dementia and other biomarkers. *Brain*, *142*(4), 1063–1076.

Guo, J., Wang, Z., Liu, R., Huang, Y., Zhang, N., & Zhang, R. (2020). Memantine, donepezil, or combination therapy—What is the best therapy for Alzheimer's disease? A network meta-analysis. *Brain and Behavior, 10*(11).

Hardy, J., & Selkoe, D. J. (2002). The amyloid hypothesis of Alzheimer's disease: Progress and problems on the road to therapeutics. *Science, 297*(5580), 353–356.

Heneka, M. T., Carson, M. J., Khoury, J. El, Landreth, G. E., Brosseron, F., Feinstein, D. L., Jacobs, A. H., Wyss-Coray, T., Vitorica, J., Ransohoff, R. M., Herrup, K., Frautschy, S. A., Finsen, B., Brown, G. C., Verkhratsky, A., Yamanaka, K., Koistinaho, J., Latz, E., Halle, A., . . . & Kummer, M. P. (2015). Neuroinflammation in Alzheimer's disease. *The Lancet Neurology, 14*(4), 388–405.

Hernandez-Rodriguez, M., Correa-Basurto, J., Martínez-Ramos, F., Padilla-Martínez, I. I., Benítez-Cardoza, C. G., Mera-Jiménez, E., & Rosales-Hernández, M. C. (2014). Design of multi-target compounds as AChE, BACE1, and amyloid-β 1–42 oligomerization inhibitors: In silico and in vitro studies. *Journal of Alzheimer's Disease, 41*(4), 1073–1085.

Ionescu-Tucker, A., & Cotman, C. W. (2021). Emerging roles of oxidative stress in brain aging and Alzheimer's disease. *Neurobiology of Aging, 107*, 86–95.

Jack, C. R., Bennett, D. A., Blennow, K., Carrillo, M. C., Dunn, B., Haeberlein, S. B., Holtzman, D. M., Jagust, W., Jessen, F., Karlawish, J., Liu, E., Molinuevo, J. L., Montine, T., Phelps, C., Rankin, K. P., Rowe, C. C., Scheltens, P., Siemers, E., Snyder, H. M., . . . & Silverberg, N. (2018). NIA-AA Research Framework: Toward a biological definition of Alzheimer's disease. *Alzheimer's & Dementia, 14*(4), 535–562.

Janelidze, S., Mattsson, N., Palmqvist, S., Smith, R., Beach, T. G., Serrano, G. E., Chai, X., Proctor, N. K., Eichenlaub, U., Zetterberg, H., Blennow, K., Reiman, E. M., Stomrud, E., Dage, J. L., & Hansson, O. (2020). Plasma P-tau181 in Alzheimer's disease: Relationship to other biomarkers, differential diagnosis, neuropathology and longitudinal progression to Alzheimer's dementia. *Nature Medicine, 26*(3), 379–386.

Johnson, K. A., Sperling, R. A., Gidicsin, C. M., Carmasin, J. S., Maye, J. E., Coleman, R. E., Reiman, E. M., Sabbagh, M. N., Sadowsky, C. H., Fleisher, A. S., Murali Doraiswamy, P., Carpenter, A. P., Clark, C. M., Joshi, A. D., Lu, M., Grundman, M., Mintun, M. A., Pontecorvo, M. J., & Skovronsky, D. M. (2013). Florbetapir (F18-AV-45) PET to assess amyloid burden in Alzheimer's disease dementia, mild cognitive impairment, and normal aging. *Alzheimer's & Dementia, 9*(5S).

Karch, C. M., & Goate, A. M. (2015). Alzheimer's disease risk genes and mechanisms of disease pathogenesis. *Biological Psychiatry, 77*(1), 43–51.

Khalsa, D. S. (2015). Stress, meditation, and Alzheimer's disease prevention: Where the evidence stands. *Journal of Alzheimer's Disease, 48*(1), 1–12.

Khokale, R., Kang, A., Buchanan-Peart, K.-A. R., Nelson, M. L., Awolumate, O. J., & Cancarevic, I. (2020). Alzheimer's gone viral: Could herpes simplex virus type-1 be stealing your memories? *Cureus, 12*(11), e11726.

Kraeuter, A. K., Guest, P. C., & Sarnyai, Z. (2019). The Y-maze for assessment of spatial working and reference memory in mice. *Preclinical Models: Techniques and Protocols*, 105–111.

Kwon, S. H., Lee, H. K., Kim, J. A., Hong, S. I., Kim, H. C., Jo, T. H., Park, Y.-I., Lee, C.-K., Kim, Y.-B., Lee, S.-Y., & Jang, C. G. (2010). Neuroprotective effects of chlorogenic acid on scopolamine-induced amnesia via anti-acetylcholinesterase and anti-oxidative activities in mice. *European Journal of Pharmacology, 649*(1–3), 210–217.

LaFerla, F. M., Green, K. N., & Oddo, S. (2007). Intracellular amyloid-β in Alzheimer's disease. *Nature Reviews Neuroscience, 8*(7), 499–509. https://doi.org/10.1038/nrn2168

Lee, H. E., Lim, D., Lee, J. Y., Lim, S. M., & Pae, A. N. (2019). Recent tau-targeted clinical strategies for the treatment of Alzheimer's disease. *Future Medicinal Chemistry, 11*(15), 1845–1848.

Li, Y., Tan, M.-S., Jiang, T., & Tan, L. (2014). Microglia in Alzheimer's disease. *BioMed Research International, 2014*, 1–7.

Liang, Z., Zhang, B., Su, W. W., Williams, P. G., & Li, Q. X. (2016). C-Glycosylflavones alleviate tau phosphorylation and amyloid neurotoxicity through GSK3β inhibition. *ACS Chemical Neuroscience, 7*(7), 912–923.

Liu, C.-C., Kanekiyo, T., Xu, H., & Bu, G. (2013). Apolipoprotein E and Alzheimer disease: Risk, mechanisms and therapy. *Nature Reviews Neurology, 9*(2), 106–118.

Liu, J., Chang, L., Song, Y., Li, H., & Wu, Y. (2019). The role of NMDA receptors in Alzheimer's disease. *Frontiers in Neuroscience, 13*.

Livingston, G., Huntley, J., Sommerlad, A., Ames, D., Ballard, C., Banerjee, S., Brayne, C., Burns, A., Cohen-Mansfield, J., Cooper, C., Costafreda, S. G., Dias, A., Fox, N., Gitlin, L. N., Howard, R., Kales, H. C., Kivimäki, M., Larson, E. B., Ogunniyi, A., . . . & Mukadam, N. (2020). Dementia prevention, intervention, and care: 2020 report of the Lancet Commission. *The Lancet, 396*(10248), 413–446. https://doi.org/10.1016/S0140-6736(20)30367-6

Lunnon, K., Smith, R., Hannon, E., De Jager, P. L., Srivastava, G., Volta, M., Troakes, C., Al-Sarraj, S., Burrage, J., Macdonald, R., Condliffe, D., Harries, L. W., Katsel, P., Haroutunian, V., Kaminsky, Z., Joachim, C., Powell, J., Lovestone, S., Bennett, D. A., . . . & Mill, J. (2014). Methylomic profiling implicates cortical deregulation of ANK1 in Alzheimer's disease. *Nature Neuroscience*, *17*(9), 1164–1170.

Mallah, K., Couch, C., Borucki, D. M., Toutonji, A., Alshareef, M., & Tomlinson, S. (2020). Anti-inflammatory and neuroprotective agents in clinical trials for CNS disease and injury: Where do we go from here? *Frontiers in Immunology*, *11*, 2021.

Mandelkow, E.-M., & Mandelkow, E. (2012). Biochemistry and cell biology of tau protein in neurofibrillary degeneration. *Cold Spring Harbor Perspectives in Medicine*, *2*(7), a006247–a006247.

Mansor, N. I., Ntimi, C. M., Abdul-Aziz, N. M., Ling, K.-H., Adam, A., Rosli, R., Hassan, Z., & Nordin, N. (2020). Asymptomatic neurotoxicity of amyloid β-peptides (Aβ1–42 and Aβ25–35) on mouse embryonic stem cell-derived neural cells. *Bosnian Journal of Basic Medical Sciences*, *21*(1), 98–110.

Marston, A., Kissling, J., & Hostettmann, K. (2002). A rapid TLC bioautographic method for the detection of acetylcholinesterase and butyrylcholinesterase inhibitors in plants. *Phytochemical Analysis: An International Journal of Plant Chemical and Biochemical Techniques*, *13*(1), 51–54.

Mathew, M., & Subramanian, S. (2014). In vitro evaluation of anti-Alzheimer effects of dry ginger (Zingiber officinale Roscoe) extract. *Indian Journal of Experimental Biology*, *52*(6), 606–612.

McShane, R., Westby, M. J., Roberts, E., Minakaran, N., Schneider, L., Farrimond, L. E., Maayan, N., Ware, J., & Debarros, J. (2019). Memantine for dementia. *Cochrane Database of Systematic Reviews*, *2019*, 1–7.

Meng, Q., Lin, M.-S., & Tzeng, I.-S. (2020). Relationship between exercise and Alzheimer's disease: A narrative literature review. *Frontiers in Neuroscience*, *14*.

Miao, J., Shi, R., Li, L., Chen, F., Zhou, Y., Tung, Y. C., Hu, W., Gong, C.-X., Iqbal, K., & Liu, F. (2019). Pathological tau from Alzheimer's brain induces site-specific hyperphosphorylation and SDS- and reducing agent-resistant aggregation of tau in vivo. *Frontiers in Aging Neuroscience*, *11*.

Mobini, S., Song, Y. H., McCrary, M. W., & Schmidt, C. E. (2019). Advances in ex vivo models and lab-on-a-chip devices for neural tissue engineering. *Biomaterials*, *198*, 146–166.

Morozova, A., Zorkina, Y., Abramova, O., Pavlova, O., Pavlov, K., Soloveva, K., Volkova, M., Alekseeva, P., Andryshchenko, A., Kostyuk, G., Gurina, O., & Chekhonin, V. (2022). Neurobiological highlights of cognitive impairment in psychiatric disorders. *International Journal of Molecular Sciences*, *23*(3), 1217.

Morris, G. P., Clark, I. A., & Vissel, B. (2014). Inconsistencies and controversies surrounding the amyloid hypothesis of Alzheimer's disease. *Acta Neuropathologica Communications*, *2*(1), 135.

Murphy, M. P., & LeVine, H. (2010). Alzheimer's disease and the amyloid-β peptide. *Journal of Alzheimer's Disease*, *19*(1), 311–323.

O'Brien, R. J., & Wong, P. C. (2011). Amyloid precursor protein processing and Alzheimer's disease. *Annual Review of Neuroscience*, *34*, 185–204.

Olivares, D., Deshpande, V. K., Shi, Y., Lahiri, D. K., Greig, N. H., Rogers, J. T., & Huang, X. (2012). N-methyl D-aspartate (NMDA) receptor antagonists and memantine treatment for Alzheimer's disease, vascular dementia and Parkinson's disease. *Current Alzheimer Research*, *9*(6), 746–758.

Pang, P. S., Butler, J., Collins, S. P., Cotter, G., Davison, B. A., Ezekowitz, J. A., Filippatos, G., Levy, P. D., Metra, M., Ponikowski, P., Teerlink, J. R., Voors, A. A., Bharucha, D., Goin, K., Soergel, D. G., & Felker, G. M. (2017). Biased ligand of the angiotensin II type 1 receptor in patients with acute heart failure: A randomized, double-blind, placebo-controlled, phase IIB, dose ranging trial (BLAST-AHF). *European Heart Journal*, *38*(30), 2364–2373.

Parmar, N. S., & Prakash, S. (2006). Screening methods in pharmacology. *Alpha Science International Limited*, 140–143.

Petrasek, T., Vojtechova, I., Lobellova, V., Popelikova, A., Janikova, M., Brozka, H., Houdek, P., Sladek, M., Sumova, A., Kristofikova, Z., Vales, K., and Stuchlík, A. (2018). The McGill transgenic rat model of Alzheimer's disease displays cognitive and motor impairments, changes in anxiety and social behavior, and altered circadian activity. *Frontiers in Aging Neuroscience*, *10*, 250.

Richter, M., Vidovic, N., Biber, K., Dolga, A., Culmsee, C., & Dodel, R. (2020). The neuroprotective role of microglial cells against amyloid beta-mediated toxicity in organotypic hippocampal slice cultures. *Brain Pathology*, *30*(3), 589–602.

Rusek, M., Pluta, R., Ułamek-Kozioł, M., & Czuczwar, S. J. (2019). Ketogenic diet in Alzheimer's disease. *International Journal of Molecular Sciences*, *20*(16).

Saber, F. R., Mohsen, E., El-Hawary, S., Eltanany, B. M., Elimam, H., Sobeh, M., & Elmotayam, A. K. (2021). Chemometric-enhanced metabolic profiling of five Pinus species using HPLC-MS/MS spectrometry: Correlation to in vitro anti-aging, anti-Alzheimer and antidiabetic activities. *Journal of Chromatography B*, *1177*, 122759.

Sanchez-Mut, J. V., & Gräff, J. (2015). Epigenetic alterations in Alzheimer's disease. *Frontiers in Behavioral Neuroscience*, *9*.

Schultz, S. A., Larson, J., Oh, J., Koscik, R., Dowling, M. N., Gallagher, C. L., Carlsson, C. M., Rowley, H. A., Bendlin, B. B., Asthana, S., Hermann, B. P., Johnson, S. C., Sager, M., LaRue, A., & Okonkwo, O. C. (2015). Participation in cognitively-stimulating activities is associated with brain structure and cognitive function in preclinical Alzheimer's disease. *Brain Imaging and Behavior*, *9*(4), 729–736.

Sensi, S. L., Granzotto, A., Siotto, M., & Squitti, R. (2018). Copper and zinc dysregulation in Alzheimer's disease. *Trends in Pharmacological Sciences*, *39*(12), 1049–1063.

Sibanyoni, M. N., Chaudhary, S. K., Chen, W., Adhami, H. R., Combrinck, S., Maharaj, V., Schuster, D., & Viljoen, A. (2020). Isolation, in vitro evaluation and molecular docking of acetylcholinesterase inhibitors from South African Amaryllidaceae. *Fitoterapia*, *146*, 104650.

Singh, A., Kukreti, R., Saso, L., & Kukreti, S. (2019). Oxidative stress: A key modulator in neurodegenerative diseases. *Molecules*, *24*(8), 1583.

Slanzi, A., Iannoto, G., Rossi, B., Zenaro, E., & Constantin, G. (2020). In vitro models of neurodegenerative diseases. *Frontiers in Cell and Developmental Biology*, *8*.

Sun, W., Lee, S., Huang, X., Liu, S., Inayathullah, M., Kim, K. M., Tang, H., Ashford, J. W., & Rajadas, J. (2016). Attenuation of synaptic toxicity and MARK4/PAR1-mediated Tau phosphorylation by methylene blue for Alzheimer's disease treatment. *Scientific Reports*, *6*(1), 1–10.

Tanzi, R. E., & Bertram, L. (2005). Twenty years of the Alzheimer's disease amyloid hypothesis: A genetic perspective. *Cell*, *120*(4), 545–555.

Trichopoulou, A., Martínez-González, M. A., Tong, T. Y., Forouhi, N. G., Khandelwal, S., Prabhakaran, D., Mozaffarian, D., & de Lorgeril, M. (2014). Definitions and potential health benefits of the Mediterranean diet: Views from experts around the world. *BMC Medicine*, *12*(1), 112.

Vion, E., Page, G., Bourdeaud, E., Paccalin, M., Guillard, J., & Bilan, A. R. (2018). Trans ε-viniferin is an amyloid-β disaggregating and anti-inflammatory drug in a mouse primary cellular model of Alzheimer's disease. *Molecular and Cellular Neuroscience*, *88*, 1–6.

Vitek, G. E., Decourt, B., & Sabbagh, M. N. (2023). Lecanemab (BAN2401): An anti—beta-amyloid monoclonal antibody for the treatment of Alzheimer disease. *Expert Opinion on Investigational Drugs*, *32*(2), 89–94.

Vorhees, C. V., & Williams, M. T. (2006). Morris water maze: Procedures for assessing spatial and related forms of learning and memory. *Nature Protocols*, *1*(2), 848–858.

World Health Organization. (2023). *Dementia*. Factsheet Details. www.who.int/news-room/fact-sheets/detail/dementia

Yadikar, H., Torres, I., Aiello, G., Kurup, M., Yang, Z., Lin, F., Kobeissy, F., Yost, R., & Wang, K. K. (2020). Screening of tau protein kinase inhibitors in a tauopathy-relevant cell-based model of tau hyperphosphorylation and oligomerization. *PLoS ONE*, *15*(7), e0224952.

Yeo, C. H., Hardiman, M. J., & Glickstein, M. (1985). Classical conditioning of the nictitating membrane response of the rabbit: II. Lesions of the cerebellar cortex. *Experimental Brain Research*, *60*, 99–113.

Yokohama, M., Kobayashi, H., Tatsumi, L., & Tomita, T. (2022). Mouse models of Alzheimer's disease. *Frontiers in Molecular Neuroscience*, *15*, 1–14.

Yoshizaki, K., Asai, M., & Hara, T. (2020). High-fat diet enhances working memory in the Y-maze test in male C57BL/6J mice with less anxiety in the elevated plus maze test. *Nutrients*, *12*(7), 2036.

3 Computational Insights into the Applications of Plant-Based Therapies in Alzheimer's Disease

Dibyajyoti Das, Satyendra Deka, Biswajit Devanath, Aparoop Das, Nongmaithem Randhoni Chanu and Ramen Kalita*

3.1 INTRODUCTION: ALZHEIMER'S DISEASE AND CURRENT THERAPIES

Alzheimer's disease (AD) is a neurodegenerative disorder that affects the brain, resulting in a steady decline in memory and other cognitive abilities and, eventually, mortality within 3 to 9 years of diagnosis. Dementia is a term that refers to a group of disorders that are characterized by a deterioration in cognitive function, including changes in conduct, memory loss, and poor thinking. The underlying causes of dementia can be varied, including genetic factors, environmental factors, and a range of diseases and injuries that affect the brain. AD is the most prevalent form of dementia, accounting for 50 to 56% of cases in both autopsy and clinical studies.

The disease is characterized by the accumulation of abnormal proteins in the brain, which interferes with communication between brain cells and causes damage to nerve cells. As a result, individuals with AD may experience confusion, mood swings, difficulty with language, and a decline in their ability to perform daily activities. While treatments are available to manage symptoms, there is currently no cure for AD. According to a March 2023 WHO report, the number of people living with this condition has surpassed 55 million globally, and the prevalence of dementia is a growing concern worldwide. Shockingly, over 60% of these individuals reside in low- and middle-income countries, where access to medical care and support can be limited. Moreover, every year, there are nearly 10 million new cases of dementia, which highlights the urgent need for increased awareness and better health care resources to manage this condition. As the global population continues to age, this issue is expected to become even more critical, making it imperative to implement effective strategies to improve the quality of life for those affected by dementia. Around the world, dementia is a leading cause of disability and dependence for older individuals and currently ranks as the 7th leading cause of death. This debilitating condition can significantly impact an individual's quality of life and ability to carry out regular tasks, making it crucial to prioritize research and resources to manage and prevent the onset of dementia.

In India, around 7.4% of adults aged 60 and above are estimated to have dementia, and a staggering 8.8 million individuals in this age group are currently living with the condition. Dementia is more commonly found in women than men, and prevalence rates are higher in rural than in urban regions. Additionally, there is a significant difference in dementia prevalence across various states in the country. These findings highlight the need for targeted interventions and support to improve the quality of life and care for individuals living with dementia in India (Lee et al., 2023).

In recent years, AD has gained widespread recognition as a major medical and social issue affecting older individuals globally, regardless of whether they live in developed or underdeveloped nations. With rapidly aging populations, the incidence of AD has increased significantly,

DOI: 10.1201/9781003412069-3

and it is now considered one of the most pressing health problems facing older adults. Alzheimer's disease is a degenerative illness for which there is presently no effective cure; although there are medications available that can help manage the symptoms and slow the progression of the disease, they cannot reverse the damage already done to the brain, which can be fatal. The lack of a cure for AD means that prevention and early detection are essential to managing the disease's impact. Additionally, early detection and treatment can limit the disease's development, improving the quality of life for those living with the condition. The devastating effects of AD on individuals, families, and society as a whole highlight the need for continued research and investment in finding a cure.

3.2 NATURAL PRODUCTS AND THEIR ROLE IN TREATING ALZHEIMER'S DISEASE

Although AD has been a subject of intense research and study, there are still numerous aspects of the disease that remain elusive to modern medicine (Uddin et al., 2020). The complex and varied nature of Alzheimer's disease makes it challenging to identify its exact initial cause (Jalili-Baleh et al., 2018). Despite significant progress in understanding its underlying causes and mechanisms, there is still much to learn about the disease's pathophysiology, risk factors, and potential treatments; the intricacy of this disease and its psychological as well as physiological complications make it challenging to establish the clinical effectiveness of various treatment methods. The development of AD involves several critical biological pathways, including the dysregulation of cholinergic neurotransmitters in the brain; the build-up of extracellular Aβ plaques; the intracellular accumulation of tau microtubule protein; inflammation in the nervous system; oxidative stress; disturbances in calcium homeostasis; and imbalances in bio-metal ions such as copper, iron, and zinc (Jalili-Baleh et al., 2018).

Since time immemorial, plants have been utilized for medicine by various cultures worldwide for several thousand years, and it retains a crucial place in present-day medicine. Numerous pharmaceuticals today are sourced from plant-based substances, and the effectiveness and safety of plant medicine persistently undergo scientific examination and authentication. Plant medicine provides a diverse array of alternatives for individuals seeking natural and holistic approaches to health and wellness, extending from traditional herbal remedies to contemporary plant-based therapies. The medicines obtained from natural herbal products offer various advantages such as lower toxicity, fewer side effects than with synthetic medicines, and affordability. In fact, reports suggest that nearly 50% of treatments developed after 1994 are derived from natural products (Deshpande et al., 2019; Khatoon et al., 2018).

Studies have shown that some herbal drugs and phytochemicals have properties that could be useful in treating neurological diseases. These include anticholinesterase, anti-inflammatory, antioxidant, and neuroprotective functions. Herbal products and phytochemicals have the potential to prevent acetylcholine breakdown, reduce brain inflammation, and protect against damage caused by free radicals and toxins and might slow down the progression of neurological diseases like Alzheimer's and Parkinson's. However, further research is necessary to gain a full understanding of their benefits.

Nonetheless, these natural alternatives could provide a safe and effective alternative to conventional pharmaceuticals (Deshpande et al., 2019) for minimizing symptoms and slowing down the progression of diseases. Secondary metabolites, which are unique and important compounds found in natural products, have led to the rediscovery of natural products as a potential source of novel medicines. Given their high therapeutic potential and minimal side effects, employing natural products to treat AD can be an efficient treatment strategy; it has even been discovered that these natural compounds work well in the treatment of a variety of age-related ailments and diseases. Indeed, using natural products as a treatment for AD and other related diseases is a viable option that should be seriously considered (Table 3.1).

TABLE 3.1
Natural Products and Their Roles in Treating Alzheimer's Disease

Natural Products	Mechanism	Model	Reference
Curcuma longa	• Decreases the proinflammatory cytokine interleukin-1β • Reduces oxidative damage • Reduces plaque burden in brains	APPsw/Tg 2576 mice	(Lim et al., 2001)
Bacopa monnieri (L.)	• Blocks the deterioration of neurons in the hippocampus and the impairment of memory • Enhances cognitive function and prevents cholinergic degeneration	Male Wistar rats	(Uabundit et al., 2010)
Centella asiatica	• Prevents cognitive impairment • Reduces oxidative stress induced by brain glucose metabolism impairment	Male Wistar rats	(Nagar & Delhi, 2003)
Convolvulus pluricaulis	• Reduces AD biomarkers such as mRNA levels of tau and amyloid precursor protein (AbPP)	Male Wistar rats	(Bihaqi et al., 2012)
Ginkgo biloba	• Reduces age-related deterioration in cognitive function • Improves pathologic cognitive impairment in early-onset Alzheimer's disease and vascular dementia	Patients who exhibited amnesia and met the criteria for moderate cognitive impairment in the Diagnostic and Statistical Manual of Mental Disorders, Fourth Edition, Text Revision (DSM-IV-TR).	(Yakoot et al., 2013)
Zingiber officinaleis	• Inhibiting AChE activity • Preventing lipid peroxidation in rats' brains caused by sodium nitroprusside and quinolinic acid	Male Wistar rats In vitro acetylcholinesterase inhibition assay by Ellman's modified colorimetric method	(Oboh et al., 2012)
Allium sativum	• Improves short-term recognition memory • Reducessneuro-inflammation	Male Wistar rats	(Nillert et al., 2017)
Quercetin (loaded in zein nanoparticles)	• Improves the features of cognitive and memory deficits	Senescence-accelerated mouse-prone 8 mice	(Moreno et al., 2017)
	• Enhances cognitive performance and decreases oxidative stress, reducing extracellular Aβ	• Zebrafish AD model • 3xTg-AD mice and nontransgenic mice	(Mani et al., 2018; Sabogal-Guáqueta et al., 2015)
Epigallocatechin-3-gallate	• Decreases AChE activity and oxidative stress • Increases glutathione peroxidase activity	Male Wistar rats	(Biasibetti et al., 2013)
Berberine	• Reduces the buildup of amyloid plaque • Decreases the phenotypic pathology of AD	TgCRND8 mouse	(Durairajan et al., 2012)
Resveratrol	• Activates the hippocampus's and cortex's microglia and reduces Aβ plaque aggregation.	AbetaPP/PS-1 transgenic mouse	(Ahmed et al., 2017)
	• Reduces macromolecular damage such as lipid peroxidation, protein carbonylation, and DNA damage. • Neuroprotective effect on negatively influenced cholinergic receptors that results in memory impairment and defective locomotor behaviour	Male Wistar rats	(Karthick et al., 2016)

(*Continued*)

TABLE 3.1
(Continued)

Natural Products	Mechanism	Model	Reference
Huperzine A	• Induces secretase activity and decreases the phosphorylation of protein tau, improving memory and cognitive activities	Homozygous triple transgenic mice 3xTg-AD male mice	(Ratia et al., 2013)
Rosmarinic acid	• Reduces ROS formation, lipid peroxidation, DNA fragmentation, caspase-3 activation, and tau hyperphosphorylation, contributing a neuroprotective effect.	Cultured rat pheochromocytoma (PC12) cells.	(Iuvone et al., 2006)
Luteolin	• Decreases AChE activity, oxidative stress, and Aβ-42 peptide buildup • Inhibits ROS	AD flies (transgenic fly line expressing wild-type human Aβ42)	(Ali et al., 2019)
Cinnamaldehyde	• Enhances insulin signalling and treats cognitive impairment • Increases p-GSK3β and inhibits AChE activity	Nontransgenic rat model of Alzheimer's disease	(Madhavadas & Subramanian, 2017)
Caffeine	• Decreases tau phosphorylation, neuro-inflammation • Decreases oxidative stress	THY-Tau22 mice	(Laurent et al., 2014)
Apigenin	• Improves learning and memory • Reduces fibrillary amyloid accumulation	APP/PS1 double transgenic AD mouse	(Balez et al., 2016)
Genipin (From gardenia fruits)	• Promotes neurite development by activating neuronal nitric oxide synthase • Prevents Aβ peptide-induced neurotoxicity and guards against Aβ poisoning of the hippocampus neurons	PC12h cells Neuro2a cells	(Yamazaki & Chiba, 2005; Yamazaki et al., 2001)

3.3 COMPUTATIONAL METHODS IN DRUG DISCOVERY FOR ALZHEIMER'S DISEASE

Computer-aided drug design (CADD) is an effective technique for accelerating the drug development process. Using combinatorial chemistry techniques, libraries containing numerous chemical compounds for lead optimization and discovery can be quickly generated. CADD is considered one of the most efficient technologies for reducing costs and expediting the discovery of new drug leads.

Compared with traditional high-throughput screening methods, virtual screening approaches designed to explore large compound libraries in silico typically offer a higher success rate and generate hits more closely resembling drugs. CADD helps increase the rationality of the drug discovery process by enabling the mathematical expression and numerical analysis of all research efforts. This technique provides extensive atom-level details about small and large molecules that cannot be obtained through experimental techniques.

CADD employs extensive computer programming, mathematics, and physics to understand the chemistry and biology of molecules. The pharmacological activity of a ligand at the site of action is determined by the spatial arrangement, the electronic composition, and the interactions between the ligand's atoms and their biological counterparts. Molecular modelling and computational chemistry play crucial roles in understanding the mechanisms of drug-receptor interactions and supporting medicinal chemists in developing new therapeutic agents.

In recent years, the integration of computational methods has become an essential aspect of drug discovery. These methods play a critical role in predicting the pharmacokinetic and pharmacodynamic properties of potential drug candidates. For natural products, computational methods are particularly valuable in predicting bioactivity and identifying possible targets. There are several computational methods used in drug discovery for AD: molecular docking, pharmacophore modelling, ADME/Tox prediction, network pharmacology analysis, and in silico studies on clinical trials. These techniques allow researchers to more efficiently screen potential drug candidates and identify promising leads for further investigation. By combining computational methods with traditional experimental approaches, drug discovery for AD can be accelerated and improved, leading to the development of more effective treatments.

The use of CADD has become increasingly vital in the discovery of new drugs for AD. This approach not only conserves limited resources but also speeds up drug research and development cycles significantly. Various research groups have effectively utilized CADD to create inhibitors for beta and gamma secretases, as well as other compounds, for the purpose of AD therapy and diagnosis. In an overview of the underlying mechanism and causes of AD, we will discuss how Aβ plaques, beta-secretases, and gamma-secretases are structurally represented as well as the most recent accurate identification of drugs that combat AD and imaging agents for amyloid that utilize CADD to diagnose and treat AD.

One of the main signs of Alzheimer's disease, a common neurodegenerative ailment, is the buildup of amyloid in the brain. The clinical features of this condition include the development of tau-hyperphosphorylated neurofibrillary tangles (NFTs) and neuritic or senile plaques formed of Aβ fibrils. Defects in neurotransmitters are also common. Senile plaques and NFTs are the two primary neuropathological features of AD, as confirmed by post-mortem brain examinations of AD patients and recent research (Tiraboschi et al., 2004). The formation of extracellular amyloid fibrils, known as senile plaques, is due to the presence of Aβ, which is a metabolic byproduct of amyloid precursor protein (APP). Meanwhile, cytoskeletal lesions referred to as neurofibrillary tangles primarily consist of straight filaments (SF) and paired helical filaments (PHF) containing highly phosphorylated tau proteins as well as β-amyloid (Aβ) peptides.

In recent decades, numerous hypotheses have emerged to elucidate the complex nature of this multifactorial disorder (Kurz & Perneczky, 2011). One prominent theory that has gained traction over the past 20 years is the amyloid cascade hypothesis. According to this hypothesis, a crucial role is attributed to the amyloid precursor protein (APP). Under normal conditions, APP undergoes cleavage by α-secretase, but in the case of aberrant processing by β- and γ-secretases, there is a disruption in the equilibrium between the production and clearance of the Aβ peptide, leading to pathological consequences (Salomone et al., 2012). Due to this process, the Aβ peptides have a tendency to aggregate spontaneously, forming soluble oligomers that subsequently accumulate to form insoluble fibrils. Eventually, these fibrils are deposited in diffuse senile plaques, as reported by (Hardy, 2009; Kumar et al., 2015).

Three different kinds of enzymes, namely α-, β-, and γ-secretases, cleave APP to produce β-amyloid peptides, such as Aβ1–40 and Aβ1–42 (Savage & Gingrich, 2009). Additionally, α- and γ-secretases play a role in cleaving APP, whereas β-secretases such as Beta-secretase_1 and Beta-secretase_2 primarily contribute to Aβ production. Various genetic mutations are known to enhance Aβ generation through the activity of β- or γ-secretases in the processing of APP, as highlighted in studies by Kurz and Perneczky (2011) and Salomone et al. (2012).

Furthermore, internal APP mutations within the β-amyloid peptides facilitate the self-aggregation of Aβ, resulting in the formation of amyloid fibrils, as described by Hardy (2009). To intervene in the disease progression, numerous treatment strategies have been devised, one of which involves the development of drugs targeting Aβ peptides and β- or γ-secretases (Kumar et al., 2015). However, the efficacy of AD drug development greatly relies on understanding the structure of the protein that serve as the target and investigating the underlying pathogenic mechanisms. The subsequent sections will delve into these aspects in detail.

Computational techniques offer valuable insights by visually representing three-dimensional structures along with interactions of proteins, ligands, and their interactions. Within the realm of new drug research, the identification of an active compound serves as the fundamental basis for all subsequent investigations. Therefore, employing CADD to screen bioactive compounds assumes a pivotal role as the initial and essential step in innovative drug research. In recent times, CADD has gained widespread application in the design and discovery of novel pharmacological agents owing to its ability to identify high numbers of new compounds through targeted searches, surpassing the conventional methods of high-throughput screening and combinatorial chemistry.

In practical scenarios, CADD is frequently used to achieve 3 main goals (Sliwoski et al., 2014): (1) Reducing experimental workloads by reducing the number of predicted active compounds in large compound libraries; (2) facilitating the optimization of lead compounds by improving their affinity or refining drug metabolism and pharmacokinetics properties; and (3) creating new compounds by extending the functional groups of preexisting molecules or assembling fragments of other molecules. In this part, we discuss new CADD applications for the treatment and diagnosis of Alzheimer's disease in the context of gamma-secretase inhibitors, amyloid inhibitors, radiotracers, and beta-secretase inhibitors.

3.4 USING COMPUTATIONAL METHODS TO DISCOVER POTENTIAL DRUGS FOR TREATING ALZHEIMER'S DISEASE

3.4.1 Targeting Amyloid and Radiotracer Molecules

The utilization of CADD in the pharmaceutical field has gained significant advancements, particularly in the pursuit of potential drugs targeting amyloid-related conditions. In a study conducted by the Schein group (Liao et al., 2009), 2 lead compounds with low toxicity were identified for the treatment of AD through molecular docking, database screening, and pharmacophore modelling techniques. The researchers discovered that certain substances can prevent and repair the damaging amyloid aggregation in neurons linked to AD. These substances were chosen due to their comparable conformations to active peptides found through screening utilizing pharmacophores built from solid-state NMR structures of Aβ fibres. At low levels, 2 of the compounds demonstrated the capacity to suppress neurotoxicity brought on by AD in neuroblastoma cells as well as to prevent and reverse Aβ aggregation. Due to their low molecular weight and water solubility, these substances are intriguing candidates for leading compounds in the treatment of AD.

Additionally, it has been demonstrated that metabolites from marine sources have the potential to be used as acetylcholinesterase and amyloid-aggregation inhibitors (Stoddard et al., 2014). Researchers have employed a systematic and user-friendly computational approach to identify potential peptide inhibitors against the aggregation of Aβ in Alzheimer's disease (Eskici & Gur, 2013). This methodology involved substituting amino acids in the mutated peptide, specifically targeting the GxMxG motif known to bind Aβ40 fibrils. By employing docking techniques, the energy favourability of these mutations was evaluated, resulting in promising inhibitor candidates. This approach not only sheds light on the inhibitor binding problem but also serves as a foundation for further analysis in drug design.

Metal ions bind to the water-loving N-terminal portion of the Aβ peptide, particularly Aβ(1–16) (Tõugu et al., 2011). Although it has nothing to do with fibril formation, it is a perfect model for exploring how metal ions interact with Aβ peptides using various spectroscopic methods. By employing computational simulations, a systematic investigation was conducted to explore the diverse metal amyloid-β interactions, revealing the intricate AD aetiology involving metal-mediated events. A fresh set of AMBER force field parameters was created to simulate various Cu(II) coordination spheres of A peptides. Molecular dynamics simulations on six Cu2+-A(1–16) complexes (Xu et al., 2013) indicated that Cu(II) spontaneously bonded to A(1–16). These parameters were then verified using replica exchange.

3.4.2 Development of Novel Gamma-secretase and Beta-secretase Inhibitors

Despite extensive research spanning nearly 2 decades, the development of effective drugs targeting gamma-secretase for the treatment of AD has faced significant challenges. One significant barrier to the adoption of several computational techniques in the field of AD drug discovery has been the absence of a comprehensive three-dimensional structure of gamma-secretase. Fortunately, the quantitative structure-activity relationship (QSAR) method offers a promising solution.

Unlike other approaches, QSAR does not rely on the specific 3D structure of gamma-secretase. Instead, it focuses on establishing a quantitative correlation between the biological activity of compounds and their chemical structures using linear regression analyses. This independence from 3D structure allows QSAR to overcome the limitations imposed by the absence of detailed structural information on gamma-secretase. In the search for possible anti-Alzheimer medications, QSAR methods that use different machine learning and classification methods, including linear discriminant analysis, decision trees, support vector machines, and artificial neural networks, have been extremely effective. These methods utilize the quantitative relationships identified by QSAR to guide the discovery and design of compounds with enhanced therapeutic activity against AD. The application of QSAR in AD drug research has gained wide recognition and has been extensively explored in the scientific literature. Notable studies by Ajmani et al. (2013), Fan et al. (2010), and Ambure and Roy (2014) demonstrated the effectiveness of QSAR approaches in the discovery of promising anti-Alzheimer drugs.

Ajmani et al. (2013) constructed categorical and continuous 2D-quantitative structure-activity relationship (2D-QSAR) models utilizing a variety of chemicals supplied from the ChEMBL database to aid in the discovery of powerful gamma-secretase inhibitors. The goal was to identify the relevant descriptors that contribute to the variation in the potency of gamma-secretase inhibitors. Through the analysis of these compounds, the researchers rationalized the structure-activity relationship (SAR) based on important QSAR descriptors. They discovered that the presence of electronegative substitutions on N-substituted cyclic amines and aryl rings were notable features that influenced the potency of gamma-secretase inhibitors. The study also provided statistically meaningful QSAR model descriptions, which are expected to be helpful in the design and gamma-secretase inhibitor potential screening of larger chemical sets. The discovery of novel substances with improved gamma secretase inhibitory characteristics can be aided by these models.

Gundersen et al. (2005), to create a molecular model, utilized some piperidines as gamma-secretase inhibitors. The researchers looked for novel structures showing potency for future leads using the ROCS algorithm. The sulphonamide derivative served as an outline for the molecule for identifying similarities in structures and making comparisons against a chemical database. Compound 2 was located from the corporate database, and it scored well (low-energy conformer), in contrast to compound 1.

Further modifications were made in compound 2, which included modifications and variations at the positions of R1 (Cl, H, OMe) and R2 (OMe, H, Me). The results indicated that the structure-activity relationship was influenced by both steric and electrostatic factors. The activity of compound no. 3 was increased by substituting the hydroxamic acid with an aromatic ring and the olefin functional group with a urethane. Through radiolabelled cellular experiments, the substance's ability to function as a gamma-secretase inhibitor was also established. The design strategy for the ROCS search offers useful insights for applying this method to find innovative scaffolds for creating anti-Alzheimer medication candidates. Future research in locating possible substances with increased efficacy against AD might take a positive turn in this approach.

Beta-secretase inhibition has recently come to light as a possible therapeutic approach for AD. As a result, there is increased interest in the creation of beta-secretase inhibitors, and many beta-secretase complex crystal structures have been clarified. One of these was the original crystal structure of beta-secretase in association with the peptidomimetic inhibitor OM99-2. This inhibitor spans the P4 to P40 binding pockets of beta-secretase, providing valuable insights into its interaction with the enzyme.

OM99-2 and its derivatives have become widely utilized as key chemical structures for designing and developing beta-secretase inhibitors. The presence of OM99-2 peptidomimetics and its various analogues in the beta-secretase inhibitor repertoire highlights their significance in the field. These compounds serve as important tools for studying the structure-activity relationship and understanding the molecular mechanisms underlying beta-secretase inhibition. The availability of crystal structures, particularly the first one involving OM99-2, has facilitated the rational design and optimization of novel beta-secretase inhibitors. Researchers can leverage these structural insights to guide the development of new compounds with improved potency, selectivity, and pharmacokinetic properties. Consequently, the exploration and utilization of OM99-2 and its derivatives contribute significantly to the advancement of therapeutic approaches targeting beta-secretase for the treatment of AD.

Simulating protein flexibility presents a significant challenge in molecular docking, and there is no universally applicable solution to address this issue. However, Cosconati et al. (2012) successfully identified novel modern-day β-secretase inhibitors by considering protein flexibility and 2 two different methodologies. The researchers screened the NCI database using a collection of X-ray enzyme configurations. The 2 methodologies employed in the study demonstrated analogous predictive capacity but differed in their computational requirements. Energy-weighted virtual screening (EWVS) was better suited for screening large molecular databases due to its lower computational demands. The researchers observed a high success rate with the EWVS approach, indicating that it is an effective substitute for ensemble virtual screening, offering comparable performance while requiring fewer computational resources.

In a separate study, Moitessier et al. (2006) addressed the flexibility of the peptide-based beta-secretase inhibitors, as well as the flexible beta-secretase active site, via induced-fit docking. This method involved a new scoring function that considered various factors, such as the field of influence created by the attractive forces of the hydrogen bonds, entropy fluctuations, and inhibitor-protein interactions. The scoring function, known as rank-score, outperformed LigScore2 and PLP2 in a direct comparison.

To evaluate the predictive power of the approach, the researchers conducted a virtual screening of 80 synthetic compounds. The results demonstrated the effectiveness of the induced-fit docking approach coupled with the rank-score scoring function in identifying potential inhibitors. Overall, these studies highlight the importance of considering protein flexibility in molecular docking and showcase different methodologies that address this challenge. Both the energy-weighted virtual screening and induced-fit docking approaches, along with their respective scoring functions, offer valuable insights and contribute to the discovery of new β-secretase inhibitors.

Molecular dynamics simulations have been extensively utilized for investigating various aspects of protein folding, perturbation, and conformational changes (Koseki et al., 2012; Lindorff-Larsen et al., 2011). Prominent simulation software packages, such as NAMD (Shih et al., 2005), AMBER (Case et al., 2005), and GROMACS (Hess et al., 2008), have been employed in these studies. MD simulations were performed on an inhibitor-beta-secretase complex submerged in an explicit water environment, including more than 15,000 molecules, in order to examine the intrinsic flexibility of Beta-secretase. The duration of the simulation was around 0.3 microseconds (Gorfe et al., 2004; Neri et al., 2005). Beta-secretase was simulated in both its apo and liganded forms, with simulations lasting 1.2 nanoseconds and 4.5 nanoseconds, respectively, using CHARMM. When a substrate or inhibitor is bound, beta-secretase undergoes a significant conformational change, according to a crucial dynamics analysis of the apo trajectory (Xiong et al., 2004). Additionally, the Gaussian model was used to calculate the beta-secretase's normal vibrational modes (Neri et al., 2005) and analysed through MacroModel VBR2. These studies have shed important light on how changes in amino acids affect the active binding sites of -secretase (Wahab et al., 2009). Furthermore, virtual screening was employed to identify novel beta-secretase ligand motifs exhibiting flap-up conformation (Holloway et al., 2009; Wilson et al., 2003).

3.4.3 Molecular Docking Studies on Natural Products for Alzheimer's Disease

Molecular docking is a computational method used to predict the interactions of binding between a small molecule (ligand) and a target protein (receptor). It plays a crucial role in drug discovery and drug development by providing insights into the binding affinities and mechanisms of a potential drug candidate (Shoichet et al., 2002). In the context of AD, molecular docking studies focus on investigating the interactions between natural products and key protein targets involved in the disease's pathology (Gnanaraj et al., 2022).

Products derived from natural sources have long been a valuable source of bioactive compounds with diverse medicinal properties (Deshpande et al., 2019). They offer a vast chemical space for exploration and have shown promise in various therapeutic areas, including AD. Researchers employ molecular docking techniques to virtually screen and identify natural compounds that can potentially modulate the activity of proteins implicated in AD, such as β-amyloid plaques and tau protein tangles (Abraham et al., 2022; Pradeepkiran & Reddy, 2019). Molecular docking studies can assist in understanding the molecular mechanisms underlying the therapeutic effects of natural products in AD.

By elucidating the specific binding sites and key amino acid residues involved in ligand–receptor interactions, researchers can gain valuable information about the ligand's mode of action and potentially optimize its potency and selectivity. It can guide the design and synthesis of novel natural product derivatives with improved drug-like properties. By exploring the chemical space around known active natural compounds, researchers can generate analogues or modify existing structures to enhance their binding affinity, solubility, bioavailability, and metabolic stability. This rational approach aids in the development of more potent and selective compounds for AD treatment (García et al., 2015).

Molecular docking studies on natural products for AD offer a powerful computational tool for identifying potential drug candidates, elucidating their binding modes, and optimizing their therapeutic properties. This approach holds significant promise for discovering new treatments and advancing our understanding of the molecular basis of AD. However, it is important to note that while molecular docking studies provide valuable insights, experimental validation and further preclinical and clinical investigations are necessary to confirm the efficacy and safety of these potential natural product-based therapeutics.

Alzheimer's disease is potentially caused by the accumulation of amyloid peptides. In a research study, 342 compounds obtained from Vietnamese plants were investigated for their proliferating fibrils and these peptides' binding affinity. The surface area determined by the combined use of molecular mechanics and the Poisson–Boltzmann equation and docking were both used to examine their interactions. Through the utilization of docking and molecular dynamics (MD) simulations, the researchers identified 5 top leads that exhibited a high binding affinity to both the monomers and mature fibrils of amyloid peptides.

These compounds demonstrated the potential not only to obstruct and delay the accumulation process but also to degrade the mature fibrils of amyloid peptides. Among these leads, dracorubin and taraxerol showed greater prominence in treating AD than did curcumin, as they exhibited lower DGbind values. It is noteworthy that all 5 top leads were derived from plants. Furthermore, the researchers conducted an analysis of pharmacological characteristics, including human intestinal absorption, blood–brain barrier (BBB) permeability, toxicity, and metabolism, for both the top leads and curcumin. However, it is important to note that the researchers emphasized the consultative nature of their theoretical predictions, highlighting the necessity for careful verification through in vivo experiments (Ngo & Li, 2013).

In the context of AD, caspase 8 has been identified as a significant therapeutic target due to its involvement in neuronal death. Ahmad et al. (2020) conducted a virtual screening on 200 natural compounds using molecular docking to assess their binding capabilities with caspase 8. The formation of the caspase 8-rutaecarpine complex was found to be influenced by hydrogen bonds and hydrophobic interactions. Among the tested compounds, rutaecarpine exhibited the most powerful negative binding energy and was chosen and subjected to MD simulations for additional examination.

The interactions between caspase 8 and rutaecarpine were determined to involve 5 amino acid residues: Thr337, Lys353, Val354, Phe355, and Phe356. A:PHE355:N-ligand:N5 and ligand:H35-A:LYS353:O-ligand:H35 were 2 more hydrogen bonds that were discovered. The caspase 8-rutaecarpine complex was enhanced, its flexibility measured, and its stability examined using a 50 ns MD simulation. The results demonstrated that the complex exhibited a high level of stability. Based on these findings, rutaecarpine emerges as a promising lead compound with notable anti-Alzheimer's potential against caspase 8.

Karanjin, a furanoflavonoid primarily extracted from *Pongamia pinnata* along with various medicinal plants, has demonstrated multiple health benefits. In order to evaluate the potential neuroprotective effects of karanjin, extensive in silico studies were conducted (Gnanaraj et al., 2022). Molecular docking was performed against 5 putative targets linked to Alzheimer's illness and 4 targets linked to Parkinson's disease (PD). Using Auto Dock 4.1 and Molegro Virtual Docker software, the results were compared with those of 3 conventional medications. The Lipinski rule of 5 was used to analyse the physiochemical characteristics of karanjin, including its drug-likeness, absorption, distribution, metabolism, elimination, and toxicity profiles.

To examine the dynamic behaviour and binding free energy of 2 chosen karanjin docking complexes over a time scale of 100 ns, MD simulations were also carried out on them. The in silico findings revealed that karanjin possesses potential neuroprotective properties by virtue of its ability to bind to specific protein targets associated with AD and PD. Compared with the standard drugs, karanjin exhibited higher binding affinity against both AD and PD. Through molecular dynamics simulations, the knowledge of karanjin's thermodynamic and dynamic properties offered insightful conclusions that were in line with the docking results. This research forms the basis for the creation of a novel, inexpensive medicine for the treatment of neurodegenerative illnesses that is obtained from natural sources. Additionally, understanding the mechanism(s) by which karanjin halts the progression of AD and PD can advance the development of medications for these neurodegenerative diseases.

The main components of the Apiaceae family and the *Zosima* genus are coumarins and essential oils. A study conducted by (Karakaya et al., 2019) focuses the anticholinesterase and antioxidant properties of coumarins, essential oils, and extracts produced from *Zosima absinthifolia*'s roots, aerial parts, flowers, and fruits, particularly bergapten, imperatorin, pimpinellin, and umbelliferone. Among these compounds, umbelliferone demonstrated active inhibition against acetylcholinesterase (AChE), while pimpinellin showed activity against butyryl-cholinesterase (BuChE).

Docking studies were performed, and at the binding sites of 1-EVE and 1-P0I, which stand for AChE and BuChE, respectively, active substances were docked. AChE and BuChE's complex three-dimensional structures were downloaded from the Protein Data Bank with the PDB IDs 1EVE and 1P0I, respectively. Prior to docking, the protein was prepared in Schrödinger software, which involved removing water molecules, adding hydrogen atoms, and setting the physiological pH to 7. Subsequently, restrained minimization was performed using OPLS3e force field with the added hydrogen atoms. Ligands were prepared, and finally, docking studies were carried out using the Glide Ligand Docking Panel. The results indicated that umbelliferone and pimpinellin, isolated from *Zosima absinthifolia*, exhibited inhibitory activity against AChE and BuChE, respectively.

The researchers speculated that hydroxyl group at position C-7. of umbelliferone and the methoxy groups at positions C-5 and C-6 of pimpinellin were primarily responsible for their inhibitory activities. Additionally, the comparative analysis of essential oil constituents derived from different parts of *Zosima absinthifolia* (roots, fruits, flowers, and aerial components) revealed variations in phenolic compositions, potentially influenced by the function of the secretory canals in these plant parts. The prospects of using *Zosima absinthifolia* as a natural alternative to synthetic medications for the prevention of Alzheimer's disease has been raised by the existence of substances with strong anticholinesterase activity.

A theoretical investigation was conducted on 7 sea hare *Aplysia dactylomela*-derived halogen-based sesquiterpenes from the Persian Gulf. The study was divided into two parts: Firstly, analysis utilizing both an electronic structure theory and quantum mechanics principles of the molecules using the B3LYP level of theory, and secondly, a molecular docking study where each molecule served as a ligand and the Torpedo California Acetylcholinesterase TcAChE acted as the receptor. The results were compared with Huperzin-A (HupA), donepezil, and galanthamine, which served as reference molecules.

Among the 7 compounds examined, molecule 4 had the strongest inhibitory action against TcAChE, according to the comparative docking analysis. Molecule 4 has a slightly higher lowest binding docked energy than HupA. Both the acyl-binding pocket (Phe330) and the choline-binding site (Trp84) of TcAChE were occupied by molecule 4. Additionally, the acetate group of molecule 4 formed 2 hydrogen bonds with Asp72. Notably, these specific interactions were absent in the HupA and donepezil complexes, while galanthamine exhibited binding to these sites. The results of the molecular docking investigation were compared with the electronic characteristics found by the density functional theory study. Both techniques demonstrated that molecule 4 was more chemically reactive and had higher absolute ligand binding energy. The molecular docking investigation also showed that the molecule had higher absolute ligand binding energy than HupA, and it had similar dipole moment, inner, and frontier molecular orbital energies with HupA (Farrokhnia & Nabipour, 2014).

In another study, Choi et al. (2016) explored the therapeutic potency of 6 ginsenosides sourced from *Panax ginseng* as anti-Alzheimer's disease agents. They conducted inhibitory assays for AChE, BChE, and beta-secretase 1. Additionally, molecular docking simulations were run on Discovery Studio 4.1 and the Autodock Vina software. AChE, BChE, Beta-secretasel, ONOO- and nitrotyrosine production were all significantly inhibited by ginsenosides in the in vitro enzyme testing.

For the chosen ginsenosides and Beta-secretasel, molecular docking studies were used to accurately forecast the protein-ligand interaction geometries. In Autodock Vina, QUD served as the control ligand for validation. The docking study could not identify any unique interaction residues that were important in beta-secretase 1 inhibition and shared by the compounds. However, the ARG235 residue was found to interact with 3 compounds via a common hydrogen bond. Additionally, 2 of the compounds made hydrogen bonds with the ASN233 and THR232 residues. Surprisingly, TYR71 residues showed contact with similar compounds.

Each of the examined ginsenosides produced more than 2 hydrogen bonds with beta-secretase 1. Notably, the ginsenoside sugar molecules showed the strongest interactions with the Beta-secretase 1 active site residues. Overall, the ginsenosides formed numerous Van der Waals interactions and hydrogen bonds, contributing to the strengthening of the protein-ligand complex. This scientific study provides empirical evidence that supports the utilization of ginsenoside as a viable treatment option for individuals suffering from AD.

In vitro, enzymatic, and molecular docking investigations were used in a study by Geromichalos et al. (2012) to look into the activity of saffron extract and its ingredients. This study concentrated on saffron, a plant historically used to cure (Ferrence & Bendersky, 2004; Akhondzadeh, 1998). The results demonstrated that the aqueous methanolic crude extract from saffron had inhibitory effects that were independent of dose. Analysis of the in vitro effects of crocetin, dimethyl crocetin, and safranal on AChE inhibition indicated a dose-dependent inhibitory pattern even though the IC_{50} values were larger than those of galanthamine. The electric eel enzyme's inhibition constant (Ki) for galanthamine was 3.36 M, which was supported by kinetic studies.

The vast majority of saffron phytochemicals evaluated, however, displayed a mixed kind of inhibition and had higher Ki values. According to molecular docking studies, crocetin and dimethyl crocetin concurrently bonded to the catalytic and peripheral anionic sites of AChE rather than to the same place as the inhibitors acetylthiocholine iodide and galanthamine. Crocin, unlike acetylthiocholine iodide and galanthamine, was unable to go far into the aromatic gorge because of its hefty glycosidically linked sugar moieties and instead bonded to a separate location. The presence of these sugars might explain the relatively low in vitro inhibiting action of the extract from saffron extract. The ability of CRT to span the two distant "anionic" sites and bind to both the peripheral anionic binding site and catalytic site of AChE suggests the potential for designing novel, cholinergic transmission-altering dual-binding AChE inhibitors and amyloid-β peptide (Aβ) aggregation. In contrast, safranal exhibited greater potency as an inhibitor when compared with carotenoids, but it interacted only with the catalytic centre.

Abraham et al. (2022) focused on investigating the effectiveness of natural compounds obtained from the haustoria of the palmyra palm *Borassus flabellifer* Linn. against 2 key enzymes involved in AD, namely β-secretase and acetylcholinesterase. Through GC–MS analysis, these compounds were isolated and evaluated for their binding affinity to the target proteins, as well as their pharmacokinetic properties. The outcomes of the docking tests revealed that certain compounds derived from *Borassus flabellifer* exhibited promising potential for further research as potential treatments for AD. Among the 36 ligands tested against β-secretase and acetylcholinesterase, 3 compounds demonstrated the most favourable outcomes against both proteins.

Ligands with binding scores above −8 kcal/mol were the top performers. Additionally, 5 compounds showed strong interaction with acetylcholinesterase, while 3 compounds exhibited excellent interaction with β-secretase. These compounds also demonstrated favourable pharmacokinetic properties. The potential therapeutic use of these compounds, which target both acetylcholinesterase and β-secretase, holds significant value in the treatment of AD. Further investigations are needed to gain valuable insights into the effectiveness of these compounds in in vivo studies. Moreover, such studies could shed light on the synergistic effects of these natural compounds in combating AD.

3.4.4 Network Pharmacology Analysis of Natural Products in AD

Network pharmacology, an emerging field in pharmacology, embraces the concept of "network target, multicomponent therapeutics," which aligns with the holistic principles of Traditional Chinese Medicine (Zhang et al., 2013; Zhao et al., 2009). It offers new avenues for investigating the mechanisms of action (MOA) of herbal medicines in the treatment of AD at multiple levels. In an effort to uncover the MOA of herbs that helps in the treatment of AD, Cao's research group conducted text-based mining of the PubMed database to identify the herbs, their constituents, and their respective targets using appropriate resources from bioinformatics and methodologies (Sun et al., 2013). Although the results provided valuable insights into the MOA of herbal medicines for the treatment of AD to some extent, they were limited in capturing the complex interactions occurring at different levels.

Firstly, the literature mining, which spanned from 1995 to 2011, yielded a restricted collection of herbs and ingredients, leading to an incomplete representation of the ingredient landscape. Secondly, the identification of aims for herbal ingredients relied on HIT (Ye et al., 2011), a target-based database that encompasses only known targets of 586 herbal compounds. Consequently, many compounds lacked target information, making it challenging to gather or predict their respective targets.

Tan et al. (2017) conducted an extensive text mining analysis on PubMed to identify potential herbs with anti-Alzheimer's disease properties. The researchers aimed to summarize the key ingredients of these herbs as potential alternative treatments for AD. To accomplish this, a comprehensive list of plant constituents was obtained from internet databases such the traditional Chinese medicine systems pharmacology (TCMSP) database (Ru et al., 2014). Subsequently, a drug-likeness analysis was performed to filter the compounds for further investigation.

The researchers employed a network pharmacology approach to investigate the interactions between chemicals and proteins, as well as to identify possible therapeutic agents for treating AD. Through network analysis, they elucidated the multi-scale mechanisms of action of these herbs in managing AD. The researchers utilized a network pharmacology technique to integrate large-scale text mining, drug-likeness filtering, target prediction, and network analysis to reveal the underlying molecular processes of the most intensively researched medicinal herbs used in AD therapy.

Wang et al. (2022) aimed to explore the therapeutic potential of *Polygonati rhizoma* (PR) and identify its key active ingredients for the treatment of AD. The authors employed 2 main approaches. Firstly, to determine any probable essential components that are active in both edible and medicinal PR, comparative metabolomics was used. Secondly, network pharmacology was used to look at PR's theoretical underpinnings and therapeutic possibilities for treating AD. The researchers used medicinal PR (HJ: *Polygonatum kingianum* var. grandifolium), edible PR (SGHJ: *Polygonatum alternicirrhosum* Hand.-Mzt.), and clinical and pharmacological experimental design methods, with the former serving as the control group and the latter as the experimental group. A comparison of the groups revealed 19 overlapping differential metabolites out of a total of 335 detected metabolites.

This suggests a synergistic effect resulting from the diverse compounds, targets, and pathways characteristic of traditional Chinese medicine (TCM). The study identified the main PR therapeutic components and their potential therapeutic benefits for treating AD. The results emphasized the properties of TCM that involve many compounds, multiple targets, and multiple action pathways. Rutin, isorhamnetin, and syringetin-3-O-glucoside all specifically targeted AChE (acetylcholinesterase), according to the target network analysis of PR. These three essential active components displayed robust molecular interactions in the putative ligand-binding site of AChE, according to a molecular docking study.

All 7 of the main active components were further assessed for their inhibitory action against AChE utilizing in vitro cell validation experiments to confirm the effects of the compounds. Cafestol increased the vitality of nerve cells, rutin exhibited anti-inflammatory properties, and isorhamnetin decreased oxidative stress damage. Additionally, in vivo results obtained from AD transgenic mice indicated that cafestol, isorhamnetin, and rutin provided various benefits. These advantages included fewer A-positive regions, lower inflammatory cytokine levels, decreased AChE activity, and higher antioxidant levels. Every compound played a distinct role in the initial phases of AD.

In another recent study, Li et al. (2022) utilized Xinjiang lavender (*Lavandula angustifolia* Mill.) as a case study to investigate the material basis and mechanism of action of lavender in treating AD at various levels. The zebrafish AD model, combined with high-resolution mass spectrometry and network pharmacology theory, was employed to develop a quick screening approach for identifying the main bioactive components found in Chinese herbal remedies and understanding how they work. This study used the zebrafish AD model for the first time, and 5 different lavender extracts were given to the fish to assess how they affected their ability to move and respond. Network pharmacology was used to determine the targets for the chemicals isolated from lavender that are related to AD.

In the initial screening round, 44 potential core targets were chosen from the protein–protein interaction network using the median of degree values as the screening criterion. The research was focused on identifying luteolin-7-glucoside, apigenin-7-O-glucoside, salvianolic acid A, and rosmarinic acid as the main active ingredients for further pharmacodynamic analysis and quality control. According to the study, using a combination of drug delivery methods may increase the bioavailability and therapeutic benefits of natural small-molecule compounds, which may have low bioavailability as indicated by in vitro studies or algorithms. Therefore, it is essential to focus on the functional properties of compounds rather than just their bioavailability. This larger viewpoint can make it easier to identify important substances for pharmacodynamic analysis and quality assurance of Chinese herbal medicine. These factors led the researchers to make a preliminary determination that 2 phenolic acids and 2 glycosides are the essential components of lavender for the treatment of AD.

ENDNOTE

* Corresponding author

REFERENCES

Abraham, J. T., Maharifa, H. N. S., & Hemalatha, S. (2022). In Silico molecular docking approach against enzymes causing Alzheimer's disease using borassus flabellifer linn. *Applied Biochemistry and Biotechnology, 194*(4), 1804–1813. https://doi.org/10.1007/s12010-021-03779-3

Ahmad, S. S., Sinha, M., Ahmad, K., Khalid, M., & Choi, I. (2020). Study of caspase 8 inhibition for the management of Alzheimer's disease: A molecular docking and dynamics simulation. *Molecules, 25*(9), 1–9. https://doi.org/10.3390/molecules25092071

Ahmed, T., Javed, S., Javed, S., Tariq, A., Šamec, D., Tejada, S., Nabavi, S. F., Braidy, N., & Nabavi, S. M. (2017). Resveratrol and Alzheimer's disease: Mechanistic insights. *Molecular Neurobiology, 54*(4), 2622–2635. https://doi.org/10.1007/s12035-016-9839-9

Ajmani, S., Janardhan, S., & Viswanadhan, V. N. (2013). Toward a general predictive QSAR model for gamma-secretase inhibitors. *Molecular Diversity, 17*(3), 421–434. https://doi.org/10.1007/s11030-013-9441-2

Akhondzadeh, S. (1998). Herbal medicines in the treatment of psychiatric and neurological disorders. In L'Abate, L. (eds), *Low-Cost Approaches to Promote Physical and Mental Health* (pp. 119–138). Springer. https://doi.org/10.1007/0-387-36899-X_6

Ali, F., Rahul, Jyoti, S., Naz, F., Ashafaq, M., Shahid, M., & Siddique, Y. H. (2019). Therapeutic potential of luteolin in transgenic Drosophila model of Alzheimer's disease. *Neuroscience Letters, 692*, 90–99. https://doi.org/10.1016/j.neulet.2018.10.053

Ambure, P., & Roy, K. (2014). Advances in quantitative structure-activity relationship models of anti-Alzheimer's agents. *Expert Opinion on Drug Discovery, 9*(6), 697–723. https://doi.org/10.1517/17460441.2014.909404

Balez, R., Steiner, N., Engel, M., Muñoz, S. S., Lum, J. S., Wu, Y., Wang, D., Vallotton, P., Sachdev, P., O'Connor, M., Sidhu, K., Münch, G., & Ooi, L. (2016). Neuroprotective effects of apigenin against inflammation, neuronal excitability and apoptosis in an induced pluripotent stem cell model of Alzheimer's disease. *Scientific Reports, 6*(December 2015), 1–16. https://doi.org/10.1038/srep31450

Biasibetti, R., Tramontina, A. C., Costa, A. P., Dutra, M. F., Quincozes-Santos, A., Nardin, P., Bernardi, C. L., Wartchow, K. M., Lunardi, P. S., & Gonçalves, C. A. (2013). Green tea (-)epigallocatechin-3-gallate reverses oxidative stress and reduces acetylcholinesterase activity in a streptozotocin-induced model of dementia. *Behavioural Brain Research, 236*(1), 186–193. https://doi.org/10.1016/j.bbr.2012.08.039

Bihaqi, S., Singh, A., & Tiwari, M. (2012). Supplementation of Convolvulus pluricaulis attenuates scopolamine-induced increased tau and Amyloid precursor protein (AβPP) expression in rat brain. *Indian Journal of Pharmacology, 44*(5), 593–598. https://doi.org/10.4103/0253-7613.100383

Case, D. A., Cheatham, T. E., Darden, T., Gohlke, H., Luo, R., Merz, K. M., Onufriev, A., Simmerling, C., Wang, B., & Woods, R. J. (2005). The Amber biomolecular simulation programs. *Journal of Computational Chemistry, 26*(16), 1668–1688. https://doi.org/10.1002/jcc.20290

Choi, R. J., Roy, A., Jung, H. J., Ali, M. Y., Min, B. S., Park, C. H., Yokozawa, T., Fan, T. P., Choi, J. S., & Jung, H. A. (2016). BACE1 molecular docking and anti-Alzheimer's disease activities of ginsenosides. *Journal of Ethnopharmacology, 190*, 219–230. https://doi.org/10.1016/j.jep.2016.06.013

Cosconati, S., Marinelli, L., Di Leva, F. S., La Pietra, V., De Simone, A., Mancini, F., Andrisano, V., Novellino, E., Goodsell, D. S., & Olson, A. J. (2012). Protein flexibility in virtual screening: The BACE-1 case study. *Journal of Chemical Information and Modelling, 52*(10), 2697–2704. https://doi.org/10.1021/ci300390h

Deshpande, P., Gogia, N., & Singh, A. (2019). Exploring the efficacy of natural products in alleviating Alzheimer's disease. *Neural Regeneration Research, 14*(8), 1321–1329. https://doi.org/10.4103/1673-5374.253509

Durairajan, S. S. K., Liu, L. F., Lu, J. H., Chen, L. L., Yuan, Q., Chung, S. K., Huang, L., Li, X. S., Huang, J. D., & Li, M. (2012). Berberine ameliorates β-amyloid pathology, gliosis, and cognitive impairment in an Alzheimer's disease transgenic mouse model. *Neurobiology of Aging, 33*(12), 2903–2919. https://doi.org/10.1016/j.neurobiolaging.2012.02.016

Eskici, G., & Gur, M. (2013). Computational design of new peptide inhibitors for amyloid beta (Aβ) aggregation in Alzheimer's disease: Application of a novel methodology. *PLoS ONE, 8*(6). https://doi.org/10.1371/journal.pone.0066178

Fan, Y., Unwalla, R., Denny, R. A., Di, L., Kerns, E. H., Diller, D. J., & Humblet, C. (2010). Insights for predicting blood-brain barrier penetration of CNS targeted molecules using QSPR approaches. *Journal of Chemical Information and Modelling, 50*(6), 1123–1133. https://doi.org/10.1021/ci900384c

Farrokhnia, M., & Nabipour, I. (2014). Marine natural products as acetylcholinesterase inhibitor: Comparative quantum mechanics and molecular docking study. *Current Computer Aided-Drug Design*, *10*(1), 83–95. https://doi.org/10.2174/1573409910666140408155615

Ferrence, S. C., & Bendersky, G. (2004). Therapy with saffron and the goddess at Thera. *Perspectives in Biology and Medicine*, *47*(2), 199–226. https://doi.org/10.1353/pbm.2004.0026

García, M. E., Borioni, J. L., Cavallaro, V., Puiatti, M., Pierini, A. B., Murray, A. P., & Peñéñory, A. B. (2015). Solanocapsine derivatives as potential inhibitors of acetylcholinesterase: Synthesis, molecular docking and biological studies. *Steroids*, *104*, 95–110. https://doi.org/10.1016/j.steroids.2015.09.001

Geromichalos, G. D., Lamari, F. N., Papandreou, M. A., Trafalis, D. T., Margarity, M., Papageorgiou, A., & Sinakos, Z. (2012). Saffron as a source of novel acetylcholinesterase inhibitors: Molecular docking and in vitro enzymatic studies. *Journal of Agricultural and Food Chemistry*, *60*(24), 6131–6138. https://doi.org/10.1021/jf300589c

Gnanaraj, C., Sekar, M., Fuloria, S., Swain, S. S., Gan, S. H., Chidambaram, K., Rani, N. N. I. M., Balan, T., Stephenie, S., Lum, P. T., Jeyabalan, S., Begum, M. Y., Chandramohan, V., Thangavelu, L., Subramaniyan, V., & Fuloria, N. K. (2022). In silico molecular docking analysis of karanjin against Alzheimer's and Parkinson's diseases as a potential natural lead molecule for new drug design, development and therapy. *Molecules*, *27*(9). https://doi.org/10.3390/molecules27092834

Gorfe, A. A., Pellarin, R., & Caflisch, A. (2004). Membrane localization and flexibility of a lipidated ras peptide studied by molecular dynamics simulations. *Journal of the American Chemical Society*, *126*(46), 15277–15286. https://doi.org/10.1021/ja046607n

Gundersen, E., Fan, K., Haas, K., Huryn, D., Jacobsen, J. S., Kreft, A., Martone, R., Mayer, S., Sonnenberg-Reines, J., Sun, S. C., & Zhou, H. (2005). Molecular-modeling based design, synthesis, and activity of substituted piperidines as γ-secretase inhibitors. *Bioorganic and Medicinal Chemistry Letters*, *15*(7), 1891–1894. https://doi.org/10.1016/j.bmcl.2005.02.006

Hardy, J. (2009). The amyloid hypothesis for Alzheimer's disease: A critical reappraisal. *Journal of Neurochemistry*, *110*(4), 1129–1134. https://doi.org/10.1111/j.1471-4159.2009.06181.x

Hess, B., Kutzner, C., Van Der Spoel, D., & Lindahl, E. (2008). GRGMACS 4: Algorithms for highly efficient, load-balanced, and scalable molecular simulation. *Journal of Chemical Theory and Computation*, *4*(3), 435–447. https://doi.org/10.1021/ct700301q

Holloway, M. K., Hunt, P., & McGaughey, G. B. (2009). Structure and modeling in the design of β- and γ-secretase inhibitors. *Drug Development Research*, *70*(2), 70–93. https://doi.org/10.1002/ddr.20291

Iuvone, T., De Filippis, D., Esposito, G., D'Amico, A., & Izzo, A. A. (2006). The spice sage and its active ingredient rosmarinic acid protect PC12 cells from amyloid-β peptide-induced neurotoxicity. *Journal of Pharmacology and Experimental Therapeutics*, *317*(3), 1143–1149. https://doi.org/10.1124/jpet.105.099317

Jalili-Baleh, L., Babaei, E., Abdpour, S., Nasir Abbas Bukhari, S., Foroumadi, A., Ramazani, A., Sharifzadeh, M., Abdollahi, M., & Khoobi, M. (2018). A review on flavonoid-based scaffolds as multi-target-directed ligands (MTDLs) for Alzheimer's disease. *European Journal of Medicinal Chemistry*, *152*, 570–589. https://doi.org/10.1016/j.ejmech.2018.05.004

Karakaya, S., Koca, M., Yılmaz, S. V., Yıldırım, K., Pınar, N. M., Demirci, B., Brestic, M., & Sytar, O. (2019). Molecular docking studies of coumarins isolated from extracts and essential oils of zosima absinthifolia link as potential inhibitors for Alzheimer's disease. *Molecules*, *24*(4). https://doi.org/10.3390/molecules24040722

Karthick, C., Periyasamy, S., Jayachandran, K. S., & Anusuyadevi, M. (2016). Intrahippocampal administration of ibotenic acid induced cholinergic dysfunction via NR2A/NR2B expression: Implications of resveratrol against Alzheimer disease pathophysiology. *Frontiers in Molecular Neuroscience*, *9*(APR), 1–16. https://doi.org/10.3389/fnmol.2016.00028

Khatoon, S. S., Rehman, M., & Rahman, A. (2018). The role of natural products in Alzheimer's and Parkinson's disease. In *Studies in Natural Products Chemistry* (1st ed., Vol. 56). Elsevier B.V. https://doi.org/10.1016/B978-0-444-64058-1.00003-0

Koseki, Y., Kinjo, T., Kuroki, M., & Aoki, S. (2012). Aberrant structures of Parkinson's disease-associated ubiquitin C-terminal hydrolase L1 predicted by molecular dynamics. *Chemical Physics Letters*, *535*, 163–168. https://doi.org/10.1016/j.cplett.2012.03.083

Kumar, A., Singh, A., & Ekavali. (2015). A review on Alzheimer's disease pathophysiology and its management: An update. *Pharmacological Reports*, *67*(2), 195–203. https://doi.org/10.1016/j.pharep.2014.09.004

Kurz, A., & Perneczky, R. (2011). Novel insights for the treatment of Alzheimer's disease. *Progress in Neuro-Psychopharmacology and Biological Psychiatry*, *35*(2), 373–379. https://doi.org/10.1016/j.pnpbp.2010.07.018

Laurent, C., Eddarkaoui, S., Derisbourg, M., Leboucher, A., Demeyer, D., Carrier, S., Schneider, M., Hamdane, M., Müller, C. E., Buée, L., & Blum, D. (2014). Beneficial effects of caffeine in a transgenic model of Alzheimer's disease-like tau pathology. *Neurobiology of Aging*, *35*(9), 2079–2090. https://doi.org/10.1016/j.neurobiolaging.2014.03.027

Lee, J., Meijer, E., Langa, K. M., Ganguli, M., Varghese, M., Banerjee, J., Khobragade, P., Angrisani, M., Kurup, R., Chakrabarti, S. S., Gambhir, I. S., Koul, P. A., Goswami, D., Talukdar, A., Mohanty, R. R., Yadati, R. S., Padmaja, M., Sankhe, L., Rajguru, C., . . . Dey, A. B. (2023). Prevalence of dementia in India: National and state estimates from a nationwide study. *Alzheimer's and Dementia*, *November 2022*, 1–15. https://doi.org/10.1002/alz.12928

Li, M., Cao, X., Yan, H., Wang, M., Tashibolati, A., & Maiwulanjiang, M. (2022). Integrating zebrafish model to screen active ingredients and network pharmacology methods to explore the mechanism of lavandula angustifolia therapy for Alzheimer's disease. *ChemistrySelect*, *7*(28). https://doi.org/10.1002/slct.202201364

Liao, R., Duan, L., Lei, C., Pan, H., Ding, Y., Zhang, Q., Chen, D., Shen, B., Yu, Y., & Liu, W. (2009). Thiopeptide biosynthesis featuring ribosomally synthesized precursor peptides and conserved posttranslational modifications. *Chemistry and Biology*, *16*(2), 141–147. https://doi.org/10.1016/j.chembiol.2009.01.007

Lim, G. P., Chu, T., Yang, F., Beech, W., Frautschy, S. A., & Cole, G. M. (2001). The curry spice curcumin reduces oxidative damage and amyloid pathology in an Alzheimer transgenic mouse. *Journal of Neuroscience*, *21*(21), 8370–8377. https://doi.org/10.1523/jneurosci.21-21-08370.2001

Lindorff-Larsen, K., Piana, S., Dror, R. O., & Shaw, D. E. (2011). How fast-folding proteins fold. *Science*, *334*(6055), 517–520. https://doi.org/10.1126/science.1208351

Madhavadas, S., & Subramanian, S. (2017). Cognition enhancing effect of the aqueous extract of Cinnamomum zeylanicum on non-transgenic Alzheimer's disease rat model: Biochemical, histological, and behavioural studies. *Nutritional Neuroscience*, *20*(9), 526–537. https://doi.org/10.1080/1028415X.2016.1194593

Mani, R. J., Mittal, K., & Katare, D. P. (2018). Protective effects of quercetin in Zebrafish model of Alzheimer's disease. *Asian Journal of Pharmaceutics*, *12*(2), 660–666.

Moitessier, N., Therrien, E., & Hanessian, S. (2006). A method for induced-fit docking, scoring, and ranking of flexible ligands. Application to peptidic and pseudopeptidic β-secretase (BACE 1) inhibitors. *Journal of Medicinal Chemistry*, *49*(20), 5885–5894. https://doi.org/10.1021/jm050138y

Moreno, L. C. G. E. I., Puerta, E., Suárez-Santiago, J. E., Santos-Magalhães, N. S., Ramirez, M. J., & Irache, J. M. (2017). Effect of the oral administration of nanoencapsulated quercetin on a mouse model of Alzheimer's disease. *International Journal of Pharmaceutics*, *517*(1–2), 50–57. https://doi.org/10.1016/j.ijpharm.2016.11.061

Nagar, A., & Delhi, N. (2003). Effect of Centella asiatica on cognition and oxidative stress in an intracerebroventricular streptozotocin model of Alzheimer's disease in rats. *Clinical and Experimental Pharmacology and Physiology*, *30*(5–6), 336–342.

Neri, M., Cascella, M., & Micheletti, C. (2005). The influence of conformational fluctuations on enzymatic activity: Modelling the functional motion of β-secretase. *Journal of Physics Condensed Matter*, *17*(18). https://doi.org/10.1088/0953-8984/17/18/014

Ngo, S. T., & Li, M. S. (2013). Top-leads from natural products for treatment of Alzheimer's disease: Docking and molecular dynamics study. *Molecular Simulation*, *39*(4), 279–291. https://doi.org/10.1080/08927022.2012.718769

Nillert, N., Pannangrong, W., Welbat, J. U., Chaijaroonkhanarak, W., Sripanidkulchai, K., & Sripanidkulchai, B. (2017). Neuroprotective effects of aged garlic extract on cognitive dysfunction and neuroinflammation induced by β-amyloid in rats. *Nutrients*, *9*(1), 1–13. https://doi.org/10.3390/nu9010024

Oboh, G., Ademiluyi, A. O., & Akinyemi, A. J. (2012). Inhibition of acetylcholinesterase activities and some pro-oxidant induced lipid peroxidation in rat brain by two varieties of ginger (Zingiber officinale). *Experimental and Toxicologic Pathology*, *64*(4), 315–319. https://doi.org/10.1016/j.etp.2010.09.004

Pradeepkiran, J. A., & Reddy, P. H. (2019). Structure based design and molecular docking studies for phosphorylated tau inhibitors in alzheimer's disease. *Cells*, *8*(3). https://doi.org/10.3390/cells8030260

Ratia, M., Giménez-Llort, L., Camps, P., Muñoz-Torrero, D., Pérez, B., Clos, M. V., & Badia, A. (2013). Huprine X and huperzine a improve cognition and regulate some neurochemical processes related with

Alzheimer's disease in triple transgenic mice (3xTg-AD). *Neurodegenerative Diseases*, *11*(3), 129–140. https://doi.org/10.1159/000336427

Ru, J., Li, P., Wang, J., Zhou, W., Li, B., Huang, C., Li, P., Guo, Z., Tao, W., Yang, Y., Xu, X., Li, Y., Wang, Y., & Yang, L. (2014). TCMSP: A database of systems pharmacology for drug discovery from herbal medicines. *Journal of Cheminformatics*, *6*(1), 1–6. https://doi.org/10.1186/1758-2946-6-13

Sabogal-Guáqueta, A. M., Muñoz-Manco, J. I., Ramírez-Pineda, J. R., Lamprea-Rodriguez, M., Osorio, E., & Cardona-Gómez, G. P. (2015). The flavonoid quercetin ameliorates Alzheimer's disease pathology and protects cognitive and emotional function in aged triple transgenic Alzheimer's disease model mice. In *Neuropharmacology* (Vol. 93). Elsevier Ltd. https://doi.org/10.1016/j.neuropharm.2015.01.027

Salomone, S., Caraci, F., Leggio, G. M., Fedotova, J., & Drago, F. (2012). New pharmacological strategies for treatment of Alzheimer's disease: Focus on disease modifying drugs. *British Journal of Clinical Pharmacology*, *73*(4), 504–517. https://doi.org/10.1111/j.1365-2125.2011.04134.x

Savage, M. J., & Gingrich, D. E. (2009). Advances in the development of kinase inhibitor therapeutics for Alzheimer's disease. *Drug Development Research*, *70*(2), 125–144. https://doi.org/10.1002/ddr.20287

Shih, A. Y., Denisov, I. G., Phillips, J. C., Sugar, S. G., & Schulten, K. (2005). Molecular dynamics simulations of discoidal bilayers assembled from truncated human lipoproteins. *Biophysical Journal*, *88*(1), 548–556. https://doi.org/10.1529/biophysj.104.046896

Shoichet, B. K., McGovern, S. L., Wei, B., & Irwin, J. J. (2002). Lead discovery using molecular docking. *Current Opinion in Chemical Biology*, *6*(4), 439–446. https://doi.org/10.1016/S1367-5931(02)00339-3

Sliwoski, G. R., Meiler, J., & Lowe, E. W. (2014). Computational methods in drug discovery prediction of protein structure and ensembles from limited experimental data view project Antibody modeling, Antibody design and Antigen-Antibody interactions view project. *Computational Methods in Drug Discovery*, *66*(1), 334–395.

Stoddard, S. V., Hamann, M. T., & Wadkins, R. M. (2014). Insights and ideas garnered from marine metabolites for development of dual-function acetylcholinesterase and amyloid-β aggregation inhibitors. *Marine Drugs*, *12*(4), 2114–2131. https://doi.org/10.3390/md12042114

Sun, Y., Zhu, R., Ye, H., Tang, K., Zhao, J., Chen, Y., Liu, Q., & Cao, Z. (2013). Towards a bioinformatics analysis of anti-alzheimer's herbalmedicines from a target network perspective. *Briefings in Bioinformatics*, *14*(3), 327–343. https://doi.org/10.1093/bib/bbs025

Tan, W., Wu, T., Yang, C., Cai, H., Fang, S., Chen, Y., & Wang, Q. (2017). Network pharmacology-based study on the mechanism of action for herbal medicines in Alzheimer treatment. *Journal of Ethnopharmacology*, *196*. Elsevier. https://doi.org/10.1016/j.jep.2016.11.034

Tiraboschi, P., Hansen, L. A., Thal, L. J., & Corey-Bloom, J. (2004). The importance of neuritic plaques and tangles to the development and evolution of AD. *Neurology*, *62*(11), 1984–1989. https://doi.org/10.1212/01.WNL.0000129697.01779.0A

Tõugu, V., Tiiman, A., & Palumaa, P. (2011). Interactions of Zn(ii) and Cu(ii) ions with Alzheimer's amyloid-beta peptide. Metal ion binding, contribution to fibrillization and toxicity. *Metallomics*, *3*(3), 250–261. https://doi.org/10.1039/c0mt00073f

Uabundit, N., Wattanathorn, J., Mucimapura, S., & Ingkaninan, K. (2010). Cognitive enhancement and neuroprotective effects of Bacopa monnieri in Alzheimer's disease model. *Journal of Ethnopharmacology*, *127*(1), 26–31. https://doi.org/10.1016/j.jep.2009.09.056

Uddin, M. S., Kabir, M. T., Tewari, D., Mathew, B., & Aleya, L. (2020). Emerging signal regulating potential of small molecule biflavonoids to combat neuropathological insults of Alzheimer's disease. *Science of the Total Environment*, *700*, 134836. https://doi.org/10.1016/j.scitotenv.2019.134836

Wahab, H. A., Choong, Y. S., Ibrahim, P., Sadikun, A., & Scior, T. (2009). Elucidating isoniazid resistance using molecular modeling. *Journal of Chemical Information and Modeling*, *49*(1), 97–107. https://doi.org/10.1021/ci8001342

Wang, F., Chen, H., Hu, Y., Chen, L., & Liu, Y. (2022). Integrated comparative metabolomics and network pharmacology approach to uncover the key active ingredients of Polygonati rhizoma and their therapeutic potential for the treatment of Alzheimer's disease. *Frontiers in Pharmacology*, *13*(August), 1–18. https://doi.org/10.3389/fphar.2022.934947

Wilson, A. J., Groves, K., Jain, R. K., Park, H. S., & Hamilton, A. D. (2003). Directed denaturation: Room temperature and stoichiometric unfolding of cytochrome c by a metalloporphyrin dimer. *Journal of the American Chemical Society*, *125*(15), 4420–4421. https://doi.org/10.1021/ja028574m

Xiong, B., Huang, X. Q., Shen, L. L., Shen, J. H., Luo, X. M., Shen, X., Jiang, H. L., & Chen, K. X. (2004). Conformational flexibility of β-secretase: Molecular dynamics simulation and essential dynamics analysis. *Acta Pharmacologica Sinica*, *25*(6), 705–713.

Xu, L., Wang, X., Shan, S., & Wang, X. (2013). Characterization of the polymorphic states of copper(II)-bound Aβ(1–16) peptides by computational simulations. *Journal of Computational Chemistry*, *34*(29), 2524–2536. https://doi.org/10.1002/jcc.23416

Yakoot, M., Salem, A., & Helmy, S. (2013). Effect of Memo®, a natural formula combination, on mini-mental state examination scores in patients with mild cognitive impairment. *Clinical Interventions in Aging*, *8*, 975–981. https://doi.org/10.2147/CIA.S44777

Yamazaki, M., & Chiba, K. (2005). Neurotrophic effects of genipin on Neuro2a cells. *Journal of Health Science*, *51*(6), 687–692. https://doi.org/10.1248/jhs.51.687

Yamazaki, M., Chiba, K., Mohri, T., & Hatanaka, H. (2001). Activation of the mitogen-activated protein kinase cascade through nitric oxide synthesis as a mechanism of neuritogenic effect of genipin in PC12h cells. *Journal of Neurochemistry*, *79*(1), 45–54. https://doi.org/10.1046/j.1471-4159.2001.00533.x

Ye, H., Ye, L., Kang, H., Zhang, D., Tao, L., Tang, K., Liu, X., Zhu, R., Liu, Q., Chen, Y. Z., Li, Y., & Cao, Z. (2011). HIT: Linking herbal active ingredients to targets. *Nucleic Acids Research*, *39*(Suppl. 1), 1055–1059. https://doi.org/10.1093/nar/gkq1165

Zhang, G. B., Li, Q. Y., Chen, Q. L., & Su, S. B. (2013). Network pharmacology: A new approach for Chinese herbal medicine research. *Evidence-Based Complementary and Alternative Medicine*, *2013*. https://doi.org/10.1155/2013/621423

Zhao, J., Jiang, P., & Zhang, W. (2009). Molecular networks for the study of TCM pharmacology. *Briefings in Bioinformatics*, *11*(4), 417–430. https://doi.org/10.1093/bib/bbp063

4 Advances in Applying Computer-Aided Drug Design for Alzheimer's Disease

Biswajit Devanath, Satyendra Deka, Dibyajyoti Das, Hans Raj Bhat, Nongmaithem Randhoni Chanu and Kunal Bhattacharya*

4.1 INTRODUCTION TO ALZHEIMER'S DISEASE

Alzheimer's disease (AD) is a progressive neurodegenerative disorder that disturbs a significant number of individuals globally, leading to dementia. A progressive and permanent deterioration in memory and cognitive abilities, including judgment, language, and reasoning, are some of its signs. It also includes the loss of physical abilities and neuropsychiatric symptoms, which can hinder social and occupational functioning. Over 60–70% of dementia cases worldwide, affecting about 35 million people, are caused by AD, which is more likely to develop in people over 65 (A. Singh and Kumar 2016).

AD is one of the main causes of disability in older people and is a serious concern. Age affects the risk that AD will develop, and by 2050, dementia prevalence in low and middle-income nations is projected to grow by 68%. Symptoms of AD include progressive memory loss, difficulty in carrying out daily activities, impaired executive function, changes in thinking or unconscious behaviour, and alteration of speech and language. Early-onset symptoms may also include depressive symptoms and mood changes in 20–30% of patients. Advanced-stage AD patients experience severe memory decline, disorientation, hallucinations, and leading to death due to respiratory syndrome, infection, or fasting (Guo et al. 2020). It is essential to differentiate AD from normal age-related cognitive decline, which is gradual and less disabling. AD typically begins with minor symptoms and progresses to severe brain damage, and individuals with dementia experience a loss of abilities at varying rates (Duthey 2013).

Although significant research efforts have been devoted to understanding AD, the original mechanisms of the disease remain incompletely understood. The pathology of AD involves numerous molecular pathways, including beta-amyloid plaque and neurofibrillary tangle accumulation, inflammation, oxidative stress, and synaptic dysfunction (Guo et al. 2020). This complexity has posed significant challenges in drug discovery for AD, resulting in the failure of many promising drug candidates in clinical trials.

The pathophysiology of AD involves neuron injury and death, starting in the hippocampal region of the brain involved in learning and memory and eventually affecting the whole brain (Mount and Downton 2006). The transmembrane protein amyloid precursor protein (APP), whose function is still unknown but which is likely to be important in neural development, produces amyloid beta as an aberrant proteolytic by-product. In certain circumstances, amyloid beta monomers experience a significant conformational change that results in the creation of beta-sheet-rich tertiary structures, which eventually unite to form amyloid fibrils. These fibrils accumulate outside neurons to form

DOI: 10.1201/9781003412069-4

senile plaques, neuritic plaques, diffuse plaques, and occasionally in the walls of tiny blood vessels in the brain, a condition known as amyloid angiopathy or congophilic angiopathy (Rozemuller, Gool, and Eikelenboom 2005).

Tau, a microtubule-associated protein produced in neurons, exhibits aberrant aggregation in Alzheimer's disease. Tau, which is typically controlled by phosphorylation, functions to stabilize microtubules in the cell cytoskeleton. P-tau, or hyperphosphorylated tau, builds up in the bodies of nerve cells inside neurofibrillary tangles and dystrophic neurites connected to amyloid plaques in AD patients as paired helical filaments (Wisniewski et al. 1989).

Most persons with AD receive care from family and friends at home, which can be physically and emotionally taxing for carers. One in 8 carers gets sick or hurt, and one in 3 takes medication linked to caregiving (Duthey 2013). Carers are at risk for both physical and mental morbidity. Governments and AD organizations have made it their mission to postpone the disease's incapacitating symptoms and ultimately prevent it. As a result, government support for research has greatly grown in recent years (Maas et al. 2017).

AD is believed to develop from multiple factors, such as lifestyle, genetics, and environment, although there may be other risk variables. The biggest risk factor for AD is age, with a patient's risk of the disease doubling every 5 years after age 65. AD affects 32% of people who are 85 years of age or older. Risk genes and deterministic genes are the 2 categories of genes that affect how the disease develops.

Risk genes make an illness more likely to manifest, but they do not ensure that it will. Deterministic genes, on the other hand, result in sickness, which means that everyone who has the gene will experience the disorder. However, only a small number of extended families worldwide, roughly less than 1% of cases, are thought to be affected by rare deterministic genes that lead to AD. Individuals with these genes typically experience symptoms in their 40s or 50s (Rose, 1995).

CADD uses computational tools to find and improve potential drug candidates. CADD can greatly speed up the drug discovery process by predicting the binding affinities of small molecules to target proteins and provide knowledge about the mechanisms of action and potential adverse effects of proposed compounds. With multiple potential medication candidates found using this method, CADD has become a potent tool in the hunt for novel therapies for AD. Recent developments in CADD have shown promise in identifying new medication candidates for AD, even if current treatments only offer little symptom alleviation (Zeng and Wu 2016).

4.2 COMPUTER-AIDED DRUG DESIGN

CADD is a technique that employs computational methods and algorithms to explore, devise, and refine small molecules that have the potential to be developed into drug candidates. The primary objective of CADD is to discover molecules that can specifically bind to a particular target, such as a protein or an enzyme, with high affinity and influence its activity in an intended manner. In the realm of drug discovery, CADD has emerged as a crucial tool that enables researchers to screen vast chemical libraries and expedite the development of novel drugs while curtailing associated expenses (Baldi 2010).

CADD is a multistep process that includes target identification, molecular modelling, virtual screening, and optimization (Namitha and Velmurugan 2022). Firstly, identifying the protein or enzyme target associated with a particular disease or condition is critical. Different methods, such as high-throughput screening, bioinformatics, or structural biology, can be employed to achieve this. Next, molecular modelling techniques are used to create a 3D structure of the protein or enzyme and its binding site. The information obtained is then utilized to design or select small molecules that can fit into the binding site and interact with the target.

Virtual screening, which predicts the binding affinity of small compounds to the target protein or enzyme using computer algorithms, is another essential component of CADD (Chandra, Nayak, and

Singh 2021). Molecular docking, molecular dynamics simulations, or machine learning algorithms are commonly used in virtual screening. These techniques can forecast the binding energy, binding mode, and other properties of small molecules, which are then utilized to prioritize and select the most promising drug candidates (Maia et al. 2020). After identifying potential drug candidates, they can be optimized further using computational techniques. This involves modifying the chemical structure of the molecules to enhance their potency, selectivity, and pharmacokinetic properties, such as absorption, distribution, metabolism, and excretion. Structure-based design, ligand-based design, or de novo design are among the various methods utilized in optimization (Chandra, Nayak, and Singh 2021).

CADD provides numerous benefits over traditional drug discovery approaches, such as high-throughput screening and empirical testing. It enables researchers to explore a more extensive chemical space, discover novel drug candidates that may have been overlooked using conventional techniques, and expedite the drug development process (Tiwari and Singh 2021). CADD eliminates the need for extensive experimental testing, reducing the numbers of false positives and negatives, as well as the overall time and cost of drug development. Numerous drugs currently on the market or undergoing clinical trials have been successfully developed using CADD. CADD has also contributed to the discovery of medications for conditions such as cancer and HIV (Wang et al. 1999).

CADD is advantageous in reducing the occurrence of false positives and false negatives in drug discovery. False positives refer to molecules that are predicted to be active but are not, whereas false negatives refer to molecules that are predicted to be inactive but are actually active. These errors can result in selecting the wrong drug candidates or rejecting potentially effective ones, leading to increased costs and time in the drug development process (Lin 2022). CADD provides more precise predictions of a molecule's activity and properties, thus reducing these errors.

Furthermore, CADD speeds up the discovery of new drugs by enabling the quick screening of extensive chemical libraries to find possible drug candidates. This technology can optimize drug candidates more efficiently by predicting their pharmacokinetic and toxicological properties, thereby reducing the number of experimental tests required. Traditional drug discovery approaches can take years or even decades to develop a new drug from target identification to clinical trials, but CADD can significantly reduce the time required for drug development (Sliwoski, Meiler, and Lowe 2014).

While CADD has many advantages, there are also some limitations to this technology. One of the major challenges of CADD is the accuracy of the computational models that are used to predict the activity and properties of small molecules. These models depend on various assumptions and approximations and might not accurately represent the complex interactions that occur in living organisms (Shaker et al. 2021). Therefore, experimental validation is still necessary to verify the safety and effectiveness of potential drug candidates.

Another issue with CADD is the quality and availability of the data used to develop and validate the computational models. The accuracy and predictive power of the models can be influenced by the quality and reliability of the data used. Therefore, it is crucial to use high-quality data and to continuously refine and update the models as new data becomes available (Baldi 2010). Nevertheless, CADD has revolutionized the drug discovery process and remains an essential tool for developing new drugs. CADD has the potential to accelerate the discovery of new treatments for a variety of diseases and to improve the efficiency and success rate of clinical trials (Zhao et al. 2020).

The process of CADD integrates experimental data with computational methods to identify, design, and enhance small molecules as potential drug candidates. CADD has transformed drug discovery by enabling scientists to screen extensive chemical libraries, thereby reducing the time and cost involved in developing new drugs. CADD has also enhanced the success rate of clinical trials by providing more accurate predictions of a molecule's activity and properties (Ou-Yang et al. 2012).

4.2.1 Major Tools and Techniques Used in CADD

There are various tools and techniques used in CADD, each with its own strengths and limitations. Some of the commonly used tools and techniques in CADD include the following.

4.2.1.1 Molecular Modelling

Molecular modelling is the process of creating 3D models of molecules, including proteins, enzymes, and small molecules, using computer software. This technique can provide valuable insights into the structure and function of molecules, enabling researchers to design new molecules that can interact with them. In drug discovery, molecular modelling is an important component of CADD, as it allows for the design of molecules that are more likely to be effective in treating specific diseases. The main objective of molecular modelling is to gain a better understanding of the properties and behaviour of molecules and use this knowledge to design new molecules with desired properties (Gohlke and Klebe 2002).

Molecular modelling is a versatile tool that encompasses several methods, including homology modelling, ab initio modelling, and molecular dynamics simulations. Homology modelling predicts the structure of a related protein using a known protein structure as a template. Ab initio modelling uses computational algorithms to predict the structure of a molecule based on its chemical composition and physical properties (Berdnikova et al. 2021). Molecular dynamics can simulate the motion and interactions of molecules over time. Molecular modelling can aid drug discovery by identifying binding sites on target proteins or enzymes and designing molecules to interact with them. Molecular docking is one method that predicts the binding affinity of small molecules to target proteins or enzymes, enabling researchers to identify potential drug candidates that bind with high specificity and affinity (Manigrasso, Marcia, and De Vivo 2021). Molecular modelling can also be used to optimize the properties of existing drug candidates by designing analogues of existing drugs that have improved pharmacokinetic or pharmacodynamic properties, for example to increase a drug's solubility, stability, or selectivity (Wang et al. 2022).

Molecular modelling is a valuable tool that can provide detailed insights into the structure and function of molecules at the atomic level, allowing researchers to design new molecules and understand how they interact with biological systems. Molecular modelling is also useful for rapidly screening large chemical libraries and identifying potential drug candidates, which can reduce the time and cost of drug discovery. However, the accuracy of the computational algorithms used in molecular modelling is a major challenge, as these algorithms rely on assumptions and approximations that may not fully capture the complex interactions that occur in living organisms. Therefore, experimental validation is necessary to confirm the activity and safety of potential drug candidates (Ekins, Mestres, and Testa 2007).

4.2.1.2 Virtual Screening

Virtual screening is a technique that employs computational algorithms to predict the binding affinity of small molecules to a target protein or enzyme. Various methods can be used for virtual screening, including molecular docking, molecular dynamics simulations, and machine learning algorithms. Using virtual screening, researchers can identify potential drug candidates that can bind to the target protein with high affinity and specificity. This approach is a powerful tool in CADD that enables rapid screening of large chemical libraries to identify potential drug candidates with the desired binding characteristics. The main objective of virtual screening is to identify promising drug candidates that can interact with the target protein in a specific and potent manner (Lyne 2002).

Virtual screening employs various techniques such as molecular docking, pharmacophore modelling, and machine learning algorithms. Molecular docking uses computational algorithms to predict the binding mode and energy of small molecules to a target protein or enzyme, while pharmacophore modelling identifies the crucial molecular features required for a molecule to bind to a target protein or enzyme (Opo et al. 2021). Machine learning algorithms rely on statistical models to predict the binding affinity of small molecules based on their structural characteristics. This method can screen large chemical libraries and prioritize molecules for further testing, ultimately reducing the time and cost of drug discovery by focusing on potential drug candidates with a higher probability of being effective and having the desired pharmacological properties (Kitchen et al. 2004).

Virtual screening is a powerful method in drug discovery that offers the advantage of quickly screening vast chemical libraries to identify potential drug candidates. This can facilitate the drug discovery process and uncover novel chemical scaffolds that might have been missed using other methods. Additionally, virtual screening can help optimize existing drug candidates by identifying antilogs that possess enhanced pharmacokinetic or pharmacodynamic properties (Shen et al. 2005).

Virtual screening has some limitations despite its usefulness. The accuracy of the computational algorithms used to predict the binding affinity of small molecules is a primary challenge in virtual screening. These algorithms rely on different assumptions and approximations and may not fully represent the complex interactions that occur in living organisms. Thus, experimental validation is still necessary to confirm the activity and safety of potential drug candidates. Nonetheless, virtual screening is still a potent tool in CADD, enabling the rapid screening of extensive chemical libraries and the identification of potential drug candidates with high affinity and specificity (Srinivas Reddy et al. 2007).

4.2.1.3 Quantitative Structure-Activity Relationship (QSAR) Modelling

The process of using statistical models to anticipate a molecule's activity by examining its structural characteristics is called QSAR modelling; it is useful for screening large chemical libraries and deciding which molecules should be further evaluated. In drug discovery, QSAR modelling is a potent computational technique that predicts the biological activity of chemical compounds based on their structural properties. To accomplish this, QSAR modelling requires the creation of mathematical models that correlate a molecule's chemical structure with its biological activity. These models can then be utilized to forecast the activity of novel chemical compounds based on their structural characteristics (Verma, Khedkar, and Coutinho 2010).

The foundation of QSAR modelling is based on the concept that a chemical compound's biological activity can be quantitatively associated with its physicochemical properties. QSAR models are constructed using statistical and machine learning algorithms such as support vector machines, artificial neural networks, and multiple linear regression. The process of building QSAR models includes collecting data, calculating descriptors, building the model, and validating it (Shen et al. 2005).

The data collection step involves selecting a diverse group of chemical compounds with known biological activity. Molecular descriptors that represent the physicochemical properties of the chemical compounds are calculated during the descriptor calculation stage. The next stage, model building, entails developing a mathematical model that relates the molecular descriptors to the biological activity. Finally, the model's performance is assessed through model validation, which involves evaluating the model's accuracy using a test set of compounds that were not used in model development (Guha 2013).

In drug discovery, QSAR modelling offers several benefits. Firstly, it can predict the biological activity of novel chemical compounds before their synthesis, saving time and resources. Secondly, it can identify the essential structural characteristics that determine a compound's biological activity, aiding the development of new compounds with improved activity. Lastly, QSAR modelling can optimize the properties of existing drug candidates by discovering structural modifications that can improve their pharmacodynamic or pharmacokinetic properties (Perkins et al. 2003).

Despite its advantages, QSAR modelling has certain limitations. The precision of the models is determined by the quality and diversity of the data set used in their development, so they are only as good as the data they are built upon. Moreover, the complexity of the biological system being studied can limit the effectiveness of QSAR models, and the models are not always accurate in predicting a compound's activity in vivo. Therefore, experimental validation is still crucial to confirm the safety and activity of potential drug candidates (Miladiyah et al. 2018).

4.2.1.4 Molecular Dynamics Simulations

Molecular dynamics (MD) simulations employ computer algorithms to mimic the movement and interactions of molecules, enabling researchers to study their dynamics and anticipate their performance under varying conditions. Such simulations are widely employed in CADD to model the behaviour and motion of molecules over time. During an MD simulation, classical mechanics equations are utilized to calculate the positions and velocities of individual atoms within a molecular system at discrete time intervals (Naqvi et al. 2019).

MD simulations are a versatile computational tool that can be applied to study various molecular systems, such as proteins, nucleic acids, and small molecules. The main objective of MD simulations is to explore the dynamics, structure, and function of molecules and their behaviour under different conditions (Alonso, Bliznyuk, and Gready 2006). Through MD simulations, researchers can gain insight into the molecular properties that are hard or impossible to observe using experimental methods, for instance by investigating protein folding, ligand binding, and biological membrane dynamics. MD simulations are also useful in designing new drug candidates by predicting the selectivity and binding affinity of small molecules to a target protein or enzyme. Researchers use techniques such as free energy calculations, which rely on MD simulations to estimate the binding free energy of a ligand to a target protein or enzyme, to identify potential drug candidates that can bind to the target protein with high affinity and specificity (Ribeiro et al. 2016). The significant advantage of MD simulations is that they can model the dynamics of molecular systems realistically. They capture the effects of solvent interactions and thermal motion, which are crucial in understanding the behaviour of molecules in biological environments (Vlachakis et al. 2014).

Nonetheless, MD simulations have some limitations. A significant challenge is their computational expense, which increases exponentially with the size of the system under study and the length of the simulation. Therefore, researchers use MD simulations for relatively small systems over short time periods (Huggins, Mulholland, and Hul n.d.). Additionally, MD simulations rely on classical mechanics equations that might not accurately represent the behaviour of molecules in some instances. For example, they may not capture the behaviour of highly charged or polar molecules or quantum mechanical effects. In summary, MD simulations are a potent tool in CADD for modelling the movement and behaviour of molecules over time. Although MD simulations have some limitations, they can provide valuable insights into the structure, function, and dynamics of molecules and can be used to design new drug candidates (Friedman, Boye, and Flatmark 2013).

4.2.1.5 De Novo Design

De novo design is a powerful tool in CADD that involves the use of computer algorithms to create new molecules based on specific criteria, such as target affinity, selectivity, and pharmacokinetic properties. Unlike other methods that use existing molecules as a starting point, de novo design generates molecules from scratch (D. B. Singh and Pathak 2020). This process utilizes various computational tools and algorithms, including molecular docking, MD simulations, quantum mechanics calculations, and machine learning algorithms, to explore chemical space and generate new drug candidates. By exploring new chemical space, de novo design can help researchers to identify novel drug candidates that may have been overlooked using other methods (Hung and Chen 2014).

De novo design is a multistep process used in CADD to design new drug molecules from scratch. The process involves three main steps: Identifying a target protein or enzyme and determining its three-dimensional structure, generating a library of potential drug candidates using computational tools, and optimizing the drug candidates through further simulations and experimental testing (Giordano et al. 2022). De novo design has the advantage of creating novel drug candidates with unique properties and interactions with the target protein or enzyme. This is useful in situations where existing drugs have limited efficacy or undesirable side effects.

However, de novo design also has limitations. The computational complexity of exploring chemical space and generating new drug candidates is a major challenge, especially for larger and more complex molecules. Additionally, experimental testing and validation are often required to confirm the predicted properties and interactions of the drug candidates, which can be time-consuming and expensive. Many potential drug candidates may also be rejected (Ooms 2012).

4.2.1.6 Fragment-Based Drug Design

Fragment-based drug design (FBDD) is a method used in CADD to design new drug candidates based on small chemical fragments that can bind to a target protein or enzyme. FBDD can help researchers to identify novel chemical scaffolds and to optimize their properties for drug development. In FBDD, a library of small fragment molecules is screened against the target protein or enzyme to identify potential binding fragments, which are then linked together to form larger drug molecules (Renaud, Neumann, and Van Hijfte 2015).

FBDD is often used in cases where traditional high-throughput screening methods have been unsuccessful, as it can allow for the identification of weakly binding fragments that may not be detected using other screening methods (Congreve et al. 2008). This can lead to the discovery of novel and effective drug candidates with unique binding modes and properties (Bottegoni et al. 2012). Additionally, FBDD can help reduce the size and complexity of drug molecules, which can improve their pharmacokinetic properties and reduce the likelihood of unwanted side effects. However, FBDD also has some limitations. One of the major challenges is the complexity of fragment linking, which can be difficult and time-consuming. Additionally, the identified fragments may not always be complementary to the binding site of the target protein or enzyme, which can limit the success of FBDD in some cases (Bon 2022).

4.2.1.7 Structure-Based Drug Design

Structure-based drug design (SBDD) entails creating new compounds that can attach to a target protein or enzyme using its 3D structure as a guide. Researchers can better understand the molecular underpinnings of drug-target interactions and the features of drug candidates by using SBDD (Chaudhary and Mishra 2016), which is based on the three-dimensional structure of the target protein or enzyme. To find and improve potential drugs that can bind to the target protein or enzyme with high affinity and selectivity, SBDD uses computational tools and techniques (Gurung et al. 2021). The process of SBDD typically involves several steps:

- Protein structure determination: The three-dimensional structure of the target protein or enzyme is determined using various experimental methods such as X-ray crystallography, NMR spectroscopy, or cryo-electron microscopy (Hayet et al. 2021).
- Binding site identification: The binding site of the target protein or enzyme is identified using various computational tools such as molecular docking or site-directed mutagenesis (Hassan Baig et al. 2016).
- Ligand screening: A library of potential drug candidates is screened against the binding site of the target protein or enzyme using various computational tools such as molecular docking, MD simulations, or quantum mechanics calculations (Usha et al. 2018).
- Optimization: The identified drug candidates are optimized using various computational and experimental methods to improve their binding affinity, selectivity, and pharmacokinetic properties (Wang et al. 2018).

The ability to rationally design therapeutic candidates based on the three-dimensional structure of the target protein or enzyme is one of the key benefits of SBDD. This can result in the identification of extremely potent and selective medication candidates that have a high affinity and specificity for binding to the target protein or enzyme. Additionally, SBDD can help reduce the time and cost associated with drug discovery by allowing researchers to focus on a specific target protein or enzyme and design drug candidates that are more likely to be successful (Ooms 2012).

However, SBDD also has some limitations. One of the major challenges is the availability and quality of protein structures, which can be difficult to obtain for some targets. Additionally, SBDD may not always be effective in cases where the binding site of the target protein or enzyme is highly flexible or dynamic. SBDD is a powerful method in CADD that allows for the rational design of drug candidates based on the three-dimensional structure of the target protein or enzyme. While SBDD has its challenges, it has the potential to generate highly potent and selective drug candidates that can be developed into successful therapeutics (Araujo et al. 2021).

4.3 TARGETS FOR ALZHEIMER'S DISEASE

AD is a degenerative disorder of the nervous system that leads to dementia, characterized by memory loss and declining cognitive abilities. The condition starts with mild symptoms and progressively worsens over time, making it one of the leading causes of death worldwide. Several distinct features of AD have been identified, including the accumulation of beta-amyloid proteins around nerve cells, hyperphosphorylation of tau, oxidative stress, impaired regulation of metal ions, and reduced levels of acetylcholine (Kumar et al. 2016). Diagnosis of AD is challenging, as there is no single definitive test available. Pharmacotherapy options for AD provide limited symptomatic relief and primarily aim to improve cognitive function. Computational biology approaches, such as molecular docking, have emerged as useful tools for identifying novel therapeutic targets and ligands. Docking is widely used in computer-aided drug design and development to virtually screen large chemical libraries and propose structural hypotheses on how the ligands interact with their targets, ultimately leading to lead optimization (Lahiri et al. 2002).

The amyloid precursor protein (APP), a transmembrane protein largely present in neuronal and glial cells of the brain, is the source of the Aβ fragment. Specific enzymes cleave the protein, resulting in a variety of pieces. In healthy individuals, the enzymes α secretase and γ secretase break the protein to create an internal APP domain (AICD) and extracellular nontoxic soluble fragment (3kD) that aid in regulating synaptic plasticity, oxidative and metabolic stress resistance, and neuronal excitability (Cenini and Voos 2019). However, in pathological conditions, β-secretase (BACE) cleaves APP, followed by γ secretase, generating extracellular Aβ42 fragment monomers. The Aβ42 strands in these monomers combine to create soluble oligomers, which then arrange into β-sheet conformations to produce the ordered fibrils known as β-plaques. These sheets are turned into fibrils by polymerizing each strand in parallel with a different monomer, with contact between the Gln15 and Gly37 parts of the implicated monomers.

Plaques accumulate mostly on the surface of brain cells, causing the cell membrane to distort and the cell structure to change. Plaques are thought to be less hazardous to neurons than soluble oligomers. AD is characterized by the build-up of amyloid plaques in synapses, which impairs the connection between nearby neurons and impairs signal transmission (Biran et al. 2009).

In the neurons of the central nervous system (CNS), tau proteins are microtubule-associated proteins (MAPs). Tau proteins' main job is to keep cells' microtubule structures stable. Six isoforms of tau proteins are produced in human tissues by alternative splicing of mRNA. The quantity of binding domains in the structure of tau proteins affects how well they bind to microtubules. More effectively than those with 3 binding domains, tau proteins with 4 domains stabilize the microtubules (Lahiri et al. 2002).

Tau proteins' function is controlled by phosphorylation, and they cooperate with the microtubule protein tubulin to assist microtubule assembly. The term "tauopathies" refers to a group of neurodegenerative disorders caused by tau hyperphosphorylation. Amyloid plaques and neuroinflammation in AD lead to tau hyperphosphorylation. When tau and MAP proteins are together, hyperphosphorylated tau proteins can self-assemble into paired filaments of helical and straight filaments (Rousseaux 2008). These structures oligomerize, causing the accumulation of neurofibrillary tangles that block impulse transmission close to the axon area, impairing neuronal cell function and eventually causing cell death. As a result of these modifications, tau pathology spreads throughout nearby brain regions by multiplying into neighbouring neurons. Tau oligomers, which are another factor contributing to neurofibrillary degeneration in AD, are more harmful than tangles, and conformational changes in the structure of tau proteins including have been seen in transgenic rats (Bennett, Yu, and De Jager 2014).

Although the development of AD can be sparked by Aβ deposits, the neuroinflammation that results from this disease is more complicated and is fuelled by the activation of other brain cells. Evidence suggests that glial cells are mainly responsible for this phenomenon, responding quickly to brain injuries by initiating repair mechanisms. Glial cells are nonexcitable cells that perform many critical brain functions. While astrocytes are a crucial neuro-supportive cell type that carefully regulates the pH, environment, ion homeostasis, oxidative stress, and blood flow, microglia serve as the first line of defence for the immune system in the brain (Cao et al. 2018). They make up the neurovascular unit that maintains the integrity of the BBB along with microglia, oligodendrocytes, neurons, pericytes, and endothelial cells. Additionally, astrocytes play a significant role in synaptogenesis, control neuronal and synaptic plasticity, and maintain neurons' trophic and metabolic needs.

Astrocytes and microglia significantly alter their shape and function when exposed to disease, developing a reactive phenotype that aids in removing harmful stimuli: by reducing Aβ-mediated neurotoxicity, eliminating and decomposing Aβ, and forming a barrier around plaques, reactive astrocytes shield neurons from damage (Behl 2021). However, persistent and uncontrolled glial cell activation can result in reactive gliosis, which is characterized by the production of various cytokines and proinflammatory mediators and eventually promotes neuroinflammatory response. This state is a defining feature of AD brains. By enhancing its phagocytosis, clearance, and destruction as well as secreting the glia-derived neurotrophic factor, which is beneficial for neuronal survival, activated microglia decrease the build-up of Aβ (Piccioni et al. 2021).

In the cortex, basal ganglia, and basal forebrain of the brain, acetylcholine is a key neurotransmitter. In order for nerve impulses to be sent via the brain, acetylcholine must be produced, released, and reabsorbed. According to research, the cholinergic lesion in AD develops early and is largely presynaptic rather than postsynaptic. It can even happen during prodromal or asymptomatic stages (Behl 2021). Accordingly, the decline in cholinergic activity is brought on by the NBM's cholinergic neurons and their axons that project to the cerebral cortex.

Cholinergic lesions also affect the nicotinic and muscarinic receptors of the cerebral cortex. There is ample evidence of the loss of nicotinic receptors on cortical neurons, but there is also evidence of a presynaptic component based on the loss of nicotinic receptors found on cholinergic axons from the NBM that are deteriorating (Teleanu et al. 2022). On the other hand, muscarinic receptors exhibit a decrease in the presynaptic M2 receptors, while the mostly postsynaptic M1 receptors are unaffected. Evidence, however, points to potential malfunction in the remaining postsynaptic M1 receptors. The cerebral cortex, hippocampus, and amygdala experience anterograde cortical cholinergic deafferentation as a result of the loss of basal cholinergic neurons. Although it cannot be completely ruled out, it is unlikely that an initial degeneration of cortical cholinergic terminals will result in retrograde degeneration of NBM neurons (Cenini and Voos 2019).

The neurotransmitter glutamate plays a vital role in various brain activities, such as memory, learning, cognition, and motor behaviour, and it is also a precursor to GABA and glutathione. Proton magnetic resonance spectroscopy (1 H MRS) can measure glutamatergic activity and reflect neuronal integrity, which makes it a promising tool for investigating neurological and psychiatric disorders, including amyotrophic lateral sclerosis (ALS) and AD. Studies have shown age-related changes in 1 H MRS biomarkers, indicating lower concentrations of glutamate and GABA in the human brain as age increases. The use of 1 H MRS has also improved the accuracy of predicting the changes from less cognitive impairment to dementia (Conway 2020).

4.3.1 Acetylcholinesterase

Acetylcholinesterase inhibitors (AChEIs) have long been thought of as prospective treatments for Alzheimer's disease (AD) and other dementias since these conditions are linked to the loss and degradation of cholinergic neurons. In fact, donepezil, galantamine, and rivastigmine are 3 of the 4 AChEIs that are currently approved medications for AD. As a result, docking experiments frequently target acetylcholinesterase. For instance, studies have shown that compounds such as miltirone and salvianolic acid A, which are found in *Salvia miltiorrhiza* (red sage) extract, as well as cinerin C, which is extracted from *Prosopis cineraria* pods, bind to acetylcholinesterase using molecular docking (S. Wang et al. 1999).

4.3.2 Beta-Secretase and Gamma-Secretase Enzymes

Inhibiting the BACE and GS that cause Aβ production can stop the development of Aβ plaques and the progression of AD. In order to do this, molecular docking has been used to assess possible GS inhibitors. This methodology's top-scoring compound has been used to find structurally comparable compounds for pharmacophore mapping (Dejakaisaya, Kwan, and Jones 2021).

4.3.3 Caspases

Caspases are essential for triggering apoptosis in neurons; therefore, inhibiting them may be a way to stop neurodegeneration-related neuronal death in conditions including ALS, AD, PD, and HD. Several studies have used in silico drug design and molecular docking to target caspases for the treatment of NDs. For instance, a study tested 10 noncytotoxic nitrones to see if they could stop the death of the HT22 neuronal cell line and lower levels of active caspase-3 and oxidative stress. These nitrones may be used by medicinal chemistry as a starting point for the creation of innovative ND treatments because molecular docking experiments revealed that they bonded to a location close to the catalytic area of caspase-3 (Wang, Wang, and Wang 2017).

4.3.4 Acetylcholine (ACh) Receptors

Many studies, both in vivo and in vitro, have demonstrated that memory loss in people with AD is largely caused by decreased cholinergic activity. In light of this, the nicotinic acetylcholine receptor (nAChR) is a prospective therapeutic target for AD. Compounds that act on nicotinic receptors have been reported to improve memory, cognition, and spatial ability in animal models, according to multitarget CADD research (Papandreou et al. 2011).

4.3.5 N-Methyl-D-Aspartate Receptor

Glutamate and glycine signals are mediated by N-methyl-D-aspartate (NMDA) receptors in the central nervous system (CNS), and these signals are crucial for the synaptic plasticity required for memory and learning processes. NMDA antagonists can be used to address NMDA receptor-related neurotransmission abnormalities brought on by excessive glutamate exposure. The identification of new NMDA receptor inhibitors using CADD has been motivated by the identification of conantokins, MK-801, and memantine as NMDA receptor inhibitors. This method could lead to the discovery of novel chemicals that could be used as AD treatments (Zhang et al. 2016).

4.3.6 ROCK-I and NOX2 Enzymes

Inhibiting NOX2 and ROCK-I simultaneously has been suggested as a viable treatment for neuroinflammation and degenerative neurological illnesses such as AD. In phagocytic activities, a multi-protein complex called NOX2 catalyses the production of superoxide from oxygen, whereas ROCK-I is a key modulator of cell adhesion, proliferation, and migration. ROCK-I stimulates NOX2 integration into the NADPH oxidase complex via Rac in the early stages of neurological illness, causing microglial cells with high ROCK-I and NOX2 levels to gradually destroy neurons. A study employed in silico drug design to find novel compounds that can inhibit ROCK-I and NOX2 (Serhan et al. 2019).

4.4 RECENT ADVANCES IN CADD FOR ALZHEIMER'S DISEASE

Understanding of the aetiology, genetics, and pathophysiological mechanisms underlying AD has advanced significantly in recent years, but there are currently no effective treatments available. Acetylcholinesterase inhibitors are the mainstay of current clinical treatments, which aim to reduce symptoms and decrease cognitive loss in the near future but are unable to stop the pathological course of AD in all patients (Martinez and Castro 2006). The purpose of developing drugs has shifted to altering the underlying pathology by lowering the aggregation of Aβ peptide in brain tissue due to breakthroughs in our knowledge of the pathogenic mechanisms causing AD. Many anti-amyloid medications, including defibrillators, vaccines, cholesterol-lowering medications, oligosaccharides, metal chelators, testosterone, oestrogen, and anti-inflammatory medications, have entered preliminary clinical trials. Since more than 20 years ago, molecular neuroimaging has been utilized to aid in the early diagnosis and assessment of therapy response in AD utilizing methods such as positron emission tomography (PET) or single-photon emission computed tomography (SPECT) (Lane, Hardy, and Schott n.d.). As a result, numerous amyloid imaging agents, such as radiolabelled Aβ antibodies, peptide fragments, and tiny compounds, have been created for the diagnosis of AD. The ability of these amyloid imaging agents to predict the progression of moderate cognitive impairment into AD may be particularly valuable. However, classic empirical or pharmacologic approaches still present substantial difficulties in the development of viable AD medications (Tayeb et al. 2012).

The creation of novel AD medications has benefited greatly from CADD in recent years, which has also significantly sped up the drug development process and helped conserve scarce resources. Several research teams have found beta-secretase inhibitors, gamma-secretase inhibitors, and other substances for AD treatment and diagnosis using CADD (Guo et al. 2020). The pathophysiology of AD, Aβ plaque construction, beta-secretases, and gamma-secretases are very important to understand the mechanism of AD. CADD has successfully identified anti-AD medicines and amyloid imaging agents for AD diagnosis and therapy (Zeng and Wu 2016).

In drug development, it can be extremely helpful to visualize the three-dimensional structures of proteins, ligands, and protein–ligand interactions using computer techniques. Drug research's fundamental purpose is to identify an active leading chemical, and CADD screening of bioactive compounds has emerged as a critical component of this procedure (Tiwari and Singh 2021). Because CADD offers a greater hit rate for novel compounds and allows for a more focused search than more conventional techniques such as HTS and combinatorial chemistry, it is now often used in the design and discovery of new pharmacological drugs. CADD is frequently used in practise for 3 main tasks: filtering enormous compound libraries into more manageable groups of projected active compounds, directing lead compound optimization, and creating new compounds. The results showed recent uses of CADD in the creation of beta-secretase inhibitors, gamma-secretase inhibitors, amyloid-β inhibitors, and radiotracers for the detection and management of Alzheimer's disease (Abbasi and Alam 2016).

The use of CADD in the pharmaceutical industry for prospective amyloid drug research has grown quickly. Two low-toxic drugs that inhibit and reverse amyloid-β aggregation and neurotoxicity were found by the Schein group using pharmacophore modelling, database screening, and molecular docking techniques. Sixteen different compounds were chosen for experimental screening by creating three-dimensional pharmacophores for the NCI database using the docked active peptides' lowest energy conformations in the solid-state NMR structures of an Aβ fibre.

Thioflavin T fluorescence and ELISA experiments revealed 2 compounds to be able to stop and reverse Aβ to aggregate at concentrations of 2–3 M, as well as to block Aβ's neurotoxic effects on neuroblastoma cells. Using biological and spectroscopic experiments, vitamin K3 analogues were likewise successful in preventing the aggregation of Aβ, reducing the free radicals that Aβ created and shielding cells from Aβ-induced toxicity. VK3–10 was shown to have great potential for further development as anti-amyloidogenic medicine for the treatment of AD using MD modelling and molecular docking techniques to determine the binding affinity and binding mechanism between vitamin K3 analogue and Aβ. Similar findings on the dual inhibition of acetylcholinesterase and amyloid-β aggregation by a number of marine metabolites for the treatment of AD were obtained.

New peptide inhibitors were screened as potential treatments for Aβ aggregation in Alzheimer's disease using a structured and simple computational methodology. The approach systematically substituted amino acids with equivalent residues in the mutated peptide, then determined whether these mutations were energetically favourable through docking. This method produced computationally promising inhibitor candidates and provided insight into the inhibitor binding problem, giving direction and a foundation for further drug design analysis (Zeng and Wu 2016).

The creation of amyloid imaging agents for early detection is a developing preventative and therapeutic strategy in the area of AD study. In this in silico investigation, models for 2D-QSAR and group-based QSAR (G-QSAR) were developed using a variety of PET/SPECT imaging agents with strong binding affinity against Aβ plaques. Genetic function approximation was used to create these models, and they were subsequently validated using a variety of statistical criteria. Indicating that the models were helpful for creating new PET/SPECT imaging agents and forecasting their binding affinity towards Aβ plaques, created compounds based on both models were projected to be similar to, or more active than, anyone in the original dataset.

A number of benzyloxy benzene compounds lacking a bulky, rigid π-conjugated scaffold have been developed and evaluated as Aβ probes. The binding affinity relationship between these adaptable ligands and the hydrophobic Val18-Phe20 channel on the flat spine of Aβ fibre was ascertained by combining molecular docking and 3D-QSAR experiments. Autoradiography and biodistribution in vitro and in vivo further supported the high affinity of radioiodinated ligands to Aβ plaques in the brains of AD patients or Tg mice, as well as the high initial uptake and rapid washout feature. A new generation of Aβ-targeting ligands with a flexible framework may be designed and developed using this method, which also offers fresh insights for better localization and quantitative imaging of Aβ plaques (Papagiannopoulou and Hadjipavlou-Litina 2018; Zeng and Wu 2016).

For in vivo imaging of Aβ-plaques using PET/SPECT, a number of promising imaging agents have recently been discovered. The 3D-QSAR technique has been used to further improve and forecast the binding affinity of structurally related Thio-T analogues and stilbene compounds for Aβ plaques PET/SPECT imaging. Based on Thio-T derivatives and stilbene derivatives, Kim and group created 2 efficient CoMFA and CoMSIA models to find active molecules.

However, no researcher has conducted a molecular docking investigation on Aβ plaques to uncover the specific interactions between the chemicals and the active site because Aβ plaques do not have a crystal structure. Therefore, it is imperative to quickly ascertain the crystal structure of Aβ plaques in order to explore the quantitative structure-activity correlations between the agents and Aβ plaques as well as the interaction mechanism (Guo et al. 2020).

Beta-secretase (BACE-1) inhibition has recently gained popularity as a potential treatment for AD. Countless BACE-1 inhibitors and crystal structures of the BACE-1 complex have been published. With the help of the OM99-2 peptidomimetic inhibitor, which covers the P4 to P40 binding sites, the first crystal structure of BACE-1 was discovered. Beta-secretase inhibitors' primary chemical structures are typically OM 99-2 and their variants. The development of AD medications has utilized computational techniques such as MD modelling, molecular docking, and QSAR based on the crystallographic structures of BACE1-OM99-2 complexes (Ghosh and Osswald 2014).

The analysis of protein folding, disruption, and conformational changes has made considerable use of MD simulations; programmes include NAMD, AMBER, and GROMACS. To investigate the intrinsic flexibility of BACE-1, MD simulations of inhibitor-BACE-1 complexes were run for a total simulation time of about 0.3 μs in an explicit water bath of about 15K molecules. CHARMM was used to simulate both the apo and liganded forms of BACE-1. The apo trajectory was submitted to essential dynamics analysis, which showed that substrate/inhibitor interaction results in a significant conformational shift in BACE-1. The normal vibrational modes of BACE-1 were computed using the β-Gaussian model and MacroModel VBR2 analysis. Through virtual screening, these studies have found novel flap-up BACE-1 ligand patterns and expanded our understanding of the consequences of amino acid changes on the active binding sites of β-secretase (Emon 2021).

β-secretase activity has been linked to the essential amino acids Asp32 and Asp228. According to research using MD simulations of the β-secretase complex with the inhibitor OM99–2, only the neutral protonation state of Asp32 and the ionized state of Asp228 can explain the structural characteristics of their interactions with the hydroxy group of inhibitors found in previous X-ray crystallographic analyses. The Rajamani group used the crystal structure of the BACE-OM99–2 complex to predict the binding affinities of hydroxy ethylene derivatives as beta-secretase inhibitors, with the Asp228 protonated and Asp32 deprotonated states. The outcome was based on a linear regression between the binding affinities and electrostatic and van der Waals characteristics of those inhibitors. Although most cases are restricted to short-production MD simulations because of computer system limitations, evaluating protein dynamics, particularly doing long-timescale simulations, will play a significant role in the future development of more effective anti-Alzheimer medications (Carter and Anderton 1994).

In molecular docking, simulating protein flexibility is a challenging issue for which there is no universally applicable solution. In one investigation, novel-secretase inhibitors were found by screening the NCI database using an ensemble of static X-ray enzyme structures. The study used 2 distinct approaches to account for protein flexibility. Although the processing needs of these 2 approaches were similar, energy-weighted virtual screening (EWVS) was more appropriate for screening huge molecular databases. Given the low computing requirements and comparable performance of EWVS compared with virtual ensemble screening, the high success rate implies that it is a viable alternative (Slater and Kontoyianni 2019).

Another team, Moitessier et al., investigated highly flexible peptidic and pseudo-peptidic β-secretase inhibitors as well as the adaptable BACE-1 active site using an induced-fit docking method. They used a fresh scoring system called rank-score, which took into account the interactions between inhibitors and proteins as well as hydrogen bonds and binding-induced entropy changes. In a side-by-side comparison, this scoring algorithm outperformed LigScore2 and PLP2 and accounted for desolvation, protein entropy loss, and complex salvation. The authors virtually screened 80 synthetic chemicals to gauge the prediction potential of their method (Lagoutte-Renosi et al. 2022).

4.5 CONCLUSIONS AND FUTURE PERSPECTIVES

Drugs for Alzheimer's disease have proven effective in reducing symptoms throughout the past century, but important problems in diagnosis, prevention, and therapy still need to be resolved. Currently, there are no treatments that target the underlying causes of the condition; instead, they primarily aim to relieve symptoms. Making particular anti-Alzheimer medications that focus on the progression of the disease by lowering A generation, preventing A aggregation, and encouraging A clearance is thus a potential strategy. Only a few drugs have undergone clinical trials, despite the fact that inhibiting A synthesis by targeting β- and γ-secretase is a new method. By preventing Aβ production and aggregation in the early phases of the disease, computational approaches may speed up the discovery of AD drugs. The pathogenic process of AD can be better understood through the use of computational approaches, including QSAR, MD, QM, LIE, molecular docking, and pharmacophore modelling. These techniques can also reveal crucial details about the mechanism by which inhibitors and proteins interact.

It is possible to optimize particular anti-Alzheimer molecules by conventional chemical synthesis and thoughtful drug development. CADD can be crucial in the creation of medications for other illnesses. We must comprehend the various stages of the intricate pathological process of AD if we are to unlock it soon. Given AD's complex pathogenesis, multitarget drug discovery may be a potential approach to treating the disease. Future priorities should include the use of computational methods to find nontoxic lead compounds that are secretase and amyloid inhibitors. The improvement of early AD diagnosis and therapy evaluation would result from the reinforcement of CADD applications in the creation of novel amyloid imaging agents. The development of next-generation AD medications will finally be advanced by the use of quantitative system pharmacology, computational system polypharmacology methods, and chemogenomics knowledgebases.

Alzheimer's disease is a complex neurodegenerative disorder that poses immense challenges in developing effective treatments. However, CADD techniques have revolutionized the drug discovery process by enabling researchers to explore new avenues and accelerate the identification and optimization of potential therapeutic agents. It is very important to understand the molecular mechanisms and pathological features of AD, including the role of Aβ plaques and tau proteins in disease progression. This knowledge forms the foundation for target identification and validation, where computational models play a crucial role in identifying key protein targets implicated in AD pathology. By designing molecules that can modulate these targets, researchers can potentially slow down the progression of the disease.

Virtual screening techniques, aided by extensive virtual compound libraries and advanced algorithms, have emerged as powerful tools for identifying potential drug candidates. The ability to rapidly screen millions of compounds against target proteins allows for the identification of molecules with desirable properties, paving the way for the rational design of more potent and specific drug candidates. Structure-based drug design and molecular dynamics simulations further optimize lead compounds by predicting their binding interactions, ultimately enhancing their efficacy. The advancements in applying CADD for AD have brought us closer to the development of effective treatments for this debilitating disease. By leveraging computational models, virtual screening, machine learning, and experimental techniques, researchers are making significant progress in identifying and optimizing potential therapeutic agents. These breakthroughs hold great promise for improving the lives of individuals affected by Alzheimer's disease, offering hope for a future with better treatment options and, ultimately, a cure.

ENDNOTE

* Corresponding author

REFERENCES

Abbasi, Mohammad, and Md Alam. 2016. "Alzheimer's Disease Management: Current Therapy and Recent Drug Development." *International Neuropsychiatric Disease Journal* 7(2): 1–19.

Alonso, Hernán, Andrey A. Bliznyuk, and Jill E. Gready. 2006. "Combining Docking and Molecular Dynamic Simulations in Drug Design." *Medicinal Research Reviews* 26(5): 531–568.

Araújo, F. A., Cavalcante, A. L. Q., de Barros Braga, M., Kato, R. B., Ramos, R. T. J., Santos, E. F. F. D. L. 2021. Computational Genomics. In: Singh, V., Kumar, A. (eds), *Advances in Bioinformatics*. Springer, Singapore. 1st Ed, pp: 213–241.

Baldi, A. 2010. "Computational Approaches for Drug Design and Discovery: An Overview." *Systematic Reviews in Pharmacy* 1(1): 99–105.

Behl, Christian. 2021. "The Search for Novel Targets in Alzheimer's Disease—The 90s Redux." *Progress in Molecular Biology and Translational Science* 177. 1st ed. Elsevier Inc. http://doi.org/10.1016/bs.pmbts.2020.08.002.

Bennett, David A., Lei Yu, and Philip L. De Jager. 2014. "Building a Pipeline to Discover and Validate Novel Therapeutic Targets and Lead Compounds for Alzheimer's Disease." *Biochemical Pharmacology* 88(4): 617–630. http://doi.org/10.1016/j.bcp.2014.01.037.

Berdnikova, Daria V., Paolo Carloni, Sybille Krauß, and Giulia Rossetti. 2021. "Role and Perspective of Molecular Simulation-Based Investigation of RNA-Ligand Interaction: From Small Molecules and Peptides to Photoswitchable RNA Binding." *Molecules* 26(11).

Biran, Yif'At, et al. 2009. "Pharmacotherapeutic Targets in Alzheimer's Disease." *Journal of Cellular and Molecular Medicine* 13(1): 61–86.

Bon, M., Bilsland, A., Bower, J., & McAulay, K. 2022. Fragment-based drug discovery—The importance of high-quality molecule libraries. *Molecular Oncology*, 16(21), 3761–3777. https://doi.org/10.1002/1878-0261.13277

Bottegoni, Giovanni, Angelo D. Favia, Maurizio Recanatini, and Andrea Cavalli. 2012. "The Role of Fragment-Based and Computational Methods in Polypharmacology." *Drug Discovery Today* 17(1–2): 23–34. http://doi.org/10.1016/j.drudis.2011.08.002.

Cao, Jiqing, Jianwei Hou, Jing Ping, and Dongming Cai. 2018. "Advances in Developing Novel Therapeutic Strategies for Alzheimer's Disease 11 Medical and Health Sciences 1109 Neurosciences." *Molecular Neurodegeneration* 13(1): 1–20.

Carter, J., and B. Anderton. 1994. "Molecular Pathology of Alzheimer's Disease." *British Journal of Hospital Medicine* 51(10): 522–528.

Cenini, Giovanna, and Wolfgang Voos. 2019. "Mitochondria as Potential Targets in Alzheimer Disease Therapy: An Update." *Frontiers in Pharmacology* 10(July): 1–20.

Chandra, I., Nayak, C., Singh, S. K. 2021. Predicting Protein Folding and Protein Stability by Molecular Dynamics Simulations for Computational Drug Discovery. In: Singh, S. K. (eds), *Innovations and Implementations of Computer Aided Drug Discovery Strategies in Rational Drug Design*. Springer, Singapore. https://doi.org/10.1007/978-981-15-8936-2_7.

Chaudhary, K. K., and N. Mishra. 2016. "Central Bringing Excellence in Open Access OPEN ACCESS A Review on Molecular Docking: Novel Tool for Drug Discovery." *A Review on Molecular Docking: Novel Tool for Drug Discovery. JSM Chem* 4(3): 1029. www.pdb.org.

Congreve, Miles, Gianni Chessari, Dominic Tisi, and Andrew J. Woodhead. 2008. "Recent Developments in Fragment-Based Drug Discovery." *Journal of Medicinal Chemistry* 51(13): 3661–3680.

Conway, Myra E. 2020. "Alzheimer's Disease: Targeting the Glutamatergic System." *Biogerontology* 21(3): 257–274. https://doi.org/10.1007/s10522-020-09860-4.

Dejakaisaya, Hattapark, Patrick Kwan, and Nigel C. Jones. 2021. "Astrocyte and Glutamate Involvement in the Pathogenesis of Epilepsy in Alzheimer's Disease." *Epilepsia* 62(7): 1485–1493.

Duthey, Béatrice. 2013. "Background Paper 6.11 Alzheimer Disease and Other Dementias, Update on 2004." *World Health Organization* (February): 1–77.

Ekins, S., J. Mestres, and B. Testa. 2007. "In Silico Pharmacology for Drug Discovery: Applications to Targets and Beyond." *British Journal of Pharmacology* 152(1): 21–37.

Emon, Maek. 2021. "Mechanism-Based Stratification of Alzheimer's and Parkinson's Disease Using Artificial Intelligence." https://d-nb.info/123866332X/34.

Friedman, Ran, Kjetil Boye, and Kjersti Flatmark. 2013. "Molecular Modelling and Simulations in Cancer Research." *Biochimica et Biophysica Acta—Reviews on Cancer* 1836(1): 1–14.

Giordano, Deborah, Carmen Biancaniello, Maria Antonia Argenio, and Angelo Facchiano. 2022. "Drug Design by Pharmacophore and Virtual Screening Approach." *Pharmaceuticals* 15(5): 1–16.

Gohlke, Holger, and Gerhard Klebe. 2002. "Approaches to the Description and Prediction of the Binding Affinity of Small-Molecule Ligands to Macromolecular Receptors." *Angewandte Chemie—International Edition* 41(15): 2644–2676.

Guha, R. 2013. On Exploring Structure–Activity Relationships. In: Kortagere, S. (eds), In Silico Models for Drug Discovery. *Methods in Molecular Biology* (Vol. 93). Humana Press, Totowa, NJ. https://doi.org/10.1007/978-1-62703-342-8_6

Guo, Tiantian, et al. 2020. "Molecular and Cellular Mechanisms Underlying the Pathogenesis of Alzheimer's Disease." *Molecular Neurodegeneration* 15: 1–37.

Gurung, Arun Bahadur, et al. 2021. "An Updated Review of Computer-Aided Drug Design and Its Application to COVID-19." *BioMed Research International* 2021.

Hassan Baig, Mohammad, et al. 2016. "Computer Aided Drug Design: Success and Limitations." *Current Pharmaceutical Design* 22(5): 572–581.

Hayet, S., K. M. Sujan, A. Mustari, and M. A. Miah. 2021. "Hemato-Biochemical Profile of Turkey Birds Selected from Sherpur District of Bangladesh." *International Journal of Advanced Research in Biological Sciences* 8(6): 1–5. http://doi.org/10.22192/ijarbs.2021.08.06.001.

Huggins, D. J. et al. 2019. Biomolecular simulations: From dynamics and mechanisms to computational assays of biological activity. Wiley Interdisciplinary Reviews: *Computational Molecular Science*, 9(3), e1393.

Hung, Che Lun, and Chi Chun Chen. 2014. "Computational Approaches for Drug Discovery." *Drug Development Research* 75(6): 412–418.

Ghosh, A. K., & Osswald, H. L. (2014). BACE1 (β-Secretase) Inhibitors for the Treatment of Alzheimer's Disease. *Chemical Society Reviews*, 43(19), 6765. https://doi.org/10.1039/c3cs60460h

Kitchen, Douglas B., Hélène Decornez, John R. Furr, and Jürgen Bajorath. 2004. "Docking and Scoring in Virtual Screening for Drug Discovery: Methods and Applications." *Nature Reviews Drug Discovery* 3(11): 935–949.

Kumar, Ashwini, et al. 2016. "Current and Novel Therapeutic Molecules and Targets in Alzheimer's Disease." *Journal of the Formosan Medical Association* 115(1): 3–10. http://doi.org/10.1016/j.jfma.2015.04.001.

Lagoutte-Renosi, Jennifer, et al. 2022. "Molecular Modelling in Cardiovascular Pharmacology: Current State of the Art and Perspectives." *Drug Discovery Today* 27(4): 985–1007.

Lahiri, Debomoy K., Martin R. Farlow, Nigel H. Greig, and Kumar Sambamurti. 2002. "Current Drug Targets for Alzheimer's Disease Treatment." *Drug Development Research* 56(3): 267–281.

Lane, C. A., Hardy, J., & Schott, J. M. 2017. Alzheimer's disease. *European Journal of Neurology*, 25(1), 59–70. https://doi.org/10.1111/ene.13439.

Lidia, Ana, Queiroz Cavalcante, Marcus De Barros Braga, and Rodrigo B. Kato. 2021. "Advances in Bioinformatics." *Advances in Bioinformatics*: 1–446

Lin, Yipeng. 2022. "Review of Modern Computer-Aided Drug Design Methods." *International Journal of Biology and Life Sciences* 1(1): 47–50.

Lyne, Paul D. 2002. "Structure-Based Virtual Screening: An Overview." *Drug Discovery Today* 7(20): 1047–1055.

Maas, Andrew I. R., et al. 2017. "Traumatic Brain Injury—Integrated Approaches to Improving Clinical Care and Research." *Elsevier*: 9–126.

Maia, Eduardo Habib Bechelane, et al. 2020. "Structure-Based Virtual Screening: From Classical to Artificial Intelligence." *Frontiers in Chemistry* 8(April).

Manigrasso, Jacopo, Marco Marcia, and Marco De Vivo. 2021. "Computer-Aided Design of RNA-Targeted Small Molecules: A Growing Need in Drug Discovery." *Chem* 7(11): 2965–2988. https://doi.org/10.1016/j.chempr.2021.05.021.

Martinez, Ana, and Ana Castro. 2006. "Novel Cholinesterase Inhibitors as Future Effective Drugs for the Treatment of Alzheimer's Disease." *Expert Opinion on Investigational Drugs* 15(1): 1–12.

Miladiyah, Isnatin, Jumina Jumina, Sofia Mubarika Haryana, and Mustofa Mustofa. 2018. "Biological Activity, Quantitative Structure—Activity Relationship Analysis, and Molecular Docking of Xanthone Derivatives as Anticancer Drugs." *Drug Design, Development and Therapy* 12: 149–158. https://doi.org/10.2147/DDDT.S149973.

Mount, Claire, and Christian Downton. 2006. "Alzheimer: Progress or Profit." *Nature Medicine* 12(7): 780–784.

Namitha, K. N., and V. Velmurugan. 2022. "Review of Bioinformatic Tools Used in Computer Aided Drug Design (CADD)." *World Journal of Advanced Research and Reviews* 14(2): 453–465.

Naqvi, Ahmad Abu Turab, Taj Mohammad, Gulam Mustafa Hasan, and Md. Imtaiyaz Hassan. 2019. "Advancements in Docking and Molecular Dynamics Simulations Towards Ligand-Receptor Interactions and Structure-Function Relationships." *Current Topics in Medicinal Chemistry* 18(20): 1755–1768.

Ooms, F. 2012. "Molecular Modelling and Computer Aided Drug Design. Examples of Their Applications in Medicinal Chemistry." *Current Medicinal Chemistry* 7(2): 141–158.

Opo, Firoz A. Dain Md, et al. 2021. "Structure Based Pharmacophore Modelling, Virtual Screening, Molecular Docking and ADMET Approaches for Identification of Natural Anti-Cancer Agents Targeting XIAP Protein." *Scientific Reports* 11(1): 1–18. https://doi.org/10.1038/s41598-021-83626-x.

Ou-Yang, Si Sheng, et al. 2012. "Computational Drug Discovery." *Acta Pharmacologica Sinica* 33(9): 1131–1140. http://doi.org/10.1038/aps.2012.109.

Papagiannopoulou, Dionysia, and Dimitra Hadjipavlou-Litina. 2018. "Computational Modelling of Diagnostic Imaging Agents for Alzheimer's Disease: Molecular Imaging Agents for the in Vivo Detection of Amyloid Plaques in Alzheimer's Disease." *Neuromethods* 132: 463–479.

Papandreou, Magdalini A., et al. 2011. "Memory Enhancing Effects of Saffron in Aged Mice Are Correlated with Antioxidant Protection." *Behavioural Brain Research* 219(2): 197–204.

Perkins, Roger, Hong Fang, Weida Tong, and William J. Welsh. 2003. "Quantitative Structure-Activity Relationship Methods: Perspectives on Drug Discovery and Toxicology." *Environmental Toxicology and Chemistry* 22(8): 1666–1679.

Piccioni, G., Mango, D., Saidi, A., Corbo, M., & Nisticò, R. 2021. Targeting Microglia-Synapse Interactions in Alzheimer's Disease. *International Journal of Molecular Sciences*, 22(5):2342.

Reddy, A. Srinivas, et al. 2007. "Virtual Screening in Drug Discovery—A Computational Perspective." *Current Protein & Peptide Science* 8(4): 329–351.

Renaud, Jean Paul, Thomas Neumann, and Luc Van Hijfte. 2015. "Fragment-Based Drug Discovery." *Small Molecule Medicinal Chemistry: Strategies and Technologies* 47(14): 221–249.

Ribeiro, João V., et al. 2016. "QwikMD—Integrative Molecular Dynamics Toolkit for Novices and Experts." *Scientific Reports* 6(May): 1–14.

Rose, S. 1995. "The Rise of Neurogenetic Determinism." *Nature* 373(6513): 380–382. https://doi.org/10.1038/373380a0.

Rousseaux, Colin G. 2008. "A Review of Glutamate Receptors I: Current Understanding of Their Biology." *Journal of Toxicologic Pathology* 21(1): 25–51.

Rozemuller, Annemieke J. M., Willem A. Van Gool, and Piet Eikelenboom. 2005. "The Neuroinflammatory Response in Plaques and Amyloid Angiopathy in Alzheimer's Disease: Therapeutic Implications." *Current Drug Targets – CNS & Neurological Disorders*: 4(3): 223–233.

Serhan, Michael, et al. 2019, November. "Total Iron Measurement in Human Serum with a Smartphone." *AIChE Annual Meeting, Conference Proceedings*. https://doi.org/10.1109/jtehm.2020.3005308

Shaker, Bilal, et al. 2021. "In Silico Methods and Tools for Drug Discovery." *Computers in Biology and Medicine* 137(September): 104851. https://doi.org/10.1016/j.compbiomed.2021.104851.

Shen, Jianhua, et al. 2005. "Virtual Screening on Natural Products for Discovering Active Compounds and Target Information." *Current Medicinal Chemistry* 10(21): 2327–2342.

Singh, Arti, and Anil Kumar. 2016. "Comparative Analysis of Intrahippocampal Amyloid Beta (1–42) and Intracerbroventricular Streptozotocin Models of Alzheimer's Disease: Possible Behavioral, Biochemical, Mitochondrial, Cellu Lar and Histopathological Evidences." *Journal of Alzheimer's Disease & Parkinsonism* 6(1).

Singh, D. B., Pathak, R. K. 2020. Computational Approaches in Drug Designing and Their Applications. In: Gupta, N., Gupta, V. (eds), *Experimental Protocols in Biotechnology*. Springer Protocols Handbooks. Humana, New York, NY. https://doi.org/10.1007/978-1-0716-0607-0_6

Slater, Olivia, and Maria Kontoyianni. 2019. "The Compromise of Virtual Screening and Its Impact on Drug Discovery." *Expert Opinion on Drug Discovery* 14(7): 619–637. https://doi.org/10.1080/17460441.2019.1604677.

Sliwoski, Gregory R., Jens Meiler, and Edward Willard Lowe. 2014. "Computational Methods in Drug Discovery Prediction of Protein Structure and Ensembles from Limited Experimental Data View Project Antibody Modelling, Antibody Design and Antigen-Antibody Interactions View Project." *Computational Methods in Drug Discovery* 66(1): 334–395.

Tayeb, Haythum O., Hyun Duk Yang, Bruce H. Price, and Frank I. Tarazi. 2012. "Pharmacotherapies for Alzheimer's Disease: Beyond Cholinesterase Inhibitors." *Pharmacology and Therapeutics* 134(1): 8–25. http://doi.org/10.1016/j.pharmthera.2011.12.002.

Teleanu, Raluca Ioana, et al. 2022. "Neurotransmitters—Key Factors in Neurological and Neurodegenerative Disorders of the Central Nervous System." *International Journal of Molecular Sciences* 23(11).

Tiwari, Anshul, and Sakshi Singh. 2021. "Computational Approaches in Drug Designing." *Bioinformatics: Methods and Applications*: 207–217.

Usha, Talambedu, et al. 2018. "Recent Updates on Computer-Aided Drug Discovery: Time for a Paradigm Shift." *Current Topics in Medicinal Chemistry* 17(30): 3296–3307.

Verma, Jitender, Vijay Khedkar, and Evans Coutinho. 2010. "3D-QSAR in Drug Design—A Review." *Current Topics in Medicinal Chemistry* 10(1): 95–115.

Vlachakis, Dimitrios, Elena Bencurova, Nikitas Papangelopoulos, and Sophia Kossida. 2014. "Current State-of-the-Art Molecular Dynamics Methods and Applications." *Advances in Protein Chemistry and Structural Biology* 94: 269–313.

Wang, Li, Xiaodi Wang, and Qunhui Wang. 2017. "Mulberry Seeds Perform High Hypoglycaemic Effect Partially by Inhibition of α-Glucosidase Activity." *Biomedical Research (India)* 28(8): 3568–3573.

Wang, Shaomeng, et al. 1999. "Probing the Binding of Indolactam-V to Protein Kinase C through Site-Directed Mutagenesis and Computational Docking Simulations." *Journal of Medicinal Chemistry* 42(18): 3436–3446.

Wang, Xin, Ke Song, Li Li, and Lijiang Chen. 2018. "Structure-Based Drug Design Strategies and Challenges." *Current Topics in Medicinal Chemistry* 18(12): 998–1006.

Wang, Yan, et al. 2022. "A Review on Application of Molecular Simulation Technology in Food Molecules Interaction." *Current Research in Food Science* 5(October): 1873–1881. https://doi.org/10.1016/j.crfs.2022.10.012.

Wisniewski, H. M., et al. 1989. "Spectrum of Morphological Appearance of Amyloid Deposits in Alzheimer's Disease." *Acta Neuropathologica* 78(4): 337–347.

Zeng, Huahui, and Xiangxiang Wu. 2016. "Alzheimer's Disease Drug Development Based on Computer-Aided Drug Design." *European Journal of Medicinal Chemistry* 121. Elsevier Ltd. http://doi.org/10.1016/j.ejmech.2015.08.039.

Zhang, Yan, Peiyao Li, Jianbo Feng, and Minghua Wu. 2016. "Dysfunction of NMDA Receptors in Alzheimer's Disease." *Neurological Sciences* 37(7): 1039–1047.

Zhao, Linlin, Heather L. Ciallella, Lauren M. Aleksunes, and Hao Zhu. 2020. "Advancing Computer-Aided Drug Discovery (CADD) by Big Data and Data-Driven Machine Learning Modelling." *Drug Discovery Today* 25(9): 1624–1638. https://doi.org/10.1016/j.drudis.2020.07.005.

5 Biomarker Analysis for Alzheimer's Disease

*Shraddha M. Jadhav and Karan Kumar**

5.1 INTRODUCTION

An individual's memory recall, thinking, language, and other vital cognitive functions are impaired by Alzheimer's disease (AD), a prevalent form of dementia in the aging population. AD is a progressive disorder significantly affecting daily activities (Humpel, 2011). In 1907, Alois Alzheimer made the initial discovery of "military bodies" and "dense bundles of fibrils" in the brain of a woman in her early 50s with a progressive dementing illness. These findings have been recognized as significant pathological features of AD, characterized by amyloid plaques and neurofibrillary tangles (NFT). The paper in which Dr Alzheimer described his findings, titled "Über eine eigenartige Erkrankung der Hirnrinde" ("On a peculiar disease of the cerebral cortex"), is widely regarded as seminal work in the field of dementia research (Alzheimer et al., 1995).

AD primarily presents as sporadic cases, and only a few result from genetic factors. Mutations in specific genes, including presenilin genes (APP, PSEN2, PSEN1), ε4 allele of apolipoprotein E gene (APOE), and chromosome 11 (amyloid precursor protein, APP), as referenced by Masters et al. (1985) and Kang et al. (1987), play a critical role in the development of AD. The future use of predictive biomarkers in AD clinical trials may minimize the variability of study subjects, improve patient outcomes, and save expenses incurred from recruiting, enrolling, and retaining participants with incipient AD who may not develop clinical AD over a lengthy course of time. Identifying early-stage AD and subjects at risk of AD is an essential but difficult task in clinical trials. Various biomarkers for AD prediction are currently under research, including those from cerebrospinal fluid (CSF), imaging, and genetics (Pozueta et al., 2013; Ridge et al., 2013). Recent findings have revealed that women are at a higher risk of AD than men, as early menopause increases their risk of developing the disease. Furthermore, these pathologies worsen as the disease progresses (Brun & Dictor, 1981; Crystal et al., 1988). The development of AD is hypothesized to be complex and multi-faceted, with concomitant and limiting risk factors and the presence of various pathological lesions that contribute to the disease's progression.

5.1.1 Pathways in AD

AD causes severe cognitive decline and is a progressively worsening disorder. The primary clinical classification of AD based on the extent of cognitive impairment is preclinical, mild, moderate, and late-stage. Posthumously examining AD patients' central nervous system (CNS) exposed several defining features, including synaptic loss, abnormal neuritic plaque accumulation, and neurofibrillary tangles. A hypothesis is that these pathological lesions progressively gather for 10–15 years before the onset of cognitive symptoms.

Neuritic plaques, small, round clusters found in the brains of those with AD, are a defining characteristic of the disease. The accumulation of amyloid β peptide (Aβ), a product of the APP, is responsible for forming these plaques. In healthy individuals, α-secretase breaks down APP, thereby preventing the production of Aβ. Conversely, individuals with AD often exhibit higher β-secretase activity, which cleaves APP leading to the release of sAPPβ (O'Brien & Wong, 2011). Working in tandem with γ-secretase, β-secretase creates the amino acid peptides Aβ40, Aβ42, and C99, with

DOI: 10.1201/9781003412069-5

Aβ42 being the primary culprit responsible for the buildup of toxic clusters within brain cells (De Strooper et al., 2010). These clusters lodge in blood vessels and grey matter, forming the unique neuritic plaques associated with AD.

Meanwhile, NFTs comprise twisted filamentous structures that form within nerve cells composed of a protein known as tau. In normal circumstances, tau stabilizes microtubules within neurons' axons and has a specific number of phosphate molecules attached (Alonso et al., 1994). However, the impact of tau phosphorylation on the formation of neurofibrillary tangles in AD is yet to be fully understood. Nevertheless, in AD, the hyperphosphorylation of tau proteins tends to make them detach from microtubules and clump together, forming the twisted filamentous structures known as paired helical filaments. Over time, these structures form neurofibrillary tangles within the brains of individuals with AD (Alonso et al., 1994; Buerger, Teipel, et al., 2002).

Two approved drug categories for AD treatment are cholinesterase inhibitors, such as rivastigmine, tacrine, galantamine, and donepezil, and partial N-methyl D-aspartate (NMDA) antagonists, such as memantine (Molinuevo et al., 2011). However, these drugs only temporarily relieve symptoms, as the disease can reoccur with the discontinuation of medications. Moreover, as the severity and type of symptoms vary significantly among individuals, these treatments may not be universally effective. Recent developments in AD treatment include new strategies, such as β- or γ-secretase inhibitors, Aβ immunotherapies, and Aβ aggregation inhibitors. Currently, the diagnostic criteria for AD rely on recognizing clinical symptoms that may not manifest until the disease has progressed significantly, as per research findings (Hussein et al., 2018). As a result, individuals in the prodromal (mild cognitive impairment, MCI) or preclinical stage may go undetected. Hence, there is an increasing demand for tools capable of detecting patients in the early stages.

Importantly, it is worth noting that lifestyle interventions, such as healthy eating, physical exercise, mental exercises, and social engagement, play a significant role in AD management by reducing the risk and slowing progression (Gómez-Pinilla, 2008). These interventions serve as an additional option for patients with Alzheimer's. They may work well when used alongside approved drug categories or by persons who have not yet started showing signs of Alzheimer's. Hence, there is an emerging need to explore lifestyle interventions as an additional option for dealing with the early stage of AD symptoms.

5.1.1.1 What Is a Biomarker?

Biomarkers have emerged as a potential solution to detect AD, particularly before the onset of clinical symptoms (Henley et al., 2005). They offer an objective means of measuring biological or pathological processes, such as changes in concentration or activity, compared with the mere identification of clinical symptoms that may appear only when the disease has already progressed significantly (Hussein et al., 2018; Blennow et al., 2010; Frey et al., 2005; Henley et al., 2005). The biomarkers for AD is divided into 5 main categories based on the diverse nature of the samples: biochemical, neuroanatomical, metabolic, genetic, and neuropsychological biomarkers. Despite the availability of different biomarkers for AD, developing an ideal one still faces several obstacles. Obstacles in diagnosing AD include uncertainty about the disease's progression and response to treatment and an incomplete understanding of the underlying disease process. Additionally, there is a lack of available histopathological diagnoses during an individual's lifetime. It is often challenging to differentiate AD from other forms of dementia like vascular dementia, MCI, and frontotemporal lobe dementia (Blennow et al., 2010; Frey et al., 2005; Geda, 2012; Mayeux & Schupf, 2011).

This has led to researchers investigating the potential value of biomarkers such as CSF, blood, and urine biomarkers, retinal and lens amyloid-beta accumulation, and motoric cognitive risk syndrome (Ayers & Verghese, 2014; Lönneborg, 2008; Marksteiner et al., 2011; Sweeney et al., 2015). While these biomarkers show promising potential for early detection and intervention, extensive longitudinal studies with biomarkers and pathological validation are still required to evaluate their accuracy and effectiveness, particularly for mixed dementia and non-Alzheimer's dementias (Zhu et al., 2015; Ayers & Verghese, 2014). Despite the challenges, researchers persist in developing an ideal biomarker for AD to detect the disease early, differentiate it from other forms of dementia or cognitive impairment, improve follow-up care, and facilitate disease-modifying treatment options.

5.1.2 Biomarkers Based on Multiple Stages of AD

Stage 1: Individuals without functional impairment but with possible cognitive impairment are generally assigned to the first AD stage. Individuals must undergo testing with sensitive neuropsychological instruments to detect any cognitive impairment. Biomarkers such as amyloid imaging and reduced CSF Aβ42 concentration could indicate preclinical pathologic changes, especially during the early stages of AD. That is, early detection and monitoring of CSF Aβ42 concentration and amyloid imaging could help predict AD before its onset (Sperling et al., 2011). Using biomarkers for early detection might help hasten the approval of an effective treatment for AD, leading to clinical benefits and better outcomes. Nevertheless, it requires further testing to validate clinical measurements and the treatment's efficiency.

Stage 2: In stage 2 of AD, individuals may show mild cognitive changes without noticeable functional deficits. However, applying sensitive instruments and biomarkers can aid in predicting pathophysiological changes related to AD (Sperling et al., 2011). When paired with biomarker changes, sensitive neuropsychological testing could aid in diagnosing stage 2 AD. As the disease progresses in the later phases of stage 2, neuronal dysfunction is known to initiate (Counts et al., 2017).

Stage 3 is associated with mild cognitive impairment as the patient has trouble performing daily tasks. Pathophysiological biomarkers are also present during this stage, while the first two are preclinical. Cortical thinning and hippocampal atrophy are seen and validated with sMRI (Counts et al., 2017).

Stage 4: The cognitive impairment worsens along with the increase in pathophysiological markers resulting in severe AD (Table 5.1).

TABLE 5.1
Types of Biomarkers and Tests to Detect Them

Biomarkers category	Tests to identify biomarkers	References
Biochemical	CSF (cell count, serum albumin ratio, and neurogranin) Blood-based (NFL)	(Blennow et al., 2010; Formichi et al., 2006; Lascelles, 1981; Li et al., 2007)
Metabolic	PET scan SPECT scan	Kaddurah-Daouk et al., 2011; Kiehntopf et al., 2013; Peña-Bautista et al., 2019
Neuroanatomical	CT scan MRI scan	Apostolova & Thompson, 2008; Desikan et al., 2009; Querbes et al., 2009
Neuropsychological	Episodic memory	(Albert et al., 2001; Small et al., 2000; Twamley et al., 2006)
Genetic	APP PSEN1 PSEN2 APOE4	(Barber, 2012; Bellenguez et al., 2022; García-Ayllón et al., 2013)

5.2 BIOCHEMICAL MARKERS

5.2.1 CSF Biomarkers

The CSF is a vital fluid that functions as a cushioning agent by filling the ventricular system and subarachnoid space surrounding the brain. It reflects the biochemical and molecular alterations within the brain parenchyma (Lascelles, 1981). The fusion of proteins or peptides in brain-specific activities into the CSF makes it an ideal biomarker for potential AD diagnosis. Earlier research on AD was limited to exploring the role of amyloid plaques, phosphorylated tau proteins, and intraneuronal neurofibrillary tangles. However, recent insights into the role of mitochondrial dysfunction and oxidative stress have broadened the scope of biomarker evaluation (Lascelles, 1981; Perrin et al., 2009).

The primary CSF biomarkers are serum albumin ratio, cell counts, and intrathecal immunoglobulin production. Insightful information on the status of the inflammation in the brain and blood–brain barrier (BBB) is provided by CSF biomarkers, and this information can assist in ruling out other conditions that might mimic AD (García-Ayllón et al., 2013).

The amyloidogenic pathway in AD generates insoluble Aβ peptides from APP, one of AD's most frequently studied biomarkers. Biomarkers for AD usually investigate amyloid plaques, axonal degradation, and neurofibrillary tangles (Wang et al., 2017). Studies have found that CSF $A\beta_{1-42}$ inversely correlates with amyloid plaque deposition in the brain. Furthermore, the CSF $A\beta_{1-42}/A\beta_{1-40}$ ratio has demonstrated greater precision in predicting cortical amyloid deposition when compared to CSF $A\beta_{1-42}$ (McKhann et al., 2011; Thijssen et al., 2020), allowing the differentiation between AD and other forms of dementia.

Generally, levels of Aβ differ based on disease; APP can be cleaved by either the nonamyloidogenic or amyloidogenic pathway. The nonamyloidogenic pathway prevents the formation of Aβ peptides, while the amyloidogenic pathway produces insoluble Aβ peptides. The most studied CSF biomarkers for AD are Aβ1–42, total tau (t-tau), and phosphorylated tau (p-tau). Tau is a significant constituent of neurofibrillary tangles. When hyperphosphorylated at sites such as 199, 231, 235, 396, and 404, tau experiences a loss of function and impairs axonal transport (Cedazo-Minguez & Winblad, 2010). Researchers have also observed that phospho-tau-231 and phospho-tau-181 are biomarkers that differentiate AD from other forms of dementia or major depressive disorders (Hampel et al., 2010).

It has also been suggested that a combination of CSF markers is more successful in diagnosis. For instance, high CSF concentrations of t-tau and low concentrations of Aβ42 can detect AD with greater specificity (Maddalena et al., 2003). Furthermore, the combination of CSF t-tau, p-tau, Aβ42, and the t-tau/Aβ42 ratio also assists in distinguishing AD from other types of dementia (Li et al., 2007).

AD pathophysiology involves the release of inflammatory intermediates. Therefore, neuroinflammatory markers such as TREM2, MCP-1, and YKL-40 are currently extensively investigated for their diagnostic potential in AD detection (Brosseron et al., 2023; Del Prete et al., 2020; Liu et al., 2022). Notably, GSK-3 is one of the vital tau kinases that phosphorylate the tau in neurofibrillary tangles, rendering it a potential therapeutic target for AD treatment (Blesso et al., 2022).

These several proteins involved in the cascade of inflammatory mechanisms can be detected using ELISA and other proteomics, genomics, and transcriptomics approaches. Antichymotrypsin is increased in CSF samples with AD (Pirttila et al., 1994). The presence of cytokines such as CSF interleukin-6 (IL-6), Gp130, and tumour necrosis factor (TNF-α) is also observed during the last few years while elucidating pathways of inflammation (Blum-Degena et al., 1995; Martínez et al., 2000; Yamada et al., 1995).

Despite the high sensitivity and specificity of CSF biomarkers in detecting clinical AD, the use of such biomarkers has both strengths and limitations worth considering: (i) CSF markers have high accuracy in diagnosis as they are in close relationship with the brain; (ii) high concentration of biomarkers; (iii) it can validate many individuals' pathophysiological markers. On the contrary, the limitations are: (i) It is invasive; (ii) it is less accepted by the population to be examined due to anxiety, fear, etc.; (iii) it is generally employed in the later stages of AD after using the blood biomarkers; and (iv) screening of patients and follow-up can take several years using CSF.

5.2.2 Peripheral Blood Biomarkers (PBB)

These are also called plasma biomarkers. Most of the time, diagnosing AD using CSF has several drawbacks, as a lumbar puncture for CSF collection is invasive. One potential early indicator of AD is elevated blood pressure, which has been linked to several disease markers, including increased senile plaques, hippocampal atrophy, and hypertension (Skoog & Gustafson, 2006). PBBs are the most convenient and least invasive for early AD detection, with Aβ being the most thoroughly researched (Blennow et al., 2010; Hampel et al., 2018).

One approach to diagnosing AD involves screening patients and conducting follow-up tests over several years using CSF. However, cross-sectional studies examining the relationship between plasma phospholipid-bound Aβ (PBB) and AD have produced varying and inconclusive results. While some studies indicate higher levels of Aβ1-40 or Aβ1-42, others demonstrate no significant difference. Therefore, relying solely on plasma biomarkers for AD diagnosis may not be accurate, and diagnosis should involve a combination of multiple factors rather than just one. These markers might also be valuable for early AD detection (Humpel, 2011; Janelidze et al., 2016).

These studies also show that the significant source of Aβ is peripheral tissue. The serum neurofilament protein also correlates with AD (Preische et al., 2019). Blood-based elevated tau levels are also investigated using ELISA. It is the most reliable method for measuring total tau, β-amyloid (1–42), and phosphorylated tau in CSF, with high selectivity, specificity, and sensitivity (Mattsson et al., 2016; Tatebe et al., 2017). Advanced techniques such as surface-enhanced laser desorption/ionization-time-of-flight-mass spectrometry can potentially enhance sensitivity and specificity in AD diagnosis. These modern tools could be helpful for improving the accuracy of AD diagnosis (Xiao et al., 2005).

Other factors such as BDNF, clusterin levels, and extracellular RNA (exRNA) have also been observed to increase in the late onset of AD. Brain-derived neurotrophic factor (BDNF) mainly shows altered functions in the pathogenesis of neurodegenerative diseases like AD (Da Mesquita et al., 2018; Gupta et al., 2016; Pláteník et al., 2014). Ongoing research has also identified complement factor H, alpha-1-antitrypsin, apolipoprotein J, alpha-2-macroglobulin, and apolipoprotein A-I as promising candidates for biomarkers in AD that warrant further exploration (Hampel et al., 2018).

There is promising research indicating that microRNAs (miRNAs) might have the potential to help diagnose AD. These small RNA molecules play a vital role in regulating proteins that relate to AD. BACE-1 and APP are 2 proteins regulated by miRNAs, and when these molecules are downregulated, it might indicate the presence of AD. Scientists have therefore focused on specific miRNAs, such as miR-1306-5p, miR-15b-3p, miR-15b, miR-223, and miR-451a, which are all present in low concentrations within the blood and CSF of people with AD and other types of dementia. By detecting altered levels of these miRNAs in blood or CSF, researchers may have the tools to diagnose AD in its early stages (Chen et al., 2018; Cheng et al., 2015; Wang & Zhang, 2020). Remarkably, miR-31, miR-146a, and miR-206 are involved in inflammation, cell apoptosis, and cognitive decline (Dong et al., 2015; Kenny et al., 2019).

On the other hand, extracellular vesicles that can cross the blood–brain barrier are involved in cell-to-cell communication (Musunuri et al., 2016). These EVs consist of exosomes, microvesicles, and exRNAs. These are very specific to the tissue of their origin, which makes them a potential candidate for prognosis. These BEVs include synaptic proteins such as neurogranin, synaptophysin, and synaptopodin (Goetzl et al., 2016). Moreover, cerebrovascular-derived EVs are spread in AD; therefore, more research is required in this field (Badhwar & Haqqani, 2020; Murillo et al., 2019).

Insulin levels can also be considered an essential biomarker in AD as the pathways are deregulated. Precisely, an increase in phospho-Ser312-insulin receptor substrate-1 (IRS-1) and a

TABLE 5.2
Various Biochemical Biomarkers and Their Levels in AD

Biochemical Biomarkers (CSF and Plasma)	Levels seen in AD	References
CSF BACE1	Increased	(Ewers et al., 2008)
Phosphorylated-tau	Increased (Prominent marker)	Buerger, Teipel, et al., 2002; Buerger, Zinkowski, et al., 2002 Buerger, Zinkowski, et al., 2002; Frölich et al., 1991
Intrathecal immunoglobulin synthesis	Normal	(Alonso et al., 1994; Li et al., 2007)
Total-tau	Increased	(Li et al., 2007)
Truncated amyloid-β isoforms	Increased	(Davidsson et al., 1996)
Biomarkers of synaptic degeneration	Increased	(Gupta et al., 2016)
APP isoforms	Increased	(Alonso et al., 1994)
Aβ1–42	Reduced	(Klafki et al., 2022)

decrease in phospho-panTyr-IRS-1 levels are reported in blood. Therefore, IRS-1 could also be an AD biomarker (Kapogiannis et al., 2015). The advantages of blood biomarkers are as follows: (i) Noninvasive, fast, convenient, and cost-effective; (ii) simple to measure, and no prior clinician training is required; (iii) can be used for large populations and performed in various settings; and (iv) it is a first step diagnostic process (Table 5.2).

5.3 METABOLIC MARKERS

Positron emission tomography (PET) is a molecular imaging technique for evaluating metabolic activity in various brain regions. Molecular imaging techniques such as PET or SPECT entail measuring the concentration and blood flow between the arterial blood and brain tissue. This allows for determining cerebral blood flow using kinetic analysis models. Additionally, cerebral glucose metabolism can be indirectly measured by examining synaptic activity through fluorodeoxyglucose-positron emission tomography (FDG-PET) studies. These methodologies provide insight into the metabolic changes occurring in the brain, allowing for the identification of abnormalities associated with conditions such as AD.

Consequently, a reduced glucose uptake on FDG-PET indicates impaired synaptic function (Ferrando & Damian, 2021). People with AD display a particular topographical pattern of reduced metabolism in the posterior cingulate, temporal parietal, and praecuneus regions, as seen in fluorodeoxyglucose-positron emission tomography (FDG-PET) studies. Combining hippocampal hypometabolism with decreased neocortical FDG uptake enhances AD diagnosis specificity (Mosconi et al., 2008). A similar level of specificity for diagnosis can be achieved using SPECT to study regional blood flow with Tc-hexamethyl propylene amine oxime (Yuan et al., 2009). Dopamine radiotracers are also used with SPECT to distinguish between types of dementia.

According to Kiehntopf et al. (2013), people with AD experience disruptions in 10 different metabolic pathways, including alpha-linolenic acid metabolism, glycerophospholipid metabolism, linoleic acid metabolism, citrate cycle, sphingolipid metabolism, glycerolipid metabolism, proline and arginine metabolism, glutathione metabolism, fatty acid biosynthesis, and purine metabolism. These disruptions affect the proper functioning of the body's metabolism. A detailed analysis of the metabolic pathways of individuals with AD has shown that their lipid and amino acid metabolism are substantially altered, resulting in metabolic disruptions. On the other hand, purine metabolism is found to be only mildly affected in individuals with AD.

In addition to these changes, alterations in tyrosine, tocopherol, and tryptophan pathways have been observed, along with reduced levels of norepinephrine and related metabolites. These findings suggest that metabolic dysfunctions play a vital role in the development and progression of AD and may provide potential targets for intervention and management of the disease (Kaddurah-Daouk et al., 2011). As the brain ages, β-secretase and γ-secretase become active and tend to disrupt neurotrophic metabolism, forming Aβ aggregates. In a healthy brain, APP is usually broken down by α-secretase to produce sAPPα and P3, serving crucial neurotrophic functions. However, the activation of β-secretase and γ-secretase can disturb this breakdown process, forming Aβ aggregates, leading to neuritic plaques and mitochondrial dysfunction.

The phosphorylation of tau is accelerated by metabolic disruptions in lipid rafts and Ca^{2+} balance and activates β-secretase through glycogen synthase kinase three beta and protein kinase B. The mechanisms mentioned earlier eventually lead to the development of AD, neurofibrillary tangles, and brain atrophy. Altered levels of halogen-containing compounds and glutamine, along with reduced levels of piperine, have been detected in people with AD. Choline is also a promising metabolite for AD diagnosis (Peña-Bautista et al., 2019).

Metabolic biomarkers have the potential to diagnose AD, but there are limitations to their use including a lack of specificity, sensitivity, invasiveness, variability, and cost. Various factors such as age, gender, lifestyle, and other medical conditions can affect their accuracy and specificity, making them less ideal for diagnosing AD exclusively. Metabolic biomarkers alone may prove difficult to distinguish AD from other neurological disorders. Additionally, it is not always possible to detect AD in its early stages, making it more challenging to diagnose the disease before significant damage has already occurred. Therefore, diagnosing AD based on metabolic biomarkers alone is not always a reliable approach. Additionally, some metabolic biomarker tests require invasive procedures, such as lumbar punctures or brain imaging, which can be uncomfortable or risky for some patients.

5.4 GENETIC MARKERS

Much research has been conducted on genetic markers in AD over the last few years, which has led to identifying several genes that could be linked to the disease. The previous decade's research on AD genetics has revealed a few AD susceptibility genes, including APP, PS1, and PS2, which cause early-onset AD, as well as ApoE, a gene that has been found to increase the risk of late-onset AD. The genetic factors associated with AD are remarkably complex, and researchers must explore many pathways to understand them better. AD's heterogeneity arises from various mutations and polymorphisms in different genes and environmental factors. Late-onset AD is linked with common polymorphisms with low penetrance. However, there is much more to be explored in this regard.

In contrast, while less common, Autosomal-dominant mutations are responsible for early-onset familial AD and have high penetrance. Two gene categories affect an individual's chances of developing AD: (1) Risk genes and (2) deterministic genes. Whereas deterministic genes guarantee the development of a disease, risk genes only increase the possibility of its occurrence. Researchers have identified several genes that can raise the likelihood of developing AD, with APOE-e4 being the most influential and the first risk gene to be discovered. Studies indicate that 40–65% of individuals diagnosed with Alzheimer's possess the APOE-e4 gene. *APOE* has three types: APOE-e2, APOE-e4, and APOE-e3, with APOE-e4 being one of the most prevalent risk factors for AD (Lendon et al., 1997).

Deterministic genes directly cause a disease; individuals who inherit them will undoubtedly develop the disorder. However, deterministic genes only account for a small percentage of AD cases (Gerrish et al., 2014). The 3 genes that exhibit 100% penetrance with autosomal dominant inheritance patterns are APP, presenilin 1 (PSEN1), and presenilin 2 (PSEN2). Among these genes, PSEN1 has more than 120 discovered mutations. In contrast, only 8 missense mutations have been identified in PSEN2 (Schellenberg, 1995).

TABLE 5.3
Genes Involved in Familial and Sporadic AD

Name of gene	Familial/ Sporadic	Chromosomal location	References
APP	F	21q21.3—q22.05	(Grangeon et al., 2021; Yurov et al., 2019)
PS2	F	1q31—q42	(Salari et al., 2022; Soto-Ospina et al., 2021)
PS1	F	14q24.3	(Vöglein et al., 2019)
APOE	S, F	19q32.2	(Ewers et al., 2008)
ACT	S	14q32.1	(Kamboh, 2004)
5-HTT	S	17q11.1—q12	(Oliveira et al., 1998)
ACE	S	17q23	(Khurana & Goswami, 2022; Stanislav Alexandra Alina, 2020)
PSI promoter	S	14q24	(Di Resta & Ferrari, 2019; Kamboh, 2004; Van Broeckhoven et al., 1992)
catD	S	11p15.5	(Mijanovic et al., 2021; Papassotiropoulos et al., 1999)
TGF-β1	S	19q13.1—q13.3	(Bosco et al., 2013; Luedecking et al., 2000; Yang et al., 2016)
NOS3	S	7q35	(Morris et al., 2011)
CST3	S	20p11.2	(Finckh et al., 2000)

(Abbreviations—*APP*: Amyloid precursor protein gene, *PS2*: Presenilins, *PS1*: Presenilins, *APOE*: Apolipoprotein E, *ACT:* α1-Antichymotrypsin, *5-HTT*: 5-HT transporter, *ACE*: Angiotensin-converting enzyme, *PSI promoter*: Presenilin 1 promoter polymorphisms, *catD*: Cathepsin D, *TGF-β1*: Transforming growth factor-β1, *NOS3*: Nitric oxide synthase, *CST3*: Cystatin C.)

However, a small percentage of AD cases (about 5%) follow a Mendelian pattern, and only a limited number of genetic factors listed in Table 5.3 play a direct role in the disease's development. In most cases of sporadic AD, genetic factors are predisposing and increase the risk of developing the disease but they are not powerful enough to cause it on their own. Studies have found that environmental and genetic factors play a role in AD. These factors often work together synergistically to increase the risk of developing the disease (Lusis et al., 1986).

According to recent genome scans, chromosomes 9, 10, and 12 are potentially associated with additional AD loci. However, there have been no consistent results in finding genes associated with AD in these regions. Still, research has connected α2-macroglobulin (α2M), found in the bloodstream and brain, to AD due to its ability to eliminate and break down amyloid-beta (Aβ), a protein commonly found in senile plaques, which are a hallmark of AD. The presence of amyloid-beta, another hallmark of AD, is also observed in senile plaques.

Genetic linkage identified the α2M gene as a potential disease locus for AD in families with late-onset AD. This gene is located on the short arm of chromosome 12p and spans 30 cM at the telomeric end of the region (Shibata et al., 1999). While researchers have identified several genetic markers, finding genes associated with AD in these regions has proven challenging, with unpredictable results in analysing positional candidate genes. Through research, these genetic markers have given insights into the underlying mechanisms of AD and provided opportunities to develop new treatment strategies.

5.5 NEUROPSYCHOLOGICAL MARKERS

A limited number of neuropsychological markers are available for assessing cognitive function in AD patients. Standard neuropsychological measures used to assess cognitive performance in individuals with AD include the Mini-Mental State Examination (MMSE), Montreal Cognitive Assessment, AD Assessment Scale-Cognitive, clinical dementia rating (CDR), clock drawing test (CDT), and verbal fluency test. These tests provide valuable insights into the cognitive domain of AD patients and help monitor AD progression.

TABLE 5.4
Neurophysiological Tests and Their Significance in Clinical Trials

Neurophysiological tests	Advantages	Importance in Clinical Trial	References
Montreal Cognitive Assessment	evaluates several cognitive domains, including attention, memory, language, visuospatial abilities, and executive functioning	Cognitive impairment	(Albert et al., 2001; Pinto et al., 2019)
Clock Drawing	assesses visuospatial and executive functioning	In combination with MMSE in dementia screening	(Albert et al., 2001; Rentz et al., 2021; Sato et al., 2022; Twamley et al., 2006)
AD Assessment Scale	evaluates cognitive domains such as memory, language, praxis, and orientation	helps detect AD in its early stages	(Doraiswamy et al., 1997; Podhorna et al., 2016)
Dementia Rating Scale	evaluates cognitive domains such as memory, orientation, and judgment	detects MCI	(Qian et al., 2021)
MMSE	assesses several cognitive domains, including orientation, attention, memory, language, and visuospatial abilities	allows differentiation of the Lewy bodies	(Arevalo-Rodriguez et al., 2015; Chapman et al., 2016)

The neuropsychological measures mentioned earlier generally evaluate different cognitive domains, including visuospatial abilities, memory, language, orientation, and judgment. These measures are designed to evaluate a range of cognitive domains, such as memory, attention, language, and executive functioning. Various studies have shown that asymptomatic older adults perform poorly on neuropsychological measures. The results indicate that minor cognitive deficits detected during neuropsychological assessments could be a valuable indicator for identifying individuals at risk of developing AD early on.

In the early stages of AD, episodic memory loss is a prominent and meaningful feature that occurs several years before apparent changes in cognitive and behavioural functions appear. Studies have shown that the medial temporal lobe (MTL) and hippocampal formation, which play a vital role in episodic memory, are among the first brain structures affected in individuals with AD. Many results support that cognitive decline is due to the loss of neurotransmitters such as acetylcholine in the cortex or hippocampal regions. The nerve growth factor (NGF) is considered during neurogenesis studies. It has been seen that the application of NGF can avoid the degeneration of cholinergic neurons. As in AD, levels of NGF are elevated. However, the NGF cannot be considered a significant biomarker for such low-scale results (Table 5.4).

5.6 NEUROANATOMICAL MARKERS

In AD, many anatomical changes in the brain are generally seen. The locus coeruleus (LC) is a brainstem component with much neuromelanin which is believed to regulate memory and attention, and it is the primary origin of noradrenaline in the brain (Andrés-Benito et al., 2017). During the early stages of AD, when there are no symptoms, tau NFTs are visible in the reduced area LC before they can be visualized in the entorhinal cortex and neocortex (Ehrenberg et al., 2017; Stratmann et al., 2016).

Therefore, structural imaging in cognitive neuroscience research significantly predicts AD development in typically aging individuals and monitors disease progression. Neurons facilitate communication through electrical impulses, and EEG detects this electrical activity by placing small electrodes on the scalp and showing the impulses in waves (Zhang et al., 2021). These EEG waves are potent biomarkers as they can show distinct deviation from normal individuals (Vecchio et al., 2020).

Pathological changes leading to neuronal and synaptic loss can also be detected by imaging techniques such as CT and MRI, which offer good spatial and contrast resolution. MRI has advantages including better angulation, absence of bone hardening artifacts, and identification of small lesions, especially lacunes and white matter lesions, as reported by Menéndez-González (2014). AD-related brain changes are known to occur earliest in the MTL, particularly in the hippocampus, which plays a crucial role in memory function (Juottonen et al., 1999). These techniques enable brain volume and cortical thickness studies in various brain areas (Bottino et al., 2002). Research has proved that VBM allows faster and more reliable imaging and early-onset AD detection. Recent reports indicate that patterns of atrophy initially emerge in the MTL before extending to the anterior and posterior regions of the brain. This propagation pattern matches neurofibrillary tangles (Xiao et al., 2005).

Research indicates that volumetric studies successfully differentiate between AD patients and normal elderly individuals by measuring the hippocampal and entorhinal volumes. The difference in entorhinal cortex volumes between individuals with mild cognitive impairment (MCI) and normal elderly individuals is between 12–17%. As the disease advances to mild-to-moderate AD, the difference increases to 30–40%. Recent research has integrated volumetric measures with amyloid biomarkers sensitive to pre-symptomatic AD pathology, revealing abnormal degeneration in the basal forebrain before spreading to the cortex (Du et al., 2001). Optical coherence tomography angiography (OCTA) is a technique that can be used to study microvascular changes in the eye retina, which has emerged as a potential biomarker for AD (Izzo et al., 2021). Facilitating the identification of pre-symptomatic biomarkers in individuals and tracking the progression of AD, techniques such as OCTA help ensure early detection and treatment of the disease (Dubois et al., 2007).

5.7 ROLES OF BIOMARKERS IN CLINICAL TRIALS AND DRUG DEVELOPMENT

In AD clinical trials, inclusion criteria can be met through the use of specific biomarkers such as positive CSF biomarkers, structural imaging demonstrating MTL atrophy, reduction of glucose metabolism in bilateral temporal and parietal regions, increased Aβ ligand binding, or genetic mutations that cause familial AD in individuals with MCI. A combination of biomarkers can potentially enhance the identification of the advantages of drugs and potentially increase the number of study subjects with the underlying pathology of AD. Clinical trials are evaluating several AD treatment strategies, including monoclonal antibodies targeting Aβ, small molecule inhibitors targeting β-APP processing, and various multimodal and lifestyle-based interventions. None of these treatment approaches have made significant progress in their therapeutic success. However, there is still hope for using innovative biomarkers and novel treatment strategies to improve AD management and provide revised therapeutic protocols in the future.

Bard et al. (2000) conducted a study on a group of patients with AD and showed that a particular pattern of 18 inflammatory and plasma signalling proteins was unique to the disease. This proteomic approach has provided insights into developing new treatment strategies for AD. Additionally, active and passive amyloid immunization strategies have been reported to reduce cognitive impairment and alleviate memory issues. The first anti-amyloid vaccine validated in clinical trials was AN1792, as Bard et al. (2000) reported.

Alongside the active techniques, passive amyloid immunization strategies involve bapineuzumab. This humanized monoclonal antibody targets the N-terminus of Aβ. Mice treated with these antibodies exhibited reduced cortical amyloid plaque formation (Legleiter et al., 2004; Lemere, 2009). Solanezumab functions as a "peripheral sink," which involves removing amyloid from the brain (Farlow et al., 2012). Administering solanezumab peripherally to mice has been shown to cause a rapid increase in plasma Aβ of 1,000 times while also causing a significant decrease in plaque accumulation (DeMattos et al., 2001; Racke et al., 2005).

The diverse nature of AD implies that some patients might respond better to specific treatments than others. For example, patients with low CSF Aβ1–42 and high Pittsburgh compound B retention may experience a more positive response to drugs that modify amyloidogenesis. Simultaneously, individuals with elevated levels of β-site amyloid precursor protein cleaving enzyme 1 may respond more positively to β-secretase inhibitors. Given that multiple factors cause sporadic AD, a multifaceted approach to therapy is crucial, addressing genetic and environmental factors that increase the risk of AD. Biomarkers can aid in patient stratification in various drug trials and in the more precise observation of drug efficacy.

5.8 LIMITATIONS OF BIOMARKERS

Creating reliable biomarkers to track and diagnose AD progression is a crucial and formidable undertaking as the number of individuals affected by the disease continues to rise. Despite substantial advancements in AD research, including the use of biochemical, imaging, and genetic assessments for early detection, it is imperative to understand their limitations, which include:

i. Variability: Variability is a common issue in biomarker analysis that can result from technical and biological factors such as measurement error and disease heterogeneity. For instance, biomarkers such as amyloid beta or phosphor-tau proteins in CSF or plasma can differ due to age, sex, and genetic makeup, making it challenging to determine a definitive threshold for pathological values and monitor changes over time. Moreover, biomarker variability is affected by the severity of the disease. In early stages, biomarkers are less sensitive and specific, while in later stages, they are more precise and highly significant. Additionally, proper biomarker standardization is necessary to measure and interpret the data. This can help to compare the data obtained from different laboratories and establish biomarker-based diagnostic criteria. Recent research has suggested the development of standardized protocols to overcome this factor.

ii. Reliability: Reliability refers to the degree to which a biomarker accurately reflects the fundamental biological mechanism that it is designed to evaluate. Several biomarkers such as functional MRI, PET imaging, and tau protein and amyloid beta levels are used to trace AD progression. However, these factors are unreliable to a small extent as they hold limitations in their use. PET imaging can determine levels of amyloid beta but cannot distinguish between its forms. Similarly, inflammation and other neurological conditions can impact the reliability of CSF biomarkers and lead to false positive or adverse reactions.

iii. Bias and measurement inaccuracy: Bias refers to a systematic error that occurs when there is a deviation from the proper relationship between the biomarker and the disease. There are two types of bias: nondifferential and differential.

   Nondifferential bias occurs when biases occur without regard to the outcome, favouring the null hypothesis of no association.

   Differential bias occurs when there are differences in the availability of the biomarker or the specimen acquisition, storage, measurement, or ascertainment procedures between those with the disease or outcome of interest and those without. To minimize bias in biomarker studies, controls should be adequately maintained, and objective review boards should review and monitor the conduct of the study. Additionally, appropriate training of personnel who handle specimens, implementation of quality assurance and quality-control programs, and a well-organized procedures manual can help minimize errors in collecting, storing, and measuring biomarkers. The biomarkers may alter due to external as well as internal factors. Therefore, biological stability is essential.

iv. Cost: Cost is crucial when the experiments are for smaller clinical trials or more extensive epidemiologic studies. It is essential to ensure they are readily available and feasible for inclusion in the study, as larger sample sizes may reduce costs per subject.

Despite the availability of biomarkers for research purposes, their use in clinical practice remains limited. Any variation in sample collection, storage, and transportation procedures can significantly impact test results. Therefore, it is crucial to strictly adhere to these procedures' guidelines to ensure the results' accuracy. The lack of peripheral blood-based markers with high sensitivity and specificity for AD detection in patients has been a significant challenge concerning efficaciousness, invasiveness, and cost-effectiveness. As a result, identifying and validating biomarkers that can be those resources for better clinical management soon becomes the priority of research. Moreover, a multidisciplinary approach combining genetic analysis, brain imaging, neuropsychological, eye tests, and blood biomarker analysis could provide a comprehensive picture of preclinical AD. Therefore, future studies must prioritize filling this knowledge gap with accurate and reliable peripheral blood-based biomarkers.

5.9 CONCLUSION

The nature of AD is complex and has several unanswered questions, as clinical trials have been unsuccessful in treating it. Understanding aging biology and neurodegeneration is essential for improving AD treatment and identify standard ranges for biomarker levels. This research helps identify the clinical applications for early detection and intervention. AD patient stratification is critical to personalized and precision medicine for optimizing disease prevention and pharmaceutical treatment and minimizing adverse effects. It is also essential to recognize that AD diagnosis may not be appropriate, and the differential diagnosis of the disease is essential.

ENDNOTE

* Corresponding author

REFERENCES

Albert, M. S., Moss, M. B., Tanzi, R., & Jones, K. (2001). Preclinical prediction of AD using neuropsychological tests. *Journal of the International Neuropsychological Society*, *7*(5), 631–639. https://doi.org/10.1017/S1355617701755105

Alonso, A. C., Zaidi, T., Grundke-Iqbal, I., & Iqbal, K. (1994). Role of abnormally phosphorylated tau in the breakdown of microtubules in Alzheimer disease. *Proceedings of the National Academy of Sciences of the United States of America*, *91*(12), 5562–5566.

Alzheimer, A., Stelzmann, R. A., Schnitzlein, H. N., & Murtagh, F. R. (1995). An English translation of Alzheimer's 1907 paper, "Uber eine eigenartige Erkankung der Hirnrinde". *Clinical Anatomy (New York, N.Y.)*, *8*(6), 429–431. https://doi.org/10.1002/ca.980080612

Andrés-Benito, P., Fernández-Dueñas, V., Carmona, M., Escobar, L. A., Torrejón-Escribano, B., Aso, E., Ciruela, F., & Ferrer, I. (2017). Locus coeruleus at asymptomatic early and middle Braak stages of neurofibrillary tangle pathology. *Neuropathology and Applied Neurobiology*, *43*(5), 373–392. https://doi.org/10.1111/nan.12386

Apostolova, L. G., & Thompson, P. M. (2008). Mapping progressive brain structural changes in early Alzheimer's disease and mild cognitive impairment. *Neuropsychologia*, *46*(6), 1597–1612. https://doi.org/10.1016/j.neuropsychologia.2007.10.026

Arevalo-Rodriguez, I., Smailagic, N., Roqué I. Figuls, M., Ciapponi, A., Sanchez-Perez, E., Giannakou, A., Pedraza, O. L., Bonfill Cosp, X., & Cullum, S. (2015). Mini-mental state examination (MMSE) for the detection of Alzheimer's disease and other dementias in people with mild cognitive impairment (MCI). *The Cochrane Database of Systematic Reviews*, *2015*(3), CD010783. https://doi.org/10.1002/14651858.CD010783.pub2

Ayers, E., & Verghese, J. (2014). Diagnosing motoric cognitive risk syndrome to predict progression to dementia. *Neurodegenerative Disease Management*, *4*(5), 339–342. https://doi.org/10.2217/nmt.14.39

Badhwar, A., & Haqqani, A. S. (2020). Biomarker potential of brain-secreted extracellular vesicles in blood in Alzheimer's disease. *Alzheimer's & Dementia (Amsterdam, Netherlands)*, *12*(1), e12001. https://doi.org/10.1002/dad2.12001

Barber, R. C. (2012). The genetics of Alzheimer's disease. *Scientifica*, *2012*, 246210. https://doi.org/10.6064/2012/246210

Bard, F., Cannon, C., Barbour, R., Burke, R.-L., Games, D., Grajeda, H., Guido, T., Hu, K., Huang, J., Johnson-Wood, K., Khan, K., Kholodenko, D., Lee, M., Lieberburg, I., Motter, R., Nguyen, M., Soriano, F., Vasquez, N., Weiss, K., . . . & Yednock, T. (2000). Peripherally administered antibodies against amyloid β-peptide enter the central nervous system and reduce pathology in a mouse model of Alzheimer disease. *Nature Medicine*, *6*(8), Article 8. https://doi.org/10.1038/78682

Bellenguez, C., Küçükali, F., Jansen, I. E., Kleineidam, L., Moreno-Grau, S., Amin, N., Naj, A. C., Campos-Martin, R., Grenier-Boley, B., Andrade, V., Holmans, P. A., Boland, A., Damotte, V., van der Lee, S. J., Costa, M. R., Kuulasmaa, T., Yang, Q., de Rojas, I., Bis, J. C., . . . & Lambert, J.-C. (2022). New insights into the genetic etiology of Alzheimer's disease and related dementias. *Nature Genetics*, *54*(4), Article 4. https://doi.org/10.1038/s41588-022-01024-z

Blennow, K., Hampel, H., Weiner, M., & Zetterberg, H. (2010). Cerebrospinal fluid and plasma biomarkers in Alzheimer disease. *Nature Reviews Neurology*, *6*(3), Article 3. https://doi.org/10.1038/nrneurol.2010.4

Blesso, K., Sharma, V. S., & Shetty, M. V. (2022). Insilico bimolecular interaction studies of natural compounds for treatment of Alzheimers disease. *International Journal for Research in Applied Science and Engineering Technology*, *10*(9). https://doi.org/10.22214/ijraset.2022.46599

Blum-Degena, D., Müller, T., Kuhn, W., Gerlach, M., Przuntek, H., & Riederer, P. (1995). Interleukin-1 beta and interleukin-6 are elevated in the cerebrospinal fluid of Alzheimer's and de novo Parkinson's disease patients. *Neuroscience Letters*, *202*(1–2), 17–20. https://doi.org/10.1016/0304-3940(95)12192-7

Bosco, P., Ferri, R., Grazia Salluzzo, M., Castellano, S., Signorelli, M., Nicoletti, F., Di Nuovo, S., Drago, F., & Caraci, F. (2013). Role of the transforming-growth-factor-β1 gene in late-onset Alzheimer's disease: Implications for the treatment. *Current Genomics*, *14*(2), 147–156.

Bottino, C. M., Castro, C. C., Gomes, R. L., Buchpiguel, C. A., Marchetti, R. L., & Neto, M. R. (2002). Volumetric MRI measurements can differentiate Alzheimer's disease, mild cognitive impairment, and normal aging. *International Psychogeriatrics*, *14*(1), 59–72. https://doi.org/10.1017/s1041610202008281

Brosseron, F., Maass, A., Kleineidam, L., Ravichandran, K. A., Kolbe, C. C., Wolfsgruber, S., Santarelli, F., Häsler, L. M., McManus, R., Ising, C., Röske, S., Peters, O., Cosma, N. C., Schneider, L. S., Wang, X., Priller, J., Spruth, E. J., Altenstein, S., Schneider, A., Fliessbach, K., . . . & DELCODE Study Group (2023). Serum IL-6, sAXL, and YKL-40 as systemic correlates of reduced brain structure and function in Alzheimer's disease: Results from the DELCODE study. *Alzheimer's Research & Therapy*, *15*(1), 13. https://doi.org/10.1186/s13195-022-01118-0

Brun, A., & Dictor, M. (1981). Senile plaques and tangles in dialysis dementia. *Acta Pathologica Microbiologica Scandinavica Section A Pathology*, *89A*(1–6), 193–198. https://doi.org/10.1111/j.1699-0463.1981.tb00208.x

Buerger, K., Teipel, S. J., Zinkowski, R., Blennow, K., Arai, H., Engel, R., Hofmann-Kiefer, K., McCulloch, C., Ptok, U., Heun, R., Andreasen, N., DeBernardis, J., Kerkman, D., Moeller, H.-J., Davies, P., & Hampel, H. (2002). CSF tau protein phosphorylated at threonine 231 correlates with cognitive decline in MCI subjects. *Neurology*, *59*(4), 627–629. https://doi.org/10.1212/WNL.59.4.627

Buerger, K., Zinkowski, R., Teipel, S. J., Tapiola, T., Arai, H., Blennow, K., Andreasen, N., Hofmann-Kiefer, K., DeBernardis, J., Kerkman, D., McCulloch, C., Kohnken, R., Padberg, F., Pirttilä, T., Schapiro, M. B., Rapoport, S. I., Möller, H.-J., Davies, P., & Hampel, H. (2002). Differential diagnosis of alzheimer disease with cerebrospinal fluid levels of tau protein phosphorylated at threonine 231. *Archives of Neurology*, *59*(8), 1267–1272. https://doi.org/10.1001/archneur.59.8.1267

Cedazo-Minguez, A., & Winblad, B. (2010). Biomarkers for Alzheimer's disease and other forms of dementia: Clinical needs, limitations and future aspects. *Experimental Gerontology*, *45*(1), 5–14. https://doi.org/10.1016/j.exger.2009.09.008

Chapman, K. R., Bing-Canar, H., Alosco, M. L., Steinberg, E. G., Martin, B., Chaisson, C., Kowall, N., Tripodis, Y., & Stern, R. A. (2016). Mini mental state examination and logical memory scores for entry into Alzheimer's disease trials. *Alzheimer's Research & Therapy*, *8*(1), 9. https://doi.org/10.1186/s13195-016-0176-z

Chen, J., Qi, Y., Liu, C. F., Lu, J. M., Shi, J., & Shi, Y. (2018). MicroRNA expression data analysis to identify key miRNAs associated with Alzheimer's disease. *Journal of Gene Medicine*, *20*(6), e3014. https://doi.org/10.1002/jgm.3014

Cheng, L., Doecke, J. D., Sharples, R. A., Villemagne, V. L., Fowler, C. J., Rembach, A., Martins, R. N., Rowe, C. C., Macaulay, S. L., Masters, C. L., Hill, A. F., & Australian Imaging, Biomarkers and Lifestyle (AIBL) Research Group (2015). Prognostic serum miRNA biomarkers associated with Alzheimer's disease shows concordance with neuropsychological and neuroimaging assessment. *Molecular Psychiatry*, *20*(10), 1188–1196. https://doi.org/10.1038/mp.2014.127

Counts, S. E., Ikonomovic, M. D., Mercado, N., Vega, I. E., & Mufson, E. J. (2017). Biomarkers for the early detection and progression of Alzheimer's disease. *Neurotherapeutics: The Journal of the American Society for Experimental NeuroTherapeutics, 14*(1), 35–53. https://doi.org/10.1007/s13311-016-0481-z

Crystal, H., Dickson, D. W., Fuld, P. A., Masur, D., Scott, R., Mehler, M. F., Masdeu, J. C., Kawas, C. H., Aronson, M., & Wolfson, L. (1988). Clinico-pathologic studies in dementia: Nondemented subjects with pathologically confirmed Alzheimer's disease. *Neurology, 38*(11). https://doi.org/10.1212/wnl.38.11.1682

Da Mesquita, S., Louveau, A., Vaccari, A., Smirnov, I., Cornelison, R. C., Kingsmore, K. M., Contarino, C., Onengut-Gumuscu, S., Farber, E., Raper, D., Viar, K. E., Powell, R. D., Baker, W., Dabhi, N., Bai, R., Cao, R., Hu, S., Rich, S. S., Munson, J. M., Lopes, M. B., . . . & Kipnis, J. (2018). Functional aspects of meningeal lymphatics in ageing and Alzheimer's disease. *Nature, 560*(7717), 185–191. https://doi.org/10.1038/s41586-018-0368-8

Davidsson, P., Jahn, R., Bergquist, J., Ekman, R., & Blennow, K. (1996). Synaptotagmin, a synaptic vesicle protein, is present in human cerebrospinal fluid. *Molecular and Chemical Neuropathology, 27*(2), 195–210. https://doi.org/10.1007/BF02815094

De Strooper, B., Vassar, R., & Golde, T. (2010). The secretases: Enzymes with therapeutic potential in Alzheimer disease. *Nature Reviews Neurology, 6*(2), 99–107. https://doi.org/10.1038/nrneurol.2009.218

Del Prete, E., Beatino, M. F., Campese, N., Giampietri, L., Siciliano, G., Ceravolo, R., & Baldacci, F. (2020). Fluid candidate biomarkers for Alzheimer's Disease: A precision medicine approach. *Journal of Personalized Medicine, 10*(4), 221. https://doi.org/10.3390/jpm10040221

DeMattos, R. B., Bales, K. R., Cummins, D. J., Dodart, J. C., Paul, S. M., & Holtzman, D. M. (2001). Peripheral anti-A beta antibody alters CNS and plasma A beta clearance and decreases brain A beta burden in a mouse model of Alzheimer's disease. *Proceedings of the National Academy of Sciences of the United States of America, 98*(15), 8850–8855. https://doi.org/10.1073/pnas.151261398

Desikan, R. S., Cabral, H. J., Hess, C. P., Dillon, W. P., Glastonbury, C. M., Weiner, M. W., Schmansky, N. J., Greve, D. N., Salat, D. H., Buckner, R. L., Fischl, B., & Alzheimer's Disease Neuroimaging Initiative (2009). Automated MRI measures identify individuals with mild cognitive impairment and Alzheimer's disease. *Brain: A Journal of Neurology, 132*(Pt 8), 2048–2057. https://doi.org/10.1093/brain/awp123

Di Resta, C., & Ferrari, M. (2019). New molecular approaches to Alzheimer's disease. *Clinical Biochemistry, 72*, 81–86. https://doi.org/10.1016/j.clinbiochem.2019.04.010

Dong, H., Li, J., Huang, L., Chen, X., Li, D., Wang, T., Hu, C., Xu, J., Zhang, C., Zen, K., Xiao, S., Yan, Q., Wang, C., & Zhang, C. Y. (2015). Serum MicroRNA profiles serve as novel biomarkers for the diagnosis of Alzheimer's disease. *Disease Markers, 2015*, 625659. https://doi.org/10.1155/2015/625659

Doraiswamy, P. M., Bieber, F., Kaiser, L., Krishnan, K. R., Reuning-Scherer, J., & Gulanski, B. (1997). The Alzheimer's disease assessment scale: Patterns and predictors of baseline cognitive performance in multicenter Alzheimer's disease trials. *Neurology, 48*(6), 1511–1517. https://doi.org/10.1212/WNL.48.6.1511

Du, A. T., Schuff, N., Amend, D., Laakso, M. P., Hsu, Y. Y., Jagust, W. J., Yaffe, K., Kramer, J. H., Reed, B., Norman, D., Chui, H. C., & Weiner, M. W. (2001). Magnetic resonance imaging of the entorhinal cortex and hippocampus in mild cognitive impairment and Alzheimer's disease. *Journal of Neurology, Neurosurgery, and Psychiatry, 71*(4), 441–447. https://doi.org/10.1136/jnnp.71.4.441

Dubois, B., Burn, D., Goetz, C., Aarsland, D., Brown, R. G., Broe, G. A., Dickson, D., Duyckaerts, C., Cummings, J., Gauthier, S., Korczyn, A., Lees, A., Levy, R., Litvan, I., Mizuno, Y., McKeith, I. G., Olanow, C. W., Poewe, W., Sampaio, C., Tolosa, E. and Emre, M. (2007), Diagnostic procedures for Parkinson's disease dementia: Recommendations from the movement disorder society task force. *Movement Disorders, 22*(16), 2314–2324. https://doi.org/10.1002/mds.21844

Ehrenberg, A. J., Nguy, A. K., Theofilas, P., Dunlop, S., Suemoto, C. K., Di Lorenzo Alho, A. T., Leite, R. P., Diehl Rodriguez, R., Mejia, M. B., Rüb, U., Farfel, J. M., de Lucena Ferretti-Rebustini, R. E., Nascimento, C. F., Nitrini, R., Pasquallucci, C. A., Jacob-Filho, W., Miller, B., Seeley, W. W., Heinsen, H. and Grinberg, L. T. (2017). Quantifying the accretion of hyperphosphorylated tau in the locus coeruleus and dorsal raphe nucleus: the pathological building blocks of early Alzheimer's disease. *Neuropathology and Applied Neurobiology, 43*(5), 393–408. https://doi.org/10.1111/nan.12387

Ewers, M., Zhong, Z., Bürger, K., Wallin, A., Blennow, K., Teipel, S. J., Shen, Y., & Hampel, H. (2008). Increased CSF-BACE 1 activity is associated with ApoE-ε4 genotype in subjects with mild cognitive impairment and Alzheimer's disease. *Brain, 131*(5), 1252–1258. https://doi.org/10.1093/brain/awn034

Farlow, M., Arnold, S. E., van Dyck, C. H., Aisen, P. S., Snider, B. J., Porsteinsson, A. P., Friedrich, S., Dean, R. A., Gonzales, C., Sethuraman, G., DeMattos, R. B., Mohs, R., Paul, S. M., & Siemers, E. R. (2012). Safety and biomarker effects of solanezumab in patients with Alzheimer's disease. *Alzheimer's & Dementia: The Journal of the Alzheimer's Association, 8*(4), 261–271. https://doi.org/10.1016/j.jalz.2011.09.224

Ferrando, R., & Damian, A. (2021). Brain SPECT as a biomarker of neurodegeneration in dementia in the era of molecular imaging: Still a valid option?. *Frontiers in Neurology, 12*, 629442. https://doi.org/10.3389/fneur.2021.629442

Finckh, U., von der Kammer H., Velden J., Michel T., Andresen B., Deng A., Zhang J., Müller-Thomsen T., Zuchowski K., Menzer G., Mann U., Papassotiropoulos A., Heun R., Zurdel J., Holst F., Benussi L., Stoppe G., Reiss J., Miserez A. R., Staehelin H. B., Rebeck G. W., Hyman B. T., Binetti G., Hock C., Growdon J. H., Nitsch R. M. Genetic association of a cystatin C gene polymorphism with late-onset Alzheimer disease. *Archives of Neurology*, 2000 Nov; *57*(11), 1579–1583. https://doi.org/10.1001/archneur.57.11.1579. PMID: 11074789.

Formichi, P., Battisti, C., Radi, E., & Federico, A. (2006). Cerebrospinal fluid tau, Aß, and phosphorylated tau protein for the diagnosis of Alzheimer's disease. *Journal of Cellular Physiology, 208*(1), 39–46. https://doi.org/10.1002/jcp.20602

Frey, H. J., Mattila, K. M., Korolainen, M. A., & Pirttilä, T. (2005). Problems associated with biological markers of Alzheimer's disease. *Neurochemical Research, 30*(12), 1501–1510. https://doi.org/10.1007/s11064-005-8827-7

Frölich, L., Kornhuber, J., Ihl, R., Fritze, J., Maurer, K., & Riederer, P. (1991). Integrity of the blood-CSF barrier in dementia of Alzheimer type: CSF/serum ratios of albumin and IgG. *European Archives of Psychiatry and Clinical Neuroscience, 240*(6), 363–366. https://doi.org/10.1007/BF02279767

García-Ayllón, M.-S., Campanari, M.-L., Brinkmalm, G., Rábano, A., Alom, J., Saura, C. A., Andreasen, N., Blennow, K., & Sáez-Valero, J. (2013). CSF Presenilin-1 complexes are increased in Alzheimer's disease. *Acta Neuropathologica Communications, 1*, 46. https://doi.org/10.1186/2051-5960-1-46

Geda Y. E. (2012). Mild cognitive impairment in older adults. *Current Psychiatry Reports, 14*(4), 320–327. https://doi.org/10.1007/s11920-012-0291-x

Gerrish, A., Sims, R., & Williams, J. (2014). Genetics and genomics of neuro-psychiatric diseases, III: The common dementias. In D. Kumar & C. Eng (eds), *Genomic Medicine: Principles and Practice* (2nd edn). Oxford Academic, New York (Online edn, 1 February 2015). https://doi.org/10.1093/med/9780199896028.003.0033

Goetzl, E. J., Kapogiannis, D., Schwartz, J. B., Lobach, I. V., Goetzl, L., Abner, E. L., Jicha, G. A., Karydas, A. M., Boxer, A., & Miller, B. L. (2016). Decreased synaptic proteins in neuronal exosomes of frontotemporal dementia and Alzheimer's disease. *FASEB Journal: Official Publication of the Federation of American Societies for Experimental Biology, 30*(12), 4141–4148. https://doi.org/10.1096/fj.201600816R

Gómez-Pinilla, F. (2008). Brain foods: The effects of nutrients on brain function. *Nature Reviews Neuroscience, 9*(7), 568–578. https://doi.org/10.1038/nrn2421

Grangeon, L., Cassinari, K., Rousseau, S., Croisile, B., Formaglio, M., Moreaud, O., Boutonnat, J., Meur, N. L., Miné, M., Coste, T., Pipiras, E., Tournier-Lasserve, E., Rovelet-Lecrux, A., Campion, D., Wallon, D., & Nicolas, G. (2021). Early-onset cerebral amyloid angiopathy and Alzheimer disease related to an APP locus triplication. *Neurology Genetics, 7*(5). https://doi.org/10.1212/NXG.0000000000000609

Gupta, V. B., Doecke, J. D., Hone, E., Pedrini, S., Laws, S. M., Thambisetty, M., Bush, A. I., Rowe, C. C., Villemagne, V. L., Ames, D., Masters, C. L., Macaulay, S. L., Rembach, A., Rainey-Smith, S. R., Martins, R. N., & Group, A. R. (2016). Plasma apolipoprotein J as a potential biomarker for Alzheimer's disease: Australian imaging, biomarkers and lifestyle study of aging. *Alzheimer's & Dementia: Diagnosis, Assessment & Disease Monitoring, 3*(1), 18–26. https://doi.org/10.1016/j.dadm.2015.12.001

Hampel, H., Blennow, K., Shaw, L. M., Hoessler, Y. C., Zetterberg, H., & Trojanowski, J. Q. (2010). Total and phosphorylated tau protein as biological markers of Alzheimer's disease. *Experimental Gerontology, 45*(1), 30–40. https://doi.org/10.1016/j.exger.2009.10.010

Hampel, H., O'Bryant, S. E., Molinuevo, J. L., Zetterberg, H., Masters, C. L., Lista, S., Kiddle, S. J., Batrla, R., & Blennow, K. (2018). Blood-based biomarkers for Alzheimer disease: Mapping the road to the clinic. *Nature Reviews Neurology, 14*(11), 639–652. https://doi.org/10.1038/s41582-018-0079-7

Henley, S. M., Bates, G. P., & Tabrizi, S. J. (2005). Biomarkers for neurodegenerative diseases. *Current Opinion in Neurology, 18*(6), 698–705. https://doi.org/10.1097/01.wco.0000186842.51129.cb

Humpel, C. (2011). Identifying and validating biomarkers for Alzheimer's disease. *Trends in Biotechnology, 29*(1), 26–32. https://doi.org/10.1016/j.tibtech.2010.09.007

Hussein, M. A., Woo, T., Yun, I. S., Park, H., & Kim, Y. O. (2018). Analysis of the correlation between deformational plagiocephaly and neurodevelopmental delay. *Journal of Plastic, Reconstructive & Aesthetic Surgery: JPRAS, 71*(1), 112–117. https://doi.org/10.1016/j.bjps.2017.08.015

Izzo, N. J., Yuede, C. M., LaBarbera, K. M., Limegrover, C. S., Rehak, C., Yurko, R., Waybright, L., Look, G., Rishton, G., Safferstein, H., Hamby, M. E., Williams, C., Sadlek, K., Edwards, H. M., Davis, C. S., Grundman, M., Schneider, L. S., DeKosky, S. T., Chelsky, D., Pike, I., Henstridge, C., Blennow, K., Zetterberg, H., LeVine III, H., Spires-Jones, T. L., Cirrito, J. R. & Catalano, S. M. (2021). Preclinical and clinical biomarker studies of CT1812: A novel approach to Alzheimer's disease modification. *Alzheimer's & Dementia: The Journal of the Alzheimer's Association, 17*(8), 1365–1382. https://doi.org/10.1002/alz.12302

Janelidze, S., Stomrud, E., Palmqvist, S., Zetterberg, H., van Westen, D., Jeromin, A., Song, L., Hanlon, D., Tan Hehir, C. A., Baker, D., Blennow, K., & Hansson, O. (2016). Plasma β-amyloid in Alzheimer's disease and vascular disease. *Scientific Reports, 6,* 26801. https://doi.org/10.1038/srep26801

Juottonen, K., Laakso, M. P., Partanen, K., & Soininen, H. (1999). Comparative MR analysis of the entorhinal cortex and hippocampus in diagnosing Alzheimer disease. *AJNR. American Journal of Neuroradiology, 20*(1), 139–144

Kaddurah-Daouk, R., Rozen, S., Matson, W., Han, X., Hulette, C. M., Burke, J. R., Doraiswamy, P. M., & Welsh-Bohmer, K. A. (2011). Metabolomic changes in autopsy confirmed Alzheimer's disease. *Alzheimer's & Dementia : The Journal of the Alzheimer's Association*, *7*(3), Article 3. https://doi.org/10.1016/j.jalz.2010.06.001

Kamboh, M. I. (2004). Molecular genetics of late-onset Alzheimer's disease. *Annals of Human Genetics, 68*(4), 381–404. https://doi.org/10.1046/j.1529-8817.2004.00110.x

Kang, J., Lemaire, H. G., Unterbeck, A., Salbaum, J. M., Masters, C. L., Grzeschik, K. H., Multhaup, G., Beyreuther, K., & Müller-Hill, B. (1987). The precursor of Alzheimer's disease amyloid A4 protein resembles a cell-surface receptor. *Nature, 325*(6106), Article 6106. https://doi.org/10.1038/325733a0

Kapogiannis, D., Boxer, A., Schwartz, J. B., Abner, E. L., Biragyn, A., Masharani, U., Frassetto, L., Petersen, R. C., Miller, B. L., & Goetzl, E. J. (2015). Dysfunctionally phosphorylated type 1 insulin receptor substrate in neural-derived blood exosomes of preclinical Alzheimer's disease. *FASEB Journal: Official Publication of the Federation of American Societies for Experimental Biology, 29*(2), 589–596. https://doi.org/10.1096/fj.14-262048

Kenny, A., McArdle, H., Calero, M., Rabano, A., Madden, S. F., Adamson, K., Forster, R., Spain, E., Prehn, J. H. M., Henshall, D. C., Medina, M., Jimenez-Mateos, E. M., & Engel, T. (2019). Elevated plasma microRNA-206 levels predict cognitive decline and progression to dementia from mild cognitive impairment. *Biomolecules, 9*(11), 734. https://doi.org/10.3390/biom9110734

Khurana, V., & Goswami, B. (2022). Angiotensin converting enzyme (ACE). *Clinica Chimica Acta, 524,* 113–122. https://doi.org/10.1016/j.cca.2021.10.029

Kiehntopf, M., Nin, N., & Bauer, M. (2013). Metabolism, metabolome, and metabolomics in intensive care: Is it time to move beyond monitoring of glucose and lactate? *American Journal of Respiratory and Critical Care Medicine, 187*(9), Article 9. https://doi.org/10.1164/rccm.201303-0414ED

Klafki, H.-W., Wirths, O., Mollenhauer, B., Liepold, T., Rieper, P., Esselmann, H., Vogelgsang, J., Wiltfang, J., & Jahn, O. (2022). Detection and quantification of Aβ–3–40 (APP669–711) in cerebrospinal fluid. *Journal of Neurochemistry, 160*(5), 578–589. https://doi.org/10.1111/jnc.15571

Lascelles, P. (1981). Cerebrospinal fluid in diseases of the nervous system. *Journal of Clinical Pathology, 34*(4), 453.

Legleiter, J., Czilli, D. L., Gitter, B., DeMattos, R. B., Holtzman, D. M., & Kowalewski, T. (2004). Effect of different anti-Abeta antibodies on Abeta fibrillogenesis as assessed by atomic force microscopy. *Journal of Molecular Biology, 335*(4), 997–1006. https://doi.org/10.1016/j.jmb.2003.11.019

Lemere, C. A. (2009). Developing novel immunogens for a safe and effective Alzheimer's disease vaccine. *Progress in Brain Research, 175,* 83–93. https://doi.org/10.1016/S0079-6123(09)17506-4

Lendon, C. L., Talbot, C. J., Craddock, N. J., Han, S. W., Wragg, M., Morris, J. C., & Goate, A. M. (1997). Genetic association studies between dementia of the Alzheimer's type and three receptors for apolipoprotein E in a Caucasian population. *Neuroscience Letters, 222*(3), 187–190. https://doi.org/10.1016/s0304-3940(97)13381-x

Li, G., Sokal, I., Quinn, J. F., Leverenz, J. B., Brodey, M., Schellenberg, G. D., Kaye, J. A., Raskind, M. A., Zhang, J., Peskind, E. R., & Montine, T. J. (2007). CSF tau/Aβ42 ratio for increased risk of mild cognitive impairment: A follow-up study. *Neurology, 69*(7), 631–639. https://doi.org/10.1212/01.wnl.0000267428.62582.aa

Liu, P., Wang, Y., Sun, Y., & Peng, G. (2022). Neuroinflammation as a potential therapeutic target in Alzheimer's disease. *Clinical Interventions in Aging, 17*, 665–674. https://doi.org/10.2147/CIA.S357558

Lönneborg, A. (2008). Biomarkers for Alzheimer disease in cerebrospinal fluid, urine, and blood. *Molecular Diagnosis & Therapy, 12*(5), 307–320. https://doi.org/10.1007/BF03256296

Luedecking, E. K., DeKosky, S. T., Mehdi, H., Ganguli, M., & Kamboh, M. I. (2000). Analysis of genetic polymorphisms in the transforming growth factor-β1 gene and the risk of Alzheimer's disease. *Human Genetics, 106*(5), 565–569. https://doi.org/10.1007/s004390000313

Lusis, A. J., Heinzmann, C., Sparkes, R. S., Scott, J., Knott, T. J., Geller, R., Sparkes, M. C., & Mohandas, T. (1986). Regional mapping of human chromosome 19: Organization of genes for plasma lipid transport (APOC1, -C2, and -E and LDLR) and the genes C3, PEPD, and GPI. *Proceedings of the National Academy of Sciences of the United States of America, 83*(11), 3929–3933. https://doi.org/10.1073/pnas.83.11.3929

Maddalena, A., Papassotiropoulos, A., Müller-Tillmanns, B., Jung, H. H., Hegi, T., Nitsch, R. M., & Hock, C. (2003). Biochemical diagnosis of Alzheimer disease by measuring the cerebrospinal fluid ratio of phosphorylated tau protein to beta-amyloid peptide42. *Archives of Neurology, 60*(9), 1202–1206. https://doi.org/10.1001/archneur.60.9.1202

Marksteiner, J., Kemmler, G., Weiss, E. M., Knaus, G., Ullrich, C., Mechtcheriakov, S., Oberbauer, H., Auffinger, S., Hinterhölzl, J., Hinterhuber, H., & Humpel, C. (2011). Five out of 16 plasma signaling proteins are enhanced in plasma of patients with mild cognitive impairment and Alzheimer's disease. *Neurobiology of Aging, 32*(3), 539–540. https://doi.org/10.1016/j.neurobiolaging.2009.03.011

Martínez, M., Fernández-Vivancos, E., Frank, A., De la Fuente, M., & Hernanz, A. (2000). Increased cerebrospinal fluid fas (Apo-1) levels in Alzheimer's disease. Relationship with IL-6 concentrations. *Brain Research, 869*(1–2), 216–219. https://doi.org/10.1016/s0006-8993(00)02363-5

Masters, C. L., Simms, G., Weinman, N. A., Multhaup, G., McDonald, B. L., & Beyreuther, K. (1985). Amyloid plaque core protein in Alzheimer disease and down syndrome. *Proceedings of the National Academy of Sciences of the United States of America, 82*(12), Article 12. https://doi.org/10.1073/pnas.82.12.4245

Mattsson, N., Zetterberg, H., Janelidze, S., Insel, P. S., Andreasson, U., Stomrud, E., Palmqvist, S., Baker, D., Tan Hehir, C. A., Jeromin, A., Hanlon, D., Song, L., Shaw, L. M., Trojanowski, J. Q., Weiner, M. W., Hansson, O., Blennow, K., & ADNI Investigators (2016). Plasma tau in Alzheimer disease. *Neurology, 87*(17), 1827–1835. https://doi.org/10.1212/WNL.0000000000003246

Mayeux, R., & Schupf, N. (2011). Blood-based biomarkers for Alzheimer's disease: Plasma Aβ40 and Aβ42, and genetic variants. *Neurobiology of Aging, 32*(Suppl 1), S10–S19. https://doi.org/10.1016/j.neurobiolaging.2011.09.004

McKhann, G. M., Knopman, D. S., Chertkow, H., Hyman, B. T., Jack, C. R., Jr, Kawas, C. H., Klunk, W. E., Koroshetz, W. J., Manly, J. J., Mayeux, R., Mohs, R. C., Morris, J. C., Rossor, M. N., Scheltens, P., Carrillo, M. C., Thies, B., Weintraub, S., & Phelps, C. H. (2011). The diagnosis of dementia due to Alzheimer's disease: Recommendations from the National Institute on Aging-Alzheimer's Association workgroups on diagnostic guidelines for Alzheimer's disease. *Alzheimer's & Dementia: The Journal of the Alzheimer's Association, 7*(3), 263–269. https://doi.org/10.1016/j.jalz.2011.03.005

Menéndez-González, M. (2014). Biomarkers in neurodegenerative disorders: Translating research into clinical practice. *Frontiers in Aging Neuroscience, 6*, 281. https://doi.org/10.3389/fnagi.2014.00281

Mijanovic, O., Petushkova, A. I., Brankovic, A., Turk, B., Solovieva, A. B., Nikitkina, A. I., Bolevich, S., Timashev, P. S., Parodi, A., & Zamyatnin, A. A. (2021). Cathepsin D—managing the delicate balance. *Pharmaceutics, 13*(6), Article 6. https://doi.org/10.3390/pharmaceutics13060837

Molinuevo, J. L., Rabin, L. A., Amariglio, R., Buckley, R., Dubois, B., Ellis, K. A., Ewers, M., Hampel, H., Klöppel, S., Rami, L., Reisberg, B., Saykin, A. J., Sikkes, S., Smart, C. M., Snitz, B. E., Sperling, R., van der Flier, W. M., Wagner, M., Jessen, F., & Subjective Cognitive Decline Initiative (SCD-I) Working Group (2017). Implementation of subjective cognitive decline criteria in research studies. *Alzheimer's & Dementia: The Journal of the Alzheimer's Association, 13*(3), 296–311. https://doi.org/10.1016/j.jalz.2016.09.012

Morris, C. M., Ballard, C. G., Allan, L., Rowan, E., Stephens, S., Firbank, M., Ford, G. A., Kenny, R. A., O'Brien, J. T., & Kalaria, R. N. (2011). NOS3 gene rs1799983 polymorphism and incident dementia in elderly stroke survivors. *Neurobiology of Aging, 32*(3), 554.e1–554.e6. https://doi.org/10.1016/j.neurobiolaging.2010.06.012

Mosconi, L., Tsui, W. H., Herholz, K., Pupi, A., Drzezga, A., Lucignani, G., Reiman, E. M., Holthoff, V., Kalbe, E., Sorbi, S., Diehl-Schmid, J., Perneczky, R., Clerici, F., Caselli, R., Beuthien-Baumann, B., Kurz, A.,

Minoshima, S., & de Leon, M. J. (2008). Multicenter standardized 18F-FDG PET diagnosis of mild cognitive impairment, Alzheimer's disease, and other dementias. *Journal of Nuclear Medicine: Official Publication, Society of Nuclear Medicine, 49*(3), 390–398. https://doi.org/10.2967/jnumed.107.045385

Murillo, O. D., Thistlethwaite, W., Rozowsky, J., Subramanian, S. L., Lucero, R., Shah, N., Jackson, A. R., Srinivasan, S., Chung, A., Laurent, C. D., Kitchen, R. R., Galeev, T., Warrell, J., Diao, J. A., Welsh, J. A., Hanspers, K., Riutta, A., Burgstaller-Muehlbacher, S., Shah, R. V., Yeri, A., . . . & Milosavljevic, A. (2019). exRNA atlas analysis reveals distinct extracellular RNA cargo types and their carriers present across human biofluids. *Cell, 177*(2), 463–477.e15. https://doi.org/10.1016/j.cell.2019.02.018

Musunuri, S., Khoonsari, P. E., Mikus, M., Wetterhall, M., Häggmark-Mänberg, A., Lannfelt, L., Erlandsson, A., Bergquist, J., Ingelsson, M., Shevchenko, G., Nilsson, P., & Kultima, K. (2016). Increased levels of extracellular microvesicle markers and decreased levels of endocytic/exocytic proteins in the Alzheimer's disease brain. *Journal of Alzheimer's Disease: JAD, 54*(4), 1671–1686. https://doi.org/10.3233/JAD-160271

O'Brien, R. J., & Wong, P. C. (2011). Amyloid precursor protein processing and Alzheimer's disease. *Annual Review of Neuroscience, 34*, 185–204. https://doi.org/10.1146/annurev-neuro-061010-113613

Oliveira, J. R. M., Gallindo, R. M., Maia, L. G. S., Brito-Marques, P. R., Otto, P. A., Passos-Bueno, M. R., Jr, M. M., & Zatz, M. (1998). The short variant of the polymorphism within the promoter region of the serotonin transporter gene is a risk factor for late onset Alzheimer's disease. *Molecular Psychiatry, 3*(5), Article 5. https://doi.org/10.1038/sj.mp.4000417

Papassotiropoulos, A., Bagli, M., Feder, O., Jessen, F., Maier, W., Rao, M. L., Ludwig, M., Schwab, S. G., & Heun, R. (1999). Genetic polymorphism of cathepsin D is strongly associated with the risk for developing sporadic Alzheimer's disease. *Neuroscience Letters, 262*(3), 171–174. https://doi.org/10.1016/S0304-3940(99)00071-3

Peña-Bautista, C., Roca, M., Hervás, D., Cuevas, A., López-Cuevas, R., Vento, M., Baquero, M., García-Blanco, A., & Cháfer-Pericás, C. (2019). Plasma metabolomics in early Alzheimer's disease patients diagnosed with amyloid biomarker. *Journal of Proteomics, 200*, 144–152. https://doi.org/10.1016/j.jprot.2019.04.008

Perrin, R. J., Fagan, A. M., & Holtzman, D. M. (2009). Multimodal techniques for diagnosis and prognosis of Alzheimer's disease. *Nature, 461*(7266), 916–922. https://doi.org/10.1038/nature08538

Pinto, T. C. C., Machado, L., Bulgacov, T. M., Rodrigues-Júnior, A. L., Costa, M. L. G., Ximenes, R. C. C., & Sougey, E. B. (2019). Is the Montreal Cognitive Assessment (MoCA) screening superior to the Mini-Mental State Examination (MMSE) in the detection of mild cognitive impairment (MCI) and Alzheimer's disease (AD) in the elderly? *International Psychogeriatrics, 31*(4), 491–504. https://doi.org/10.1017/S1041610218001370

Pirttila, T., Mehta, P. D., Frey, H., & Wisniewski, H. M. (1994). Alpha 1-antichymotrypsin and IL-1 beta are not increased in CSF or serum in Alzheimer's disease. *Neurobiology of Aging, 15*(3), 313–317. https://doi.org/10.1016/0197-4580(94)90026-4

Pláteník, J., Fišar, Z., Buchal, R., Jirák, R., Kitzlerová, E., Zvěřová, M., & Raboch, J. (2014). GSK3β, CREB, and BDNF in peripheral blood of patients with Alzheimer's disease and depression. *Progress in Neuro-Psychopharmacology & Biological Psychiatry, 50*, 83–93. https://doi.org/10.1016/j.pnpbp.2013.12.001

Podhorna, J., Krahnke, T., Shear, M., E Harrison, J., & for the Alzheimer's Disease Neuroimaging Initiative. (2016). Alzheimer's disease assessment scale—cognitive subscale variants in mild cognitive impairment and mild Alzheimer's disease: Change over time and the effect of enrichment strategies. *Alzheimer's Research & Therapy, 8*(1), 8. https://doi.org/10.1186/s13195-016-0170-5

Pozueta, J., Lefort, R., Ribe, E. M., Troy, C. M., Arancio, O., & Shelanski, M. (2013). Caspase-2 is required for dendritic spine and behavioural alterations in J20 APP transgenic mice. *Nature Communications, 4*, 1939. https://doi.org/10.1038/ncomms2927

Preische, O., Schultz, S. A., Apel, A., Kuhle, J., Kaeser, S. A., Barro, C., Gräber, S., Kuder-Buletta, E., LaFougere, C., Laske, C., Vöglein, J., Levin, J., Masters, C. L., Martins, R., Schofield, P. R., Rossor, M. N., Graff-Radford, N. R., Salloway, S., Ghetti, B., Ringman, J. M., Noble, J. M., Chhatwal, J., Goate, A. M., Benzinger, T. L. S., Morris, J. C., Bateman, R. J., Wang, G., Fagan, A. M., McDade, E. M., Gordon, B. A., Jucker, M., & Dominantly Inherited Alzheimer Network (2019). Serum neurofilament dynamics predicts neurodegeneration and clinical progression in presymptomatic Alzheimer's disease. *Nature Medicine, 25*(2), 277–283. https://doi.org/10.1038/s41591-018-0304-3

Qian, S., Chen, K., Guan, Q., & Guo, Q. (2021). Performance of Mattis dementia rating scale-Chinese version in patients with mild cognitive impairment and Alzheimer's disease. *BMC Neurology, 21*(1), 172. https://doi.org/10.1186/s12883-021-02173-0

Querbes, O., Aubry, F., Pariente, J., Lotterie, J. A., Démonet, J. F., Duret, V., Puel, M., Berry, I., Fort, J. C., Celsis, P., & Alzheimer's Disease Neuroimaging Initiative (2009). Early diagnosis of Alzheimer's disease using cortical thickness: Impact of cognitive reserve. *Brain: A Journal of Neurology, 132*(Pt 8), 2036–2047. https://doi.org/10.1093/brain/awp105

Racke, M. M., Boone, L. I., Hepburn, D. L., Parsadainian, M., Bryan, M. T., Ness, D. K., Piroozi, K. S., Jordan, W. H., Brown, D. D., Hoffman, W. P., Holtzman, D. M., Bales, K. R., Gitter, B. D., May, P. C., Paul, S. M., & DeMattos, R. B. (2005). Exacerbation of cerebral amyloid angiopathy-associated microhemorrhage in amyloid precursor protein transgenic mice by immunotherapy is dependent on antibody recognition of deposited forms of amyloid beta. *The Journal of Neuroscience: The Official Journal of the Society for Neuroscience, 25*(3), 629–636. https://doi.org/10.1523/JNEUROSCI.4337-04.2005

Rentz, D. M., Papp, K. V., Mayblyum, D. V., Sanchez, J. S., Klein, H., Souillard-Mandar, W., Sperling, R. A., & Johnson, K. A. (2021). Association of digital clock drawing with PET amyloid and tau pathology in normal older adults. *Neurology, 96*(14), e1844–e1854. https://doi.org/10.1212/WNL.0000000000011697

Ridge, P. G., Mukherjee, S., Crane, P. K., Kauwe, J. S., & Alzheimer's Disease Genetics Consortium (2013). Alzheimer's disease: Analyzing the missing heritability. *PloS One, 8*(11), e79771. https://doi.org/10.1371/journal.pone.0079771

Salari, S. A., Miresmaeili, S. M., & Mazaheri Naeini, M. (2022). Mutational screening in exon 6 of the PSEN2 gene in iranian patients with late-onset Alzheimer's disease. *International Journal of Medical Laboratory, 9*(4), 304–310. https://doi.org/10.18502/ijml.v9i4.11625

Sato, K., Niimi, Y., Mano, T., Iwata, A., & Iwatsubo, T. (2022). Automated evaluation of conventional clock-drawing test using deep neural network: Potential as a mass screening tool to detect individuals with cognitive decline. *Frontiers in Neurology, 13*. www.frontiersin.org/articles/10.3389/fneur.2022.896403

Schellenberg, G. D. (1995). Progress in Alzheimer's disease genetics. *Current Opinion in Neurology, 8*(4), 262–267.

Shibata, N., Ohnuma, T., Takahashi, T., Ohtsuka, E., Ueki, A., & Arai, H. (1999). Genetic association between alpha-2 macroglobulin and Japanese sporadic Alzheimer's disease. *Neuroscience Letters, 271*(2), 132–134. https://doi.org/10.1016/s0304-3940(99)00541-8

Skoog, I., & Gustafson, D. (2006). Update on hypertension and Alzheimer's disease. *Neurological Research, 28*(6), 605–611. https://doi.org/10.1179/016164106X130506

Small, B. J., Fratiglioni, L., Viitanen, M., Winblad, B., & Bäckman, L. (2000). The course of cognitive impairment in preclinical Alzheimer disease: Three- and 6-year follow-up of a population-based sample. *Archives of Neurology, 57*(6), 839–844. https://doi.org/10.1001/archneur.57.6.839

Soto-Ospina, A., Araque Marín, P., Bedoya, G. de J., & Villegas Lanau, A. (2021). Structural predictive model of presenilin-2 protein and analysis of structural effects of familial Alzheimer's disease mutations. *Biochemistry Research International, 2021*, e9542038. https://doi.org/10.1155/2021/9542038

Sperling, R. A., Aisen, P. S., Beckett, L. A., Bennett, D. A., Craft, S., Fagan, A. M., Iwatsubo, T., Jack, C. R., Jr., Kaye, J., Montine, T. J., Park, D. C., Reiman, E. M., Rowe, C. C., Siemers, E., Stern, Y., Yaffe, K., Carrillo, M. C., Thies, B., Morrison-Bogorad, M., Wagster, M. V., & Phelps, C. H. (2011). Toward defining the preclinical stages of Alzheimer's disease: Recommendations from the National Institute on Aging-Alzheimer's Association workgroups on diagnostic guidelines for Alzheimer's disease. *Alzheimer's & Dementia: The Journal of the Alzheimer's Association, 7*(3), 280–292. https://doi.org/10.1016/j.jalz.2011.03.003

Stanislav Alexandra Alina. (2020). Association of ACE gene with metabolic syndrome, hypertension and Alzheimer disease by RFLP-PCR. *Journal of Chemistry and Chemical Engineering, 14*(4). https://doi.org/10.17265/1934-7375/2020.04.001

Stratmann, K., Heinsen, H., Korf, H. W., Del Turco, D., Ghebremedhin, E., Seidel, K., Bouzrou, M., Grinberg, L. T., Bohl, J., Wharton, S. B., den Dunnen, W., & Rüb, U. (2016). Precortical phase of Alzheimer's disease (AD)-related tau cytoskeletal pathology. *Brain Pathology (Zurich, Switzerland), 26*(3), 371–386. https://doi.org/10.1111/bpa.12289

Sweeney, C. J., Chen, Y. H., Carducci, M., Liu, G., Jarrard, D. F., Eisenberger, M., Wong, Y. N., Hahn, N., Kohli, M., Cooney, M. M., Dreicer, R., Vogelzang, N. J., Picus, J., Shevrin, D., Hussain, M., Garcia, J. A., & DiPaola, R. S. (2015). Chemohormonal therapy in metastatic hormone-sensitive prostate cancer. *The New England Journal of Medicine, 373*(8), 737–746. https://doi.org/10.1056/NEJMoa1503747

Tatebe, H., Kasai, T., Ohmichi, T., Kishi, Y., Kakeya, T., Waragai, M., Kondo, M., Allsop, D., & Tokuda, T. (2017). Quantification of plasma phosphorylated tau to use as a biomarker for brain Alzheimer pathology: Pilot case-control studies including patients with Alzheimer's disease and down syndrome. *Molecular Neurodegeneration, 12*(1), 63. https://doi.org/10.1186/s13024-017-0206-8

Thijssen, E. H., Joie, R. L., Wolf, A., Strom, A., Wang, P., Iaccarino, L., Bourakova, V., Cobigo, Y., Heuer, H., Spina, S., VandeVrede, L., Chai, X., Proctor, N. K., Airey, D. C., Shcherbinin, S., Evans, C. D., Sims, J. R., Zetterberg, H., Blennow, K., Karydas, A. M., Teunissen, C. E., Kramer, J. H., Grinberg, L. T., Seeley, W. W., Rosen, H., Boeve, B. F., Miller, B. L., Rabinovici, G. D., Dage, J. L., Rojas, J. C., Boxer, A. L. & Advancing Research and Treatment for Frontotemporal Lobar Degeneration (ARTFL) investigators (2020). Diagnostic value of plasma phosphorylated tau181 in Alzheimer's disease and frontotemporal lobar degeneration. *Nature Medicine, 26*(3), 387–397. https://doi.org/10.1038/s41591-020-0762-2

Twamley, E. W., Ropacki, S. A. L., & Bondi, M. W. (2006). Neuropsychological and neuroimaging changes in preclinical Alzheimer's disease. *Journal of the International Neuropsychological Society, 12*(5), 707–735. https://doi.org/10.1017/S1355617706060863

Van Broeckhoven, C., Backhovens, H., Cruts, M., De Winter, G., Bruyland, M., Cras, P., & Martin, J.-J. (1992). Mapping of a gene predisposing to early—onset Alzheimer's disease to chromosome 14q24.3. *Nature Genetics*, *2*(4), Article 4. https://doi.org/10.1038/ng1292-335

Vecchio, F., Miraglia, F., Alù, F., Menna, M., Judica, E., Cotelli, M., & Rossini, P. M. (2020). Classification of Alzheimer's disease with respect to physiological aging with innovative EEG biomarkers in a machine learning implementation. *Journal of Alzheimer's disease: JAD, 75*(4), 1253–1261. https://doi.org/10.3233/JAD-200171

Vöglein, J., Willem, M., Trambauer, J., Schönecker, S., Dieterich, M., Biskup, S., Giudici, C., Utz, K., Oberstein, T., Brendel, M., Rominger, A., Danek, A., Steiner, H., Haass, C., & Levin, J. (2019). Identification of a rare presenilin 1 single amino acid deletion mutation (F175del) with unusual amyloid-β processing effects. *Neurobiology of Aging*, *84*, 241.e5–241.e11. https://doi.org/10.1016/j.neurobiolaging.2019.08.034

Wang, L., & Zhang, L. (2020). Circulating exosomal miRNA as diagnostic biomarkers of neurodegenerative diseases. *Frontiers in Molecular Neuroscience, 13*, 53. https://doi.org/10.3389/fnmol.2020.00053

Wang, M. J., Yi, S., Han, J. Y., Park, S. Y., Jang, J. W., Chun, I. K., Kim, S. E., Lee, B. S., Kim, G. J., Yu, J. S., Lim, K., Kang, S. M., Park, Y. H., Youn, Y. C., An, S. S. A., & Kim, S. (2017). Oligomeric forms of amyloid-β protein in plasma as a potential blood-based biomarker for Alzheimer's disease. *Alzheimer's Research & Therapy, 9*(1), 98. https://doi.org/10.1186/s13195-017-0324-0

Xiao, Z., Prieto, D., Conrads, T. P., Veenstra, T. D., & Issaq, H. J. (2005). Proteomic patterns: Their potential for disease diagnosis. *Molecular and Cellular Endocrinology, 230*(1–2), 95–106. https://doi.org/10.1016/j.mce.2004.10.010

Yamada, K., Kono, K., Umegaki, H., Yamada, K., Iguchi, A., Fukatsu, T., Nakashima, N., Nishiwaki, H., Shimada, Y., & Sugita, Y. (1995). Decreased interleukin-6 level in the cerebrospinal fluid of patients with Alzheimer-type dementia. *Neuroscience Letters, 186*(2–3), 219–221. https://doi.org/10.1016/0304-3940(95)11318-q

Yang, Q., Wang, E., Jia, H., & Wang, Y. (2016). Association between polymorphisms in transforming growth factor-β1 and sporadic Alzheimer's disease in a Chinese population. *International Journal of Neuroscience*, *126*(11), 979–984. https://doi.org/10.3109/00207454.2015.1088849

Yuan, Y., Gu, Z. X., & Wei, W. S. (2009). Fluorodeoxyglucose-positron-emission tomography, single-photon emission tomography, and structural MR imaging for prediction of rapid conversion to Alzheimer disease in patients with mild cognitive impairment: A meta-analysis. *AJNR. American Journal of Neuroradiology, 30*(2), 404–410. https://doi.org/10.3174/ajnr.A1357

Yurov, Y. B., Vorsanova, S. G., & Iourov, I. Y. (2019). Chromosome instability in the neurodegenerating brain. *Frontiers in Genetics, 10*. www.frontiersin.org/articles/10.3389/fgene.2019.00892

Zhang, H., Geng, X., Wang, Y., Guo, Y., Gao, Y., Zhang, S., Du, W., Liu, L., Sun, M., Jiao, F., Yi, F., Li, X. & Wang, L. (2021). The Significance of EEG Alpha Oscillation Spectral Power and Beta Oscillation Phase Synchronization for Diagnosing Probable Alzheimer Disease. *Frontiers in Aging Neuroscience, 13*, 631587. https://doi.org/10.3389/fnagi.2021.631587

Zhu, X. C., Tan, L., Wang, H. F., Jiang, T., Cao, L., Wang, C., Wang, J., Tan, C. C., Meng, X. F., & Yu, J. T. (2015). Rate of early onset Alzheimer's disease: A systematic review and meta-analysis. *Annals of Translational Medicine, 3*(3), 38. https://doi.org/10.3978/j.issn.2305-5839.2015.01.19

6 Mechanistic Aspects of Neurodegeneration in Alzheimer's Disease and the Role of Phytochemicals

*Trideep Saikia, Tutumoni Kalita, Bhanu P Sahu, Prakash Rajak, Bhargab Deka and Alakesh Bharali**

6.1 INTRODUCTION

As early as 1906, Alois Alzheimer diagnosed amyloid plaques and neurofibrillary tangles (NFTs) in the cortex of 51-year-old Auguste D (Alzheimer, 1907). This led to the advent of a new era of medicine that revealed the temporary vulnerability of individuals. Soon after the publication of this pathological evidence, in the year 1910, Kraepelin devised the term Alzheimer's disease (AD) to describe this pathology (Sharifi-Rad et al., 2022). AD is among the most common diseases associated with aging. Increasing numbers of patients suffering from age-associated AD pose an oppressive challenge to the medical profession, economy, and society as a whole. In the current state, over 40 million people are afflicted with AD worldwide, which accounts for 70% of all dementia cases (X. Zhang et al., 2021; Bhatti et al., 2020).

In AD, cognitive and functional abilities deteriorate gradually along with irreversible loss of neurons (Selkoe, 2000). In its early stages, this disease is characterized by apathy, short-term memory loss, and depression. A range of later symptoms are experienced, including confusion, spatial disorientation, impaired communication, difficulty maintaining personal hygiene, impaired movement, difficulty in swallowing, and language deficit. A hallmark of the disease is the development of plaques outside of the neurons composed of beta-amyloid (Aβ) protein fragments and tangles within the neurons composed of twisted strands of the protein tau.

As a consequence of the protein accumulations, there is chronic inflammation and impaired glucose metabolism leading to irreversible neuronal loss (Amor et al., 2014). The market currently offers a limited number of anti-AD drugs, and currently available therapies do not have the capability to halt the disease's onset or progression. In general, current therapies are intended to reduce symptom severity, reduce the clinical progression of the disease, delay disability, and provide symptomatic relief in mild or moderate cases (Jeremic et al., 2021). According to the U.S. Food and Drug Administration (FDA), rivastigmine, donepezil, galantamine, memantine, and memantine combined with donepezil have been approved for the treatment of AD. All the FDA-approved drugs except memantine contain acetylcholinesterase inhibitors (AChEI), which promote synaptic modulation by increasing acetylcholine.

Memantine inhibits glutamate release by antagonistically blocking the NMDA receptor subtype of glutamate receptors (Epperly et al., 2017), and this imbalance between chronic inflammation and oxidative stress necessitates the timely implementation of novel therapies with better efficacy and fewer side effects. As such, it is imperative to explore promising novel compounds that have the potential to modify the progression of the disease as well as address the limitations of the current therapy. In response to this imperative, current research in this area focuses on identifying effective

DOI: 10.1201/9781003412069-6

and preventive bioactive neuroprotective components that are readily tolerated by patients and do not cause adverse effects. Against this background, plant-derived phytochemicals have shown considerable potential for use as adjunctive and alternative therapies for the treatment of neurological disorders such as AD (Calfio et al., 2020).

The term phytochemicals refers to bioactive secondary metabolites, including polyphenols, alkaloids, peptides, terpenoids, tannins, organosulfur compounds, sterols, limonoids, polyenes, furyl compounds, lignans, thiophenes, stilbenes and saponins (Velmurugan et al., 2018). The WHO reports that approximately 80% of the world's total population relies on plant-based products or phytomolecules isolated from plants for their well-being (Welcome, 2020). Consequently, in this chapter, a detailed discussion of the molecular mechanisms underlying AD is presented, as well as the most likely phytochemicals, in particular polyphenols and alkaloids, to combat its progression.

6.2 BIOLOGICAL MECHANISMS CONTRIBUTING TO THE PATHOGENESIS OF ALZHEIMER'S DISEASE

Pathological characteristics of patients with AD have been intricately linked to overproduction and dysregulation of the peptide Aβ, resulting in neurofibrillary tangles and neurotic plaques; nevertheless, the precise mechanisms regulating neuronal death remain ambiguous. Aβ aggregation triggers the pathogenic cascade, which includes the hyperphosphorylation of the microtubule-associated protein tau and the development of intracellular protein aggregates, resulting in synaptic dysfunction, neuronal death, and cognitive impairment (Busche & Hyman, 2020; Nisbet & Götz, 2018; Pereira et al., 2021). Progressive neurodegeneration has also been associated with impaired cholinergic neurotransmission in the brain. Elevation of ACh in cholinergic neurons synapses by inhibition of enzymes involved in hydrolysis of ACh, i.e., AChE and BuChE, is believed to improve the symptoms of AD since times immemorial (Pepeu & Grazia Giovannini, 2017; Ferreira-Vieira et al., 2016).

Current FDA-approved AChE inhibitor drugs such as donepezil, tacrine, rivastigmine, and galantamine attempt to temporarily ameliorate cognitive symptoms with inadequate short-term effect (Rodda & Carter, 2012). With the availability of only NMDA receptor antagonists and cholinesterase inhibitors as pharmacological treatments for AD, there is a pressing requirement for the discovery of novel drugs that could potentially open novel therapeutic opportunities for the therapeutic management of AD (Raina et al., 2008).

6.2.1 Beta-amyloid

Glenner and Wong extracted "beta-amyloid" from the meningeal arteries of AD patients with significant identification of the peptide sequence, and consecutively, it was identified as the key constituent of neuritic plaques (Masters et al., 1985; Glenner & Wong, 1984). The fundamental biology of this peptide, as well as its impact in AD pathogenesis, has been the target of extensive research throughout the subsequent decades.

The successive hydrolysis of the APP along the amyloidogenic pathway results in the production of Aβ. The BACE1 initially cleaves APP at the N terminus of the Aβ sequence, resulting in membrane-bound CTFβ or C99 fragment and sAPPβ (Coronel et al., 2018). Following cleavage by γ-secretase, full-length Aβs and AICD with varying numbers of amino acids are generated, the most prevalent of which are those with 40 amino acids (Aβ40) and 42 amino acids (Aβ42) (Haass et al., 2012). When released as monomers because of synaptic activity, they tend to assemble into fragments of oligomers and protofibrils, followed by fibrils, eventually forming matured and ubiquitous amyloid plaques. Notwithstanding the commonalities, Aβ42 is more likely to fibrilize and aggregate, making it a more lethal Aβ peptide and a key contributor to the development of neurotoxic plaques in the basal region and hippocampus (Xia, 2019; Zhou et al., 2019).

Ten years or more prior to the onset of clinical symptoms, Aβ oligomers start to build up in vivo. They act on a variety of receptors, including α7-nACh and NMDA-type glutamate, to suppress LTP and promote LTD. Under physiological conditions, the production of Aβ peptides is in equilibrium with their clearance from the brain (Rhein & Eckert, 2007). At present, ApoE and the IDE facilitate clearance by inhibiting their aggregation. If accumulation of Aβ prolongs, the occurrence of mitochondrial damage, synaptic dysfunction, onset of inflammatory cascades by activation of astrocytes and microglia, unstable homeostasis, oxidative stress, and eventually neuronal apoptosis takes place, leading to AD (Kunjathoor et al., 2004).

The amyloid cascade hypothesis, first put forth by Hardy and Higgins in 1992, is one of numerous hypotheses that explain the underlying theory of AD pathogenesis. The theory states that Aβ is the initializing step which eventually leads to subsequent tau deposition, NFT formation, and eventually cell death (Hardy & Higgins, 1992). This hypothesis has been the leading model of AD aetiology ever since it was initially put forth, although some aspects of it have been modified or added over time (Selkoe, 2019; Musiek & Holtzman, 2015). However, studies have found an insignificant association between the stages of amyloid buildup and the severity of cognitive decline, as well as weak connections between regional cerebral hypometabolism and amyloid deposition, implying that Aβ is crucial but not adequate for AD pathogenesis and that other downstream factors play a significant role (Altmann et al., 2015; Edison et al., 2007; Nelson et al., 2012).

Objectives of Aβ hypothesis-based therapeutic approaches are to decrease Aβ generation, accumulation, and clearance through BACE1 and γ-secretase modulation. Several authors have investigated BACE1 as an intriguing option for the development of anti-AD drugs because of its higher substrate specificity. Nevertheless, numerous Phase III studies with drug candidates including verubecestat, atabecestat, umibecestat, crenezumab, and lanabecestat have demonstrated minimal clinical benefit, unusual side events, and an inability to meet the primary endpoint (McConlogue et al., 2007; Chang et al., 2004). Conversely, γ-secretase inhibitors are toxic to a range of tissues and lack APP substrate specificity. Avagacestat, semagacestat, and tarenflurbil could not meet the expected outcomes and failed to complete clinical studies owing to significant adverse outcomes such as tumours, rashes, and gastrointestinal problems; as a result, associated clinical studies have been halted (D'Onofrio et al., 2012; Schneider et al., 2014; Doody et al., 2013).

Numerous clinical trials of amyloid clearance therapy have been carried out by major pharmaceutical companies worldwide, but they have all been terminated or have not shown any fruitful results. However, the US FDA's approval of aducanumab on June 7, 2021, turned out to be a turning point in the history of AD treatment (Huang et al., 2020). It is a recombinant IgG1 monoclonal antibody obtained from patients with slow or no cognitive deterioration (Khanna et al., 2023). In March 2019, it was announced that the phase 3 clinical trials of aducanumab were completed, and the futility analysis determined that the trials had achieved their main goals of slowing down the progression of the disease, as measured by the CDR Sum of Boxes. After another analysis of the data in October 2019, an application for FDA approval was submitted despite the previous conclusion that the trials would not be successful in achieving their objectives. Although aducanumab's approval represents the first drug treatment for AD targeting Aβ, approval is subject to further results (Arndt et al., 2018; Cummings et al., 2021).

6.2.2 Tau

Tau is present in the neuronal axons of the brain, along with MTs. Tau maintains and stabilizes microtubule structure, transports to the cytoplasm, and regulates neuronal signalling, as well as regulates the structure and function of synapses (Regan et al., 2017). The development of tau pathology is a complex process influenced by numerous factors. Tau undergoes several post-translation changes, such as phosphorylation, glycation, acetylation, nitration, SUMOylation, O-GlcNAcylation, truncation, and ubication (Alquezar et al., 2021). The aberrant hyperphosphorylation and aggregation of all 6 tau isoforms appear to be the exclusive cause of the hazardous behaviour of P-tau.

The two main proteins that are involved with neuronal microtubules, MAP1 A/B and MAP2, are sequestered by P-tau, and the damaged protein also sequesters healthy tau (Ramesh et al., 2020; Chen et al., 2020). Up to 40% of P-tau in the AD brain is in the cytosol, whereas the remainder is identified as insoluble PHF in polymerized form with significant hydrophobicity (H. Zhang et al., 2021). Inhibition of synaptic anchoring or glutamate receptor trafficking causes impairment in synaptic function when hyperphosphorylated tau passes from the axonal to the somatodendritic compartment. Furthermore, the truncation and acetylation of tau also enhance its accumulation, promote mitochondrial dysfunction, and cause microtubule binding disability (Guo et al., 2017).

Tau has recently been a subject of increased focus, partly due to the unsuccessful clinical trials of various treatments targeting Aβ and partly because tau pathology shows a stronger correlation with cognitive decline than Aβ lesions. Recently, there has been research into stabilizing microtubules and inhibiting tau aggregation and kinases, as well as using immunotherapy drugs as potential treatments for tau-related disorders. Several drug candidates have been eliminated in clinical trials due to inefficacy and toxicity.

One such example is LMTX (TRx0237), a potential nonspecific inhibitor of tau aggregation that failed as an additional treatment for AD (Gauthier et al., 2016). Epothilone D, the antifungal molecule, was reported to improve tau-related pathology, microtubule density, and axonal integrity in a rodent model of tauopathy; however, clinical trials have been discontinued due to adverse events. Although AADvacl, a vaccine targeting tau, has exhibited encouraging results regarding its safety and immunological response, additional research is necessary to prove its clinical effectiveness (Mantile & Prisco, 2020).

6.2.3 Neurofibrillary Tangles

The neurofibril deposits exist as insoluble PHFs and SFs because of tau–tau interactions which promote cell dysfunction by limiting the number of synapses, triggering neurotoxicity (Grundke-Iqbal et al., 1986). Degeneration of neurons begins in the medial temporal lobe's allocortex, specifically in the hippocampus and entorhinal cortex, and extends sparsely to the isocortex, sparing the major sensory, motor, and visual regions (Crary et al., 2014). The type of neuron where NFTs develop determines the tangle's morphology. They begin as pretangles or pre-NFTs containing nonpolymerized abnormal conformers of tau in the cytoplasm and dendrites of neurons and a centred nucleus. These mature into fibrillar iNFTs, which are tau polymeric aggregates in the cytoplasm that relocate the nucleus more towards the periphery of the soma and frequently spread to deformed dendrites (Tatsumi et al., 2014).

"Flame-shaped tangles" may be observed in the pyramidal neurons of the hippocampus and in layer V of the association cortices, taking the shape of intertwining swirls of fibres. In the basal nucleus of the raphe nuclei, Meynert, locus coeruleus, and substantia nigra, these tangles form as globoid tangles. Following the degeneration of the tangled neurons, the astrocytes, microglia, and extracellular proteins such as Aβ in association with insoluble filaments that are present in the extracellular spaces, form eNFTs which are also referred to as "ghost tangles" (Lasagna-Reeves et al., 2012).

6.2.4 Cerebral Amyloid Angiopathy

Aβ peptides, which are known to deposit in the brain parenchyma as plaques, also accumulate in the walls of blood vessels as CAA. In CAA, the more soluble form of Aβ, Aβ40, is the primary component and is found in the interstitial spaces between the smooth muscle cells of the tunica media. CAA is present in 85–90% of confirmed cases of Alzheimer's disease upon autopsy. The capillaries in the leptomeningeal vessels and cerebral cortex, arteries, and arterioles are all affected by CAA, in addition to being accumulated in the vein wall (Greenberg et al., 2019; Thal et al., 2019).

In accordance with the distribution of Aβ in vessel walls, its deposits develop in the basement membrane of trans-cortical capillaries and extend to leptomeningeal arteries. Pericapillary deposits are evidence that some Aβ-plaques are the result of Aβ-loaded capillaries. Others hypothesized that poor transendothelial clearance of Aβ is related to pericapillary Aβ (Attems et al., 2011). A late event in the progression of AD is the degeneration of brain mural cells. These cells are in direct proximity with endothelial cells lining the blood arteries and play an essential role in regulating cerebral blood flow and the blood–brain barrier. The aggregation of TDP-43 and accumulation of Aβ in cerebral microvessels of the parietal cortex, which links cerebrovascular anomalies to deficits in cognition, indicates that the loss of brain mural cells is a significant factor contributing to the pathogenesis of AD (Bourassa et al., 2020).

The buildup of Aβ in veins and venules causes endothelial cell loss, venous collagenosis, significant perivenular space enlargement, and concentric rings along the venular walls. This leads to vascular dysfunction in AD, which causes microinfarcts and white matter hyperintensities. In one investigation, Aβ deposition caused a 50% reduction in blood flow, which led the cerebral blood vessels to shrink by around 8%, which decreased the energy supply. Currently, endothelial degradation, BBB breakdown, and poor perivascular Aβ drainage are recognised as hallmarks of common processes in CAA and DPA, which are seen as variants along an age-related spectrum of cerebral small-artery illnesses (Hecht et al., 2018).

6.2.5 Granulovacuolar Degeneration

GVD is a degenerative alteration that occurs in AD but is not well defined. It was first identified as intraneuronal vacuoles that were reported to be an integral component of AD. It has been speculated that these granules represent late-stage autophagic vacuoles based on their specificity for phosphor-ubiquitin mitophagy markers and other autophagic markers (Hondius et al., 2021). The abundance of granulovacuolar bodies has been found to correspond to the density of NFTs, suggesting them as a cause of neuronal injury.

Moreover, the formation of a necrosome complex in GVD is also linked to neuronal damage in AD (Wiersma et al., 2019). In staging, these formations aggregate first in the hippocampus, then in the temporal neocortex and entorhinal cortex, and in the thalamus, cingulate gyrus, amygdala, and even association cortices (Hou et al., 2020). GVD bodies have been shown to be immunoreactive for cytoskeletal proteins such as neurofilament proteins, tubulin, and tau in previous immunohistochemistry investigations (Dickson & Vickers, 2001). Because of their immunoreactivity for activated caspase 3, some researchers have also proposed involvement in apoptotic cell death. Other studies have shown positivity for multiple stress kinases, suggesting that these entities are derived from the endoplasmic reticulum and constitute stress granules that exhibit the unfolded protein response (Urnowey et al., 2006).

6.2.6 Hirano Bodies

Hirano bodies are intracytoplasmic entities present along with NFTs and GVD; they are rod-like eosinophilic inclusion bodies found in the neuronal dendrites of the stratum lacunosum of the hippocampus CA1 area (Gibson & Tomlinson, 1977). On immunohistochemical examination using haematoxylin and eosin, they were characterized as rich in tropomyosin, actin, and vinculin (Gibson & Tomlinson, 1977). Inducible nitric oxide synthase, APP carboxy-terminal fragments, and advanced glycation end products constitute other immunoreactivities linked with Hirano bodies (Serrano-Pozo et al., 2011). They have a distinct morphological appearance that comprises intertwined parallel fibres in a systematic bridging pattern like that of a tweed cloth. In the brain of an AD patient, abnormally high concentrations of these bodies have been discovered which are then subsequently transferred to the neurons of the stratum pyramidale (Gibson & Tomlinson, 1977). More research study is required to completely understand their aetiology and relevance in AD. However, these inclusions were discovered to be correlated to reduced spatial working memory and synaptic loss in animal models carrying aberrant actin-binding proteins (Furgerson et al., 2014).

6.2.7 Synaptic Loss

Synapse degradation is not a characteristic of AD solely but rather a common occurrence in many dementia-related illnesses; although the degree of impairment is not completely clear, it may be profound (Subramanian et al., 2020). This loss ranges from 10 to 60%, primarily affecting mesiotemporal and frontal lobes, and it was found to be lower in LOAD than in EOAD (Scott et al., 2016). Deterioration in the integrity of the semantic system and cognitive impairment can be caused by a decrease in synaptic function in the MTLs and in the hippocampus due to axonal dysfunction and a decrease in synapses with involvement of the presynaptic termini. Extensive research has discovered proteins that differentiate the distinctive proteomic profile of postsynaptic concentrations in cognitively intact AD patients from demented AD patients and those in control (Sheng et al., 2012).

6.2.8 Neuronal Loss

Cortical atrophy is caused by neuronal loss, which occurs due to 2 mechanisms in AD. One mechanism affects neurons with tangles, which leads to extracellular ghost tangles, and the other affects tangle-free neurons through apoptosis (Goel et al., 2022). According to investigations using animal models, inflammation-induced tau protein abnormalities can destabilize the microtubule-action system and hinder axonal transport, leading to organelle buildup and microtubule accumulation (Wirths & Zampar, 2020). This can trigger further tau protein abnormalities, which in turn promote the formation of PHFs, resulting in complete axonal blockage, synaptic loss, microglia activation, and astrogliosis (Mukhin et al., 2017). Neuronal loss follows the pattern of synaptic loss, which can exceed the neuronal loss in a specific cortical area. This indicates that synaptic loss precedes neuronal loss and that the remaining neurons have weaker connections to their synaptic partners than expected based on the number of viable neurons in each circuit (Roussarie et al., 2020; J. Zhang et al., 2021).

6.2.9 Inflammatory Reaction

Microglia, which usually monitor and clear synapses, can become activated and accumulate around senile plaques in response to factors in the local environment, including protein aggregates like Aβ and tau (Navarro et al., 2016). This activation leads to an inflammatory response, with enhanced production of cytokines and T follicular helper cells (Hansen et al., 2018). However, this immune response is not sufficient to clear Aβ and instead exacerbates inflammation. Another inflammatory response that is seen in AD patients is reactive astrocytes, which are hypothesized to play a role in neuroprotection and homeostasis (Leng & Edison, 2020). Typically, astrocytes connect synapses and neurons to the blood supply to provide support; however, in AD they are frequently seen around senile plaques at a larger distance from the centre (González-Reyes et al., 2017). Reactive astrocytes are thought to react to cytokines produced by pro-inflammatory microglia and become neurotoxic. This reactive astrocyte load develops at a later stage of AD and is connected to the tau burden. It is important to note that neurons and oligodendrocytes both contribute to regulating glia (Guan et al., 2022; Hemonnot et al., 2019).

6.3 THE ROLE OF DEPENDENCE RECEPTORS IN ALZHEIMER'S DISEASE

One potential contributing factor to AD pathology is the dysfunction of dependence receptors, a type of cell-surface receptor that plays a critical role in regulating neuronal survival and death (Thibert & Fombonne, 2010). Dependence receptors are so named because they rely on

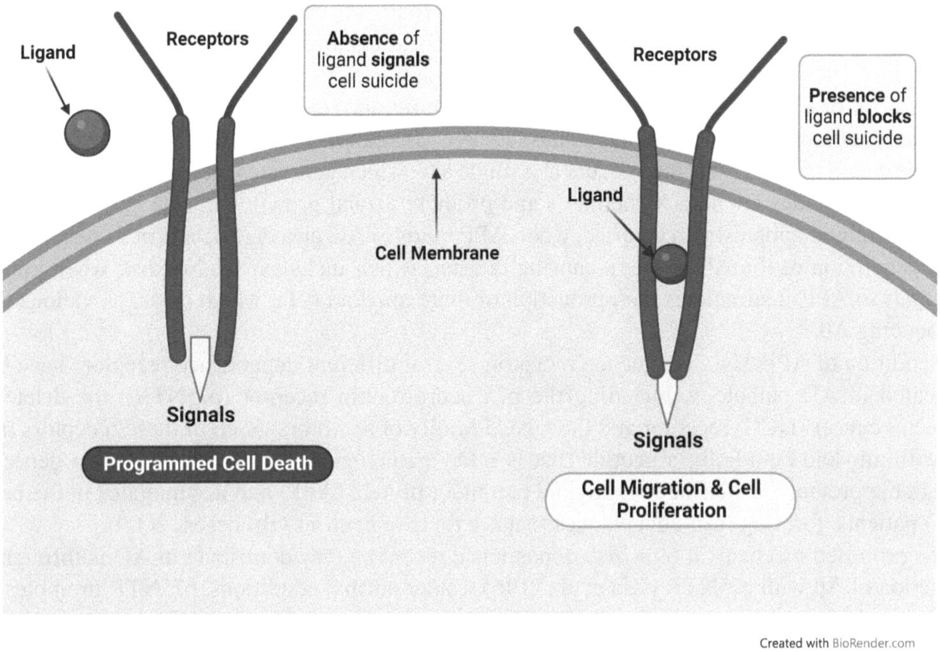

FIGURE 6.1 Role of dependence receptors in Alzheimer's disease.

extracellular signals for their survival. Under normal conditions, these receptors promote cell survival by activating pro-survival pathways in response to ligand binding. However, when the concentration of ligands falls below a certain threshold, dependence receptors can trigger apoptosis or programmed cell death. This ensures that cells that are no longer receiving the appropriate signals from their environment are eliminated, preventing their accumulation and potential deleterious effects on the organism (Guo et al., 2020). Figure 6.1 shows the role of dependence receptors in neurodegeneration related to AD.

It is noteworthy that the hallmark of Alzheimer's pathology is the accumulation of amyloid beta (Aβ), which originates from the amyloid precursor protein (APP), a type of dependence receptor. APP is a receptor of considerable size that comprises 695 amino acids. Once APP is synthesized by neurons, it is cleaved by proteases referred to as molecular scissors. Cleavage occurs at either one distinct or three specific sites, giving rise to different fragments. The cleavage at 3 sites produces 4 peptides: soluble amyloid precursor protein beta (sAPPβ), junctional caspase (Jcasp), a cleavage product of the amyloid precursor protein (C31), and amyloid beta (Aβ), all of which contribute to the underlying mechanisms of AD (Bredesen et al., 2010). These mechanisms include the loss of synapses in the brain, the shrinkage of neuronal extensions that connect with other neurons, and the activation of neuronal apoptosis. Alternatively, cleavage at one site results in the generation of 2 peptides, namely soluble amyloid precursor protein alpha (sAPPα) and alpha-cleaved fragment of the APP (αCTF). This latter process exerts effects that are completely opposite to those of the quartet of peptides mentioned earlier. These peptides maintain synaptic connections, promote neuronal growth, and inhibit apoptosis (Bredesen, 2009). Depending on how APP is cleaved, the resulting fragments will either support or destroy the cellular processes involved in memory formation and maintenance, which is referred to as neuroplastic and neuroclastic activity. This activity must be balanced for proper neuronal functioning.

Unfortunately, individuals with AD are on the wrong side of this crucial balance, as APP is frequently cleaved in a manner that generates the devastating quartet rather than the cognition-supporting duo. The mechanism of how APP is cleaved at 3 sites to produce the Alzheimer's-causing quartet, or at one site to generate the neuron-nourishing duo, is determined by numerous factors including the binding of different molecules (Zhao et al., 2020). Specifically, when APP binds to a molecule called netrin-1, cleavage occurs at a single site, which generates the sAPPα and αCTF peptides. These peptides are anti-Alzheimer's and promote axonal growth and synaptic and neuronal health and inhibit apoptosis. In contrast, when APP binds to Aβ, cleavage occurs at 3 sites, resulting in the generation of the Alzheimer's-causing quartet, which includes Aβ. Notably, when amyloid beta binds to APP, it stimulates the production of more amyloid beta, which creates a vicious cycle of producing Aβ.

In addition to APP as a dependence receptor, several different dependence receptors have been implicated in AD pathology, including the p75 neurotrophin receptor (p75NTR), the deleted in colorectal cancer (DCC) receptor and the UNC5 family of receptors. Each of these receptors interacts with amyloid-beta (Aβ), a peptide that is a key pathological feature of AD. Aβ is generated through the proteolytic cleavage of amyloid precursor protein (APP) and accumulates in the brains of AD patients, forming insoluble plaques that are toxic to neurons (Bredesen, 2009).

One proposed mechanism by which dependence receptors may contribute to AD is through the interaction of Aβ with p75NTR (Xia et al., 2016). Under normal conditions, p75NTR promotes cell survival by binding to neurotrophic factors such as nerve growth factor (NGF) and brain-derived neurotrophic factor (BDNF) (Lossi, 2022). However, when Aβ levels rise, p75NTR can bind to Aβ instead, triggering downstream signalling pathways that promote neuronal death. This may contribute to the cognitive decline associated with AD by promoting the death of neurons in critical brain regions (Rabizadeh & Bredesen, 2003). In addition to p75NTR, DCC has also been implicated in AD pathology.

DCC is a dependence receptor that plays a critical role in axon guidance during development. However, in the adult brain, DCC is expressed at low levels, and its function is less well understood. Recent studies have suggested that DCC may interact with Aβ and promote the formation of insoluble Aβ aggregates, further contributing to the toxicity of Aβ in AD (Ryan et al., 2013).

Moreover, the UNC5 family of receptors has also been implicated in AD pathology. These receptors engage in axon guidance during development and have been shown to bind to Aβ in vitro (Bredesen et al., 2004). While the precise role of UNC5 receptors in AD remains unclear, their interaction with Aβ suggests that they may also play a role in promoting the accumulation of Aβ plaques in the brain (Li et al., 2018).

6.4 THE ROLES OF DIFFERENT PHYTOCHEMICALS IN NEURODEGENERATIVE DISEASES

Phytochemicals are bioactive compounds found in plant-based foods that are accepted for potential health benefits. Several studies have indicated that certain phytochemicals have neuroprotective effects and may help prevent or slow down neurodegenerative diseases (ND). Progressive neuronal loss in particular brain regions is a hallmark of ND, leading to cognitive decline and motor impairment. Some of the most common ND include AD, Parkinson's disease (PD), and Huntington's disease (HD). Phytochemicals such as flavonoids, curcumin, resveratrol, alkaloids, and polyphenols have been shown to acquire antioxidant and anti-inflammatory properties, which can help protect neurons from oxidative stress and inflammation-induced damage. These compounds can also modulate signalling pathways involved in cell survival, differentiation, and apoptosis, thereby encouraging neuronal survival and function (Naoi et al., 2019; Kumar & Khanum, 2012; Venkatesan et al., 2015). Table 6.1 shows some common phytochemicals and their source, mechanism, and medicinal activity in terms of neurodegeneration for the treatment of Alzheimer's disease.

TABLE 6.1

Common Phytochemicals and Their Neurodegeneration Mechanisms for Treating Alzheimer's Disease

Phytochemicals	Plant Source	Structure	Mechanistic Activity	Medicinal Use	References
Curcumin	*Curcuma longa*	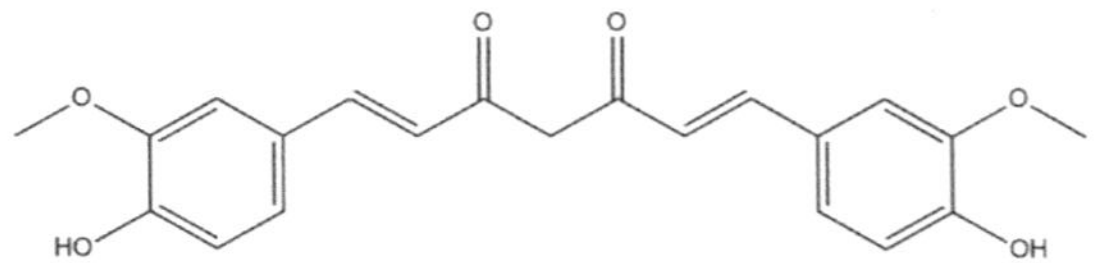 (1E,6E)-1,7-bis (4-hydroxy- 3-methoxyphenyl) -1,6- heptadiene-3,5-dione	• Curcumin binds and limits aggregation of the β-sheet conformations of amyloid characteristic to reduce the levels of cholesterol in serum and to induce BDNF secretion. • Curcumin induces antioxidant response elements, chelates iron, increases the heat shock system to improve clearance of toxic aggregates, and restores the equilibrium of the inflammatory system.	Antioxidant and anti-inflammatory	(Hu et al., 2015; Darvesh et al., 2012)
Resveratrol	*Vitis vinifera*	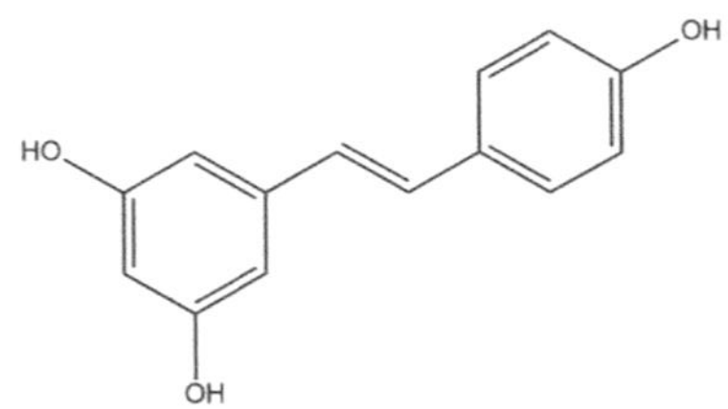 5-[(E)-2-(4-hydroxyphenyl)ethenyl]benzene-1,3-diol	• Activation of SIRT1 and vitagenes prevent the deleterious effects triggered by oxidative stress. • Stimulates the release of neurotrophy and neurite outgrowth.	Anti-Aβ deposition, antioxidant, and anti-inflammatory	(Sun et al., 2010; Yadav et al., 2022)
Quercetin	*Morus alba*	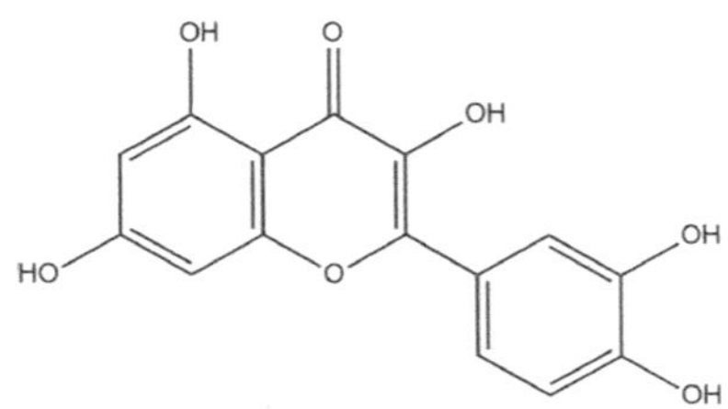 2-(3,4-dihydroxyphenyl)-3,5,7-trihydroxychromen-4-one	• Quercetin downstream signalling pathways like nuclear factor Nrf2, PON2, JNK, TNF-α, peroxisome, PGC-1α, Sirtuins, MAPK signalling cascades, CREB, and PI3K/Akt. • Inhibits acetylcholinesterase inhibitors.	Acetylcholinesterase inhibitors, antioxidant, and anti-inflammatory	(Grewal et al., 2021; Costa et al., 2016
Physostigmine	*Physostigma venenosum*	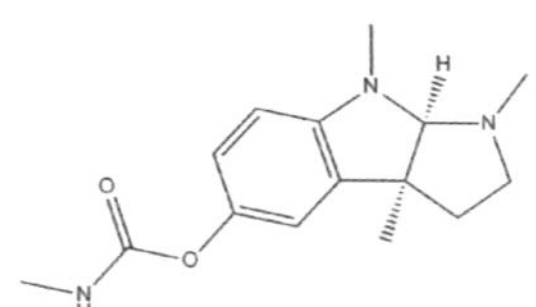 (3aS,8aR)-1,3a,8-Trimethyl-1,2,3,3a,8,8a-hexahydropyrrolo[2,3-b]indol-5-yl methylcarbamate	• Nicotinic and muscarinic receptors are stimulated due to rise in acetylcholine at the synapses.	Ach-E inhibitors	(Kalb et al., 2013)

(*Continued*)

TABLE 6.1
(Continued)

Phytochemicals	Plant Source	Structure	Mechanistic Activity	Medicinal Use	References
Caffeine	*Erigeron annuus*	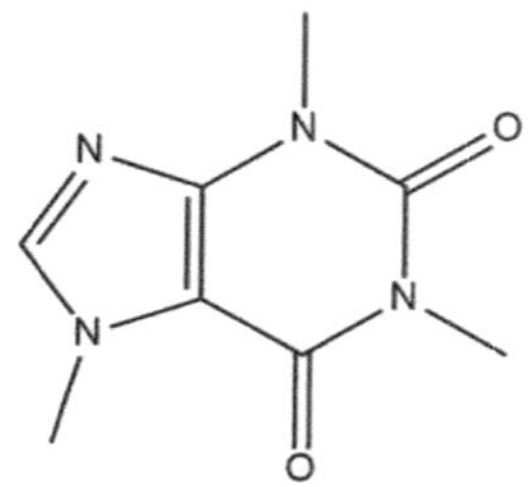 1,3,7-Trimethylpurine-2,6-dione	• Caffeine lowers increased levels of oxidative stress, prevents the activation of adenosine A2A, which controls the buildup of α-β, lower TAU hyperphosphorylation, and lowers the buildup of misfolded proteins like α-synuclein. • Caffeine shows strong antagonistic effect against adenosine A2A receptor and agonistic effect with Nrf-2 to regulate the cellular homeostasis in the brain by reducing oxidative stress.	β-Site amyloid precursor protein cleaving enzyme 1 inhibitors, anti-oxidant, and Ach-E inhibitor	(Ikram et al., 2020; Kolahdouzan & Hamadeh, 2017)
Berberin	*Coptis chinensis*	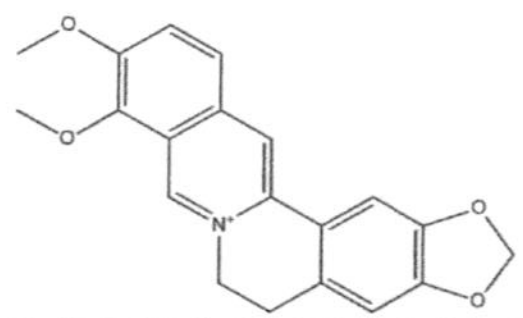 7,8,13,13a-tetradehydro-9,10-dimethoxy-2,3-[methylenebis(oxy)]berbinium	• Anti-cholinesterase and reduces Aβ and TAU protein plaques. • Hyperphosphorylation of amyloid precursor protein promotes clearance of Aβ, reduces oxidant and inflammatory damage, and increases TAU hyperphosphorylation and degradation.	Neuroinflammation and neuroprotection	(Shou & Shaw, 2022; Cheng et al., 2022)
Nicotine	*Nicotiana* sp	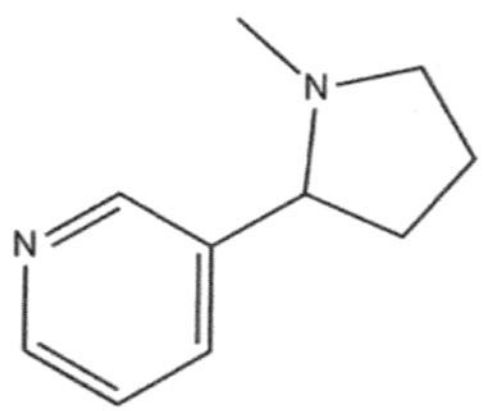 3-(1-methyl-2-pyrrolidinyl)pyridine	• Impaired Aβ-mediated neurotoxicity and NMDA-induced excitotoxicity. • Increases synaptic acetylcholine and nicotine receptor agonists. • Nicotine reduces amyloid β via activating 7nAChRs through the MAPK, NF-κB, and c-Myc pathway. • Inhibits nitric oxide production.	Acetylcholine inhibitors	(Dong et al., 2020; Piao et al., 2009)
Astaxanthin	*Haematococcus pluviali*	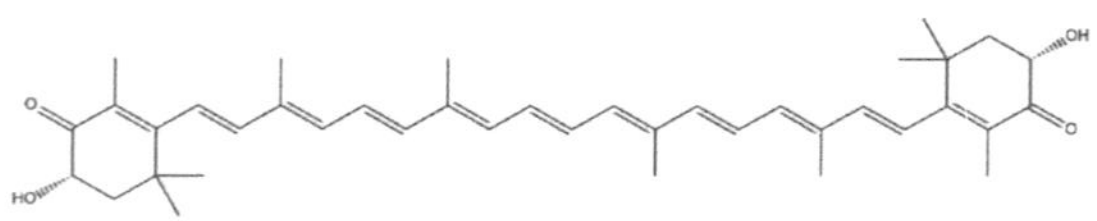 (6S)-6-Hydroxy-3-[(1E,3E,5E,7E,9E,11E,13E,15E,17E)-18-[(4S)-4-hydroxy-2,6,6-trimethyl-3-oxo-1-cyclohexenyl]-3,7,12,16-tetramethyloctadeca-1,3,5,7,9,11,13,15,17-nonaenyl]-2,4,4-trimethyl-1-cyclohex-2-enone	• Shows antioxidant effect because of ketone-bearing ionone rings by stabilizing radicals more and synergistically with polyene backbone. • Anti-inflammatory impact; inhibits the formation of NO, the NF-κB-dependent signalling pathway, and the expression of inflammatory mediators such TNF-α, PGE2, and IL-1.	Anti-inflammatory, anti-apoptotic, antioxidant effects, promotes neural plasticity	(Galasso et al., 2018; Grimmig et al., 2017)

Stilbene	*Vitis vinifera*	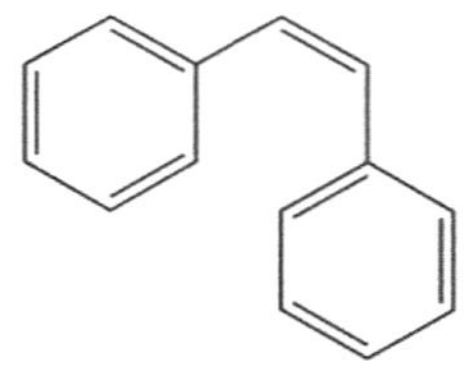 (Z)-1,2-Diphenylethene	• Inhibits Aβ fibrils.	Antioxidant and anti-inflammatory	(Rivière et al., 2007; Reinisalo et al., 2015)
Sesamin	*Sesamum indicum*	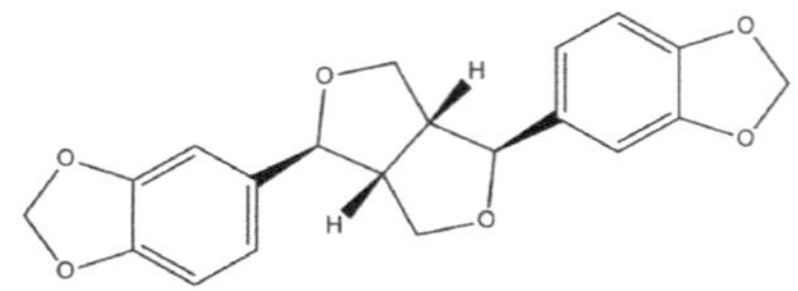 5,5′-[(1S,3aR,4S,6aR)-Tetrahydro-1H,3H-furo[3,4-c]furan-1,4-diyl]bis(2H-1,3-benzodioxole)	• Reduces Aβ toxicity	Anti-inflammatory, antioxidative, and anti-apoptotic	(Keowkase et al., 2018)
Galantamine	*Galanthus caucasicus, Galanthus woronowii*	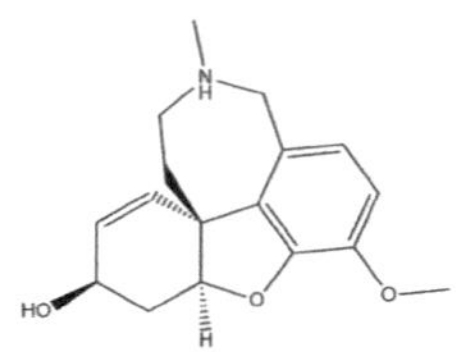 (4aS,6R,8aS)- 5,6,9,10,11,12- hexahydro- 3-methoxy- 11-methyl- 4aH- [1]benzofuro[3a,3,2-ef] [2] benzazepin-6-ol	• Galantamine improves cholinergic neuron activity and communication by inhibiting the enzyme that breaks down acetylcholine in the synaptic cleft. • Galantamine inhibits Aβ induced neurotoxicity by enhancing α7nAChR expression.	Ach-E inhibitor	(Lin et al., 2020; Lilienfeld, 2002)
Huperzine A	*Huperzia serrata*	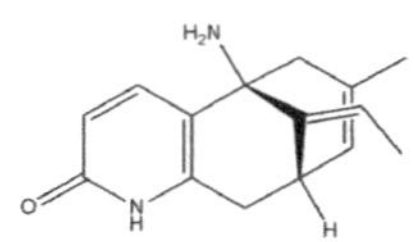 (1R,9S,13E)- 1-Amino- 13-ethylidene- 11-methyl- 6-azatricyclo[7.3.1.02,7] trideca- 2(7),3,10- trien- 5-one	• Reduces neuronal loss; decreases Aβ neurotoxicity, attenuates p53-mediated cell death, and downregulates NF-kβ signalling and cp65 translocation-related neurodegeneration. • Improved neuronal survival through PI3K/TrkB/mTOR signalling induces LTP and synaptic transmission by activation of BDNF/TrkB. • Decreased TAU phosphorylation and Aβ accumulation by GSK-3β downregulation.	Neuritogenesis, acetylcholinesterase inhibitor, neuro-inflammation and neuro-protection	(Friedli & Inestrosa, 2021; Tun & Herzon, 2012)

(Continued)

TABLE 6.1
(Continued)

Phytochemicals	Plant Source	Structure	Mechanistic Activity	Medicinal Use	References
Nobiletin	*Citrus* peel	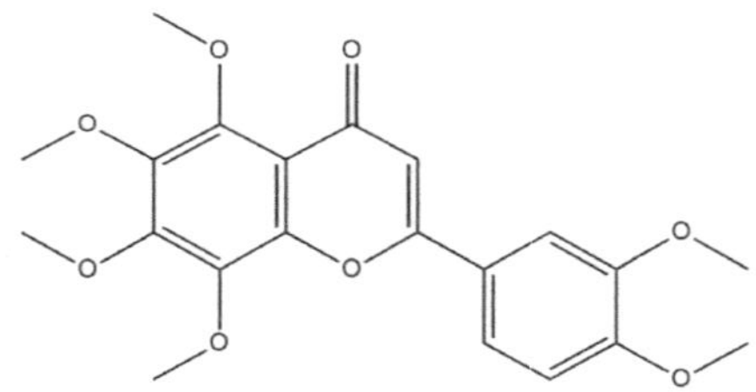 2-(3,4-dimethoxyphenyl)-5,6,7,8-tetramethoxychromen-4-one	• Increases the expression and activity of neprilysin to degrade amyloid-β. • Anti-neuroinflammatory capacity by inhibiting the lipopolysaccharide-induced production and secretion of proinflammatory mediators, such as NO, TNF-α, IL-1β, and IL-6.	Anti-amyloid-β deposition, antioxidative	(Nakajima & Ohizumi, 2019)
Sinecatechins	*Camellia sinensis*	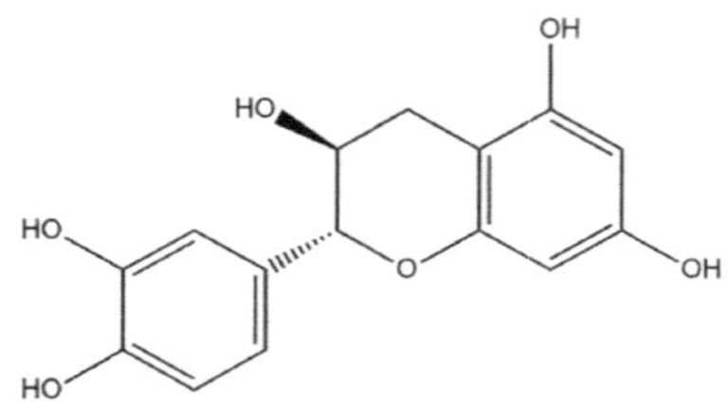 (2R,3S)-2-(3,4-dihydroxyphenyl)-3,4-dihydro-2H-chromene-3,5,7-triol	• Induces NGF secretion and neurite outgrowth	Antioxidative and anti-inflammatory	(D'Onofrio et al., 2017)
Ginkgolides A, B, C, and J	*Ginkgo biloba*	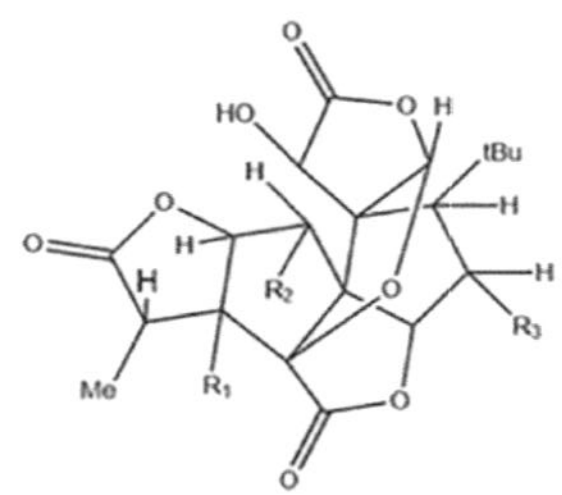	• Inhibits IL-5 and IL-13 inhibition, MAPK inhibition, TLR inhibition, Inhibits arachidonic acid inflammatory pathway, TNF receptor, and NF-kB.	Anti-inflammatory, antioxidative, and anti-apoptotic	(Gachowska et al., 2021; Singh et al., 2019)
Kaempferol, Isorhamnetin	*Ginkgo biloba*	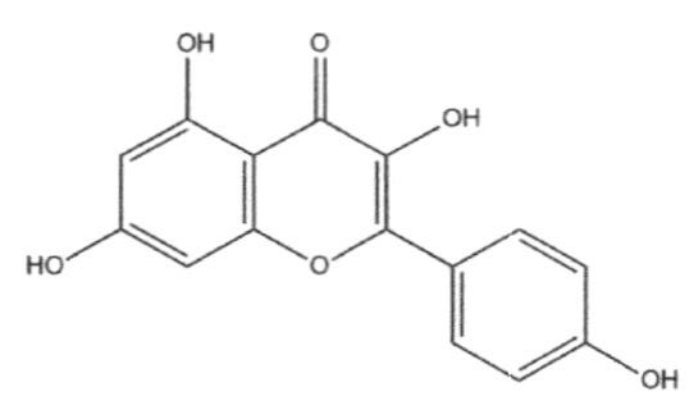 3,5,7-trihydroxy-2-(4-hydroxyphenyl)chromen-4-one	• Anti-inflammatory effects through the inhibition of pro-inflammatory mediators such as IL-6, IL-1β, NO, PGE2, and LTB4 as well as phosphorylation of NF-κB. • Reduces oxidative stress by blocking the MAPK and NF-κB pathways. • Reduces Aβ-triggered secretion of IL-6 and thereby oxidative stress	Anti-inflammatory and antioxidant	(Lim et al., 2023; Santos et al., 2020; Wei et al., 2022)

		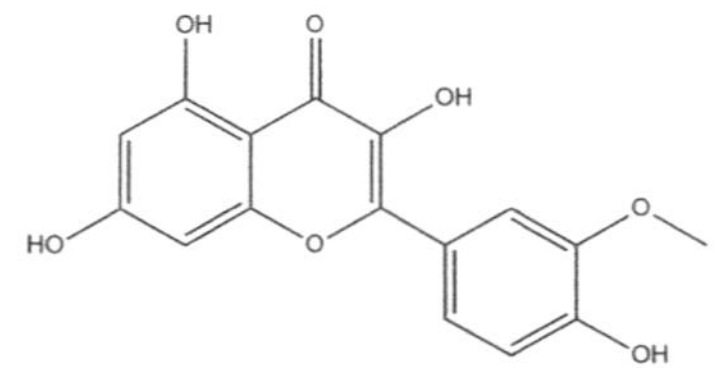 3,5,7-Trihydroxy-2-(4-hydroxy-3-methoxyphenyl)-4H-1-benzopyran-4-one			
Amentoflavone	*Selaginella tamariscina*	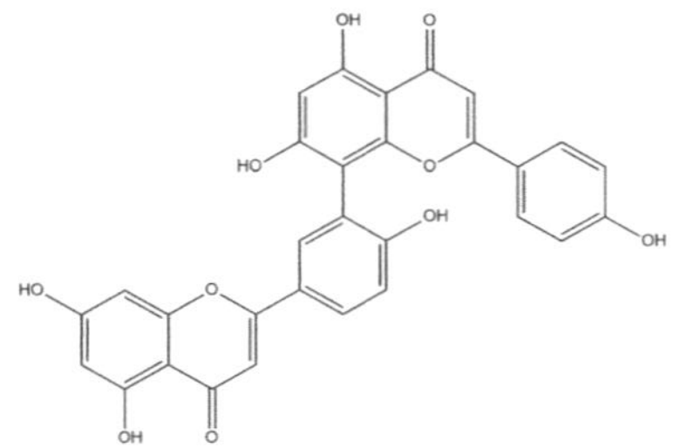 8-[5-(5,7-Dihydroxy-4-oxo-4H-1-benzopyran-2-yl)-2-hydroxyphenyl]-5,7-dihydroxy-2-(4-hydroxyphenyl)-4H-1-benzopyran-4-one	• Prevents amyloid deposition in the CNS by facilitating the cellular uptake of Aβ after binding to Aβ fibrils and destabilizes the β-sheet structure of Aβ fibrils to form amorphic Aβ aggregates.	Antioxidant and neuroprotective	(Han et al., 2022)
Morin	*Morus alba, Psidium guajava, Psidium guajava*	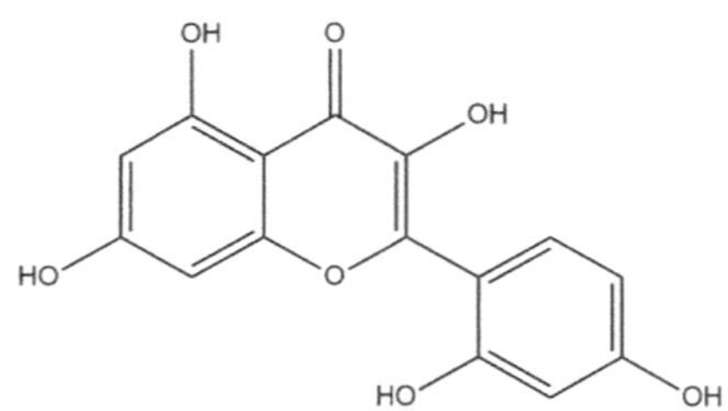 2-(2,4-Dihydroxyphenyl)-3,5,7-trihydroxy-4H-1-benzopyran-4-one	• Inhibits GRP78 expression, decreases ROS and apoptosis.	Antioxidant and anti-inflammatory	(Kataria et al., 2018; Mo et al., 2019)
Apigenin	*Matricaria chamomilla*	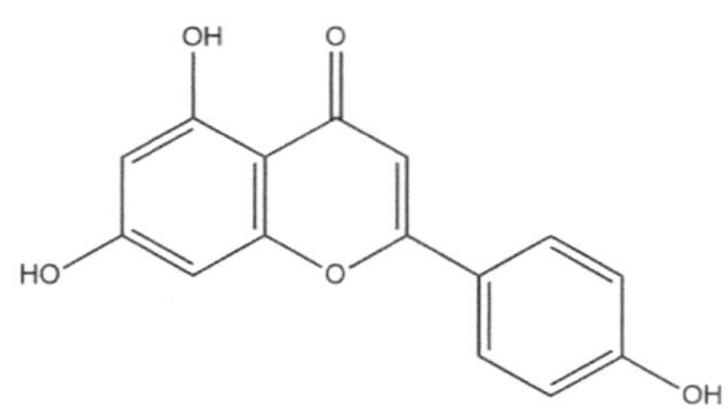 5,7-dihydroxy-2-(4-hydroxyphenyl)chromen-4-one	• Modulates inflammatory cytokine mRNA expression and reduces the expression of OX42, IL-6, and gp130. • Induces the expression of BDNF.	Neuroprotective, anti-inflammatory	(Dourado et al., 2020)

(Continued)

TABLE 6.1
(Continued)

Phytochemicals	Plant Source	Structure	Mechanistic Activity	Medicinal Use	References
Epimedium	*Epimedium grandiflorum*	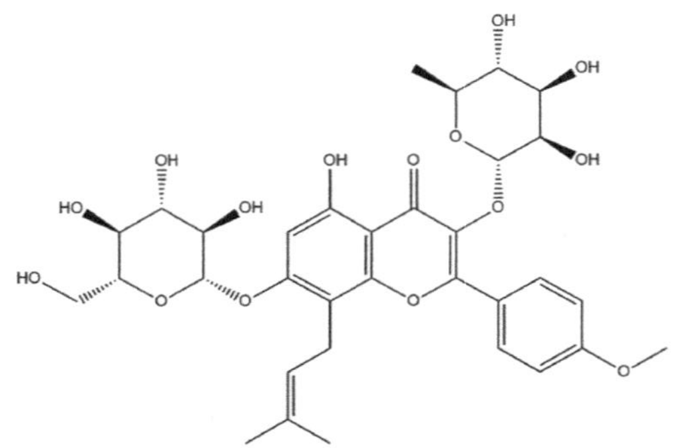 5-Hydroxy-2-(4-methoxyphenyl)-8-(3-methylbut-2-enyl)-7-[(2S,3R,4S,5S,6R)-3,4,5-trihydroxy-6-(hydroxymethyl)oxan-2-yl]oxy-3-[(2S,3R,4R,5R,6S)-3,4,5-trihydroxy-6-methyloxan-2-yl]oxychromen-4-one	• Anti-proteotoxic potency as it reduces the Aβ.	Anti-proteotoxic, antioxidant	(Cai et al., 2019)
Epimedium	*Epimedium grandiflorum*	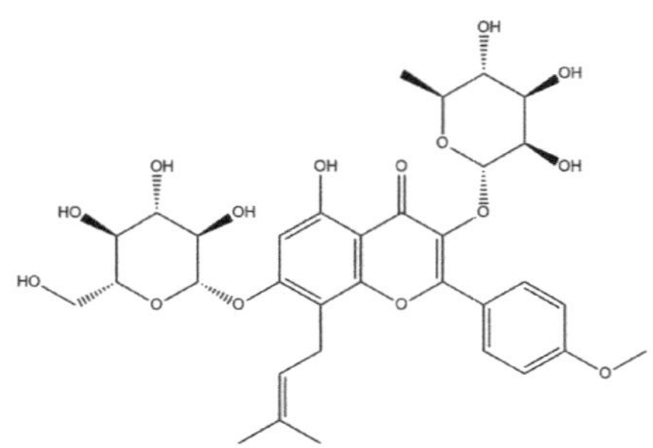 5-Hydroxy-2-(4-methoxyphenyl)-8-(3-methylbut-2-enyl)-7-[(2S,3R,4S,5S,6R)-3,4,5-trihydroxy-6-(hydroxymethyl)oxan-2-yl]oxy-3-[(2S,3R,4R,5R,6S)-3,4,5-trihydroxy-6-methyloxan-2-yl]oxychromen-4-one	• Anti-proteotoxic potency as it reduces the Aβ.	Anti-proteotoxic, antioxidant	(Cai et al., 2019)

6.4.1 Polyphenolic Molecules and Flavonoids

Plants produce polyphenolic substances called flavonoids through the phenylpropanoid pathway (Devi et al., 2021). Flavonoids are omnipresent compounds in plants such as apples, red fruits, onions, citrus fruits, nuts, and beverages such as tea, coffee, beer, and red wine. More than 6000 flavonoids are identified now, and they are produced by plants for growth and defence against stress and cold (Calderaro et al., 2022). Flavonoids can cross the blood–brain barrier to have a direct effect on the brain or in neurodegenerative disorders by exerting effects such as neuronal antioxidant, anti-inflammation, and anti-amyloidogenic properties, chelating metal, and ameliorating cognition and neuroprotection (Kelsey et al., 2010; Grosso et al., 2013). Blocking of Aβ and tau aggregation, scavenging free radicals, and sequestering metal ions are the most common mechanistic activities for flavonoids such as epicatechin-3-gallate, gossypetin, naringenin, quercetin, and myricetin (Li et al., 2022).

6.4.1.1 Polyphenol and Flavonoid Inhibition of APP Processing and Aβ Deposition

In the aromatic ring, the presence of catechol ring and OH groups in 3,5 position is responsible for the antioxidant activity of quercetin. Quercetin can indirectly prevent the synthesis of Aβ-peptides by limiting the activity of the enzyme BACE-1 that cleaves APP at its β-site. Flavonoid OH groups and phenolic rings are necessary for amyloid protein's universal β-sheet structures to interact non-covalently with one another. With the formation of hydrophobic contacts and hydrogen bonds with the created β-sheets, quercetin can prevent Aβ from aggregating (Alghamdi et al., 2022).

By activating the PI3K/Akt pathway via the oestrogen receptor alpha and the pro-ADAM10 cleavage enzyme furin via a PI3K-independent mechanism, EGCG switches APP processing into the nonamyloidogenic route (Fernandez et al., 2010). By a $GABA_A$ receptor-dependent mechanism, baicalein inhibits pro-amyloidogenic cleavage while increasing nonamyloidogenic cleavage of APP (Zhang et al., 2013). Neprilysin, a zinc-metalloprotease, is an Aβ-degrading enzyme that is diminished in affected areas of the AD. Epigallocatechin enhances neprilysin liberation from astrocytes in an Akt/PI3K-dependent way, leading to the destruction of exogenous Aβ (Yamamoto et al., 2017).

Amentoflavone, a naturally occurring bi-flavonoid, has a high affinity for Aβ-fibrils and disrupts their β-sheet structure to develop amorphous Aβ-fibril aggregates, which can be taken up by glial and N2a neuronal cells. Amentoflavone is a strong antioxidant that can chelate Cu2+, which reduces the generation of ROS and the neurotoxicity brought on by the Aβ-Cu2+ complex. By modifying the AMPK/GSK3 signalling pathway, amentoflavone shields the brain from Aβ-persuaded neurotoxicity and cognitive decline (Han et al., 2022).

6.4.1.2 Polyphenol and Flavonoid Inhibition of TAU Phosphorylation and Aggregation

Several different kinases, including Erk, Akt, p38, AMPK, GSK-3β, cdk5, and protein phosphatase 2A, regulate the phosphorylation of tau (Calderaro et al., 2022). Flavonoids can scavenge free radicals and ROS due to their antioxidant properties that are produced in the brain during aging and disease processes. These free radicals can cause oxidative destruction to cellular modules, including tau proteins, and promote its aggregation. By neutralizing these free radicals, flavonoids can prevent or slow down the aggregation of tau.

Flavonoids' capacity to regulate kinase and phosphatase activities prevents tau phosphorylation and aggregation (Hole et al., 2021). Phosphatases are enzymes that remove phosphate groups from proteins, whereas kinases add phosphate groups to proteins. In AD, tau develops hyperphosphorylation due to the upregulation of certain kinases (GSK-3β, Cdk5) and the downregulation of certain phosphatases (protein phosphatase 2A, and calcineurin). Flavonoids have been shown to inhibit the activity of these kinases and activate the activity of these phosphatases, leading to a reduction in tau phosphorylation and aggregation (Fan et al., 2022). Flavonoids also inhibit tau phosphorylation and aggregation by modulating the activity of heat shock proteins. Heat shock proteins (HSP100,

HSP90, HSP70, HSP60, HSP40 and Small HSP) are a family of proteins that help to protect cells from stress and promote protein folding and stability. Flavonoids have been shown to upregulate the expression of certain heat shock proteins, which can help to prevent the aggregation of tau and promote its clearance (Campanella et al., 2018).

Quercetin suppresses the kinase activity of GSK-3β, which reduces the hyperphosphorylation of tau proteins and contributes to the neuroprotective effect of the compound (Calderaro et al., 2022). Naringenin regulates PI3K/Akt/GSK-3K pathway to reduce hyper-phosphorylation of tau (Zhang et al., 2017). Baicalin, hesperidin, naringin, and Epigallocatechin gallate prevent GSK-3β from being phosphorylated at Ser9 by activation of PI3K/Akt (Gu et al., 2016; Wang et al., 2012). Proanthocyanidins and catechin inhibit tau phosphorylation at Ser202/Ser205 (AT8), Ser396/Ser404 (PHF1), Thr212/Ser214 (AT100) [139] and Thr181 by ERK1/2 pathway (Santa-Maria et al., 2012).

6.4.1.3 Polyphenols and Flavonoids in Neuroinflammation

Neuroinflammation is a CNS immune response where glial cells (microglia) become stimulated during neuroinflammation and release inflammatory mediators such as cytokines, chemokines, and ROS/RNS. Through interactions with neuronal receptors, flavonoids can affect the kinase signalling pathways, transcription factors, and gene and/or protein expression that control memory and learning processes in the hippocampus (Devi et al., 2021). By reducing NO generation, wogonin prevents the activation-induced death of C6 glial cells. Wogonin inhibits NF-κB mediated activation of iNOS to inhibit NO generation (Kim et al., 2022). Luteolin and apigenin increase BDNF to protect dopaminergic neurons by diminishing oxidative stress, microglial activation, and neuroinflammation. By disabling the NLRP3 (NOD-, LRR-, and pyrin domain-containing protein 3) assembly, apigenin suppresses IL-1 and caspase-1. By the activation of the PPAR-gamma, apigenin prevents the activation of the NLRP3 inflammasome to reduce oxidative stress and neuroinflammation; hence, apigenin improved psychological, biochemical, and mitochondrial enzyme function while reducing dopaminergic neuronal degeneration (Li et al., 2016; Anusha et al., 2017).

Daidzein stops the stimulation of amyloid-beta or lipopolysaccharide by downregulating the expression of TNF-α, IL-1, and IL-6 in elementary astrocytes (Anusha et al., 2017). Catechin (flavonoid) prevented tert-butyl-hydroperoxide-encouraged cell death in murine microglia N9 cells by inhibiting NF-kB activity, activating p53, and increasing extracellular signal-regulated protein kinase (ERK) (Lau et al., 2007). By lowering iNOS, NO, and TNF-α levels, as well as inhibiting p38 signalling cascades and the STAT-1 transcription factor, naringenin therapy inhibits neuronal cell death in LPS/IFN activated glial cells (Vafeiadou et al., 2009). Biochanin A shields dopaminergic neurons from LPS-induced injury by preventing microglia from becoming activated and lowering superoxide, TNF-α, and NO levels (Chen et al., 2007).

By preventing the release of TNF-α, IL-1β, ERK, c-Jun NH (2)-terminal kinase (JNK), and p38 MPAKs in LPS-stimulated BV-2 microglial cells, nobiletin reduces neuroinflammation (Cui et al., 2010). TNF-α, IL-6, and IL-1 are all reduced by naringenin, which also significantly controls the NF-κB signalling pathway, which is linked to inflammatory processes. By significantly reducing the phosphorylation and nuclear translocation of P65, a subunit of NF-κB, and by raising the levels of siSIRT1 in the hippocampus, naringenin causes a reduction in the expression of NF-κB (Vafeiadou et al., 2009; Gao et al., 2010). Rutin inhibits LPO and activates SOD, catalase, GPx, and total GSH activity to protect dopaminergic neurons from 6-OHDA-induced neurotoxicity (Khan et al., 2012).

6.4.1.4 Role of Polyphenol and Flavonoid in Oxidative Stress

The buildup of ROS, such as hydrogen peroxide (H_2O_2), hydroxyl radicals (•OH), and superoxide radicals (•O2$^-$), is linked to the death of neurons. Apoptosis in neuronal cells is caused by ROS accumulation originate, which also triggers protein oxidation, DNA damage, and lipid peroxidation (LPO). Using antioxidants such as flavonoids can help alleviate the toxicity free radicals. These flavonoids can reduce the aetiology of numerous neurological illnesses and counteract the toxicity of oxidative stress (Wen et al., 2020). Human neural SH-SY5Y neuronal cells treated with flavonoids such as quercitrin, iso-quercitrin, and afzelin showed positive effects via controlling inflammation, apoptosis, and ROS-scavenging by preventing the production of nitric oxide synthase, cyclooxygenase-2, and caspase activation (Kim et al., 2018). By boosting dopaminergic neurons, dopamine, and serotonin levels in the striatum, baicalein reduced aberrant behaviour, oxidative stress, and the astroglia response (Cheng et al., 2008).

By activating the transcription factor Nrf2 in a PI3K-dependent manner, flavonoids function as antioxidants. The expression of antioxidant-related genes, such as GPx, CAT, heme-oxygenase 1 (HO-1), and SOD, is increased when Nrf2 activates the antioxidant response element (ARE) (Thimmulappa et al., 2002). Kaempferol enhances motor coordination, increases striatal dopamine and its metabolites, boosts SOD and GSH activity, lowers LPO levels (Li & Pu, 2011). The activation of AMPK and the drug's binding to the oestrogen receptor (ER) are both factors in genistein's antioxidant actions since they encourage the production of antioxidant enzymes such as SOD, CAT, and GPx (Suzuki et al., 2002; Borras et al., 2006). Naringenin increase GSH, an endogenous antioxidant, and additionally increases SOD, CAT, GPx, and glutathione reductase (GR) activity and decrease in H_2O_2 and protein carbonyls levels (Chtourou et al., 2014). Quercetin increases catalase and SOD activity, as well as mitochondrial complex-I activity (Karuppagounder et al., 2013).

6.4.1.5 Miscellaneous Role of Polyphenols and Flavonoids

Proteotoxicity refers to the damage to cells and tissues because of the accumulation of misfolded proteins. Misfolded proteins can lead to a range of diseases, including AD, PD, and HD. Flavonoids have neuroprotective effects and can prevent or reduce misfolded protein aggregation. Flavonoids may also promote the clearance of misfolded proteins. Epigallocatechin gallate promotes the degradation of misfolded proteins by the proteasome, a cellular mechanism for removing damaged proteins (Devi et al., 2021). Flavonoids are helpful in managing endoplasmic reticulum (ER) stress due to unfolded or misfolded proteins. Flavonoids have also been shown to improve ER protein-folding capacity and enhance the degradation of misfolded proteins in the ER. This can be achieved through the upregulation of chaperone proteins, such as glucose-regulated protein 78 (GRP78) and protein disulfide isomerase (PDI), which help to fold and stabilize proteins in the ER (Takano et al., 2007; Martucciello et al., 2020). Kaempferol mitigates ER stress-stimulated cell death by targeting caspase 3/7 (Abdullah & Ravanan, 2018).

6.4.2 Alkaloids

Aside from polyphenolics, several alkaloids have received considerable attention as possible treatment options for neurodegenerative diseases, including AD. Alkaloids, primarily isoquinoline (berberine, galantamine, morphine, and related alkaloids), indole (physostigmine and related alkaloids), terpenoids, and steroidal alkaloids have been endowed with potential inhibitory effects on AChE and Aβ (D'Onofrio et al., 2017).

Many studies reported that berberine demonstrated a neuroprotective effect by decreasing extracellular Aβ production and β-secretase (Kim et al., 2001; Asai et al., 2007). Berberine pretreatment decreased the production of Aβ-induced interleukin-6 (IL-6) along with monocyte chemoattractant protein-1 (MCP-1) (Kim et al., 2014). Compared with human embryonic kidney 293 (HEK293) cells, berberine inhibited the production of Aβ and β-secretase and therefore enhanced the extracellular signal-regulated kinase 1 and 2 (p-ERK1/2) expression. This demonstrated that it can inhibit Aβ production via activation of ERK1/2-induced Aβ precursor protein cleaving enzyme.

Furthermore, pretreatment with berberine decreased TNF-α, and LDH activities by downregulation of TNF receptor 1 in neuroblastoma cells (Kim et al., 2014; Xu et al., 2013). Berberine therapy in amyloid precursor protein (APP) transgenic mice reduced the number of errors in memory tasks of the Morris water maze (MWM), demonstrating improved cognitive impairment. It allows Aβ plaque immunostaining correlated with area reduction of hippocampus and cortex sections in the brain of TgCRND8 mice. Berberine is also supposed to decrease the Aβ plaque accumulation in the brain by inhibiting the degree of microgliosis and astrocytosis in the brain (Kim et al., 2014). Henceforth, the mechanism supporting the role of berberine as a neuroprotective agent is mainly expressed for Aβ plaque inhibition, tau plaque phosphorylation and degradation, AchE inhibition, and hyperphosphorylation of APP (Shou & Shaw, 2022; Cheng et al., 2022).

Galantamine, another alkaloid, was initially used to treat poliomyelitis disease. However, because of its double action on the cholinergic system by inhibiting AChE and modulating nAChR activity, galantamine was used for cognitive function and neuroprotection (Ng et al., 2015; Scott & Goa, 2000). It improves cholinergic neuron activity and communication by inhibiting the enzyme responsible for breaking down acetylcholine in the synaptic cleft (Lin et al., 2020).

The binding of galantamine to AChE modulates neurotransmission by inhibiting the breakdown of ACh and thereby enhancing ACh levels (Ng et al., 2015). It also inhibits Aβ induced neurotoxicity by enhancing α7nAChR expression (Lin et al., 2020; Lilienfeld, 2002). Additionally, it modulates nicotinic neurotransmission by the allosteric activation of both presynaptic and postsynaptic nAChR that mediates the release of potential neurotransmitters of the brain, including dopamine, ACh, glutamate (Ng et al., 2015; Woodruff-Pak et al., 2001). Thus, galantamine clarifies its effect on cognitive function by acting as an agonist on the nAChR.

Another indole alkaloid, physostigmine, isolated from *Physostigma venenosum* (Fabaceae) seeds, is also known to have a mild effect on neurodegenerative diseases. Although physostigmine acts by inhibiting both AChE and BChE, it is more selective and known for its AChE inhibition. Due to this inhibition of the AchE, Ach levels in the synapses rise, leading to stimulation of both nicotinic and muscarinic receptors (Kalb et al., 2013). However, the use of physostigmine for the improvement of cognitive function is limited due to its gastrointestinal effects and very short half-life. To overcome these limitations associated with physostigmine, certain modifications in the physostigmine structure have been carried out, leading to the development of a new synthetic molecule, rivastigmine, with a safety and therapeutic profile. Apart from AChE and BChE inhibition, physostigmine, rivastigmine, and other derivatives also act via different mechanisms contributing to improved cognitive function in AD (Konrath et al., 2013).

Nicotine, another alkaloid isolated from *Nicotiana* sp, acts as a neuroprotective agent by binding to Aβ and activating 7nAChRs through the MAPK, NF-κB, and c-Myc pathways, resulting in inhibition of Aβ aggregation (Dong et al., 2020, Shim et al., 2008). It is also known to increase synaptic ACh and act as a nicotinic receptor agonist. It also inhibits nitric acid production suggesting its role as neuroprotectant (Piao et al., 2009). Although these studies suggested the role of nicotine in AD, it only showed enhanced attention rather than enhancing memory (Howe & Price, 2001). Owing to its short half-life, toxicity, tachyphylaxis, and high addictiveness, nicotine has been debarred from clinical use (Ferrea & Winterer, 2009).

Another alkaloid caffeine present in coffee beans is well-known for its CNS-stimulating efficacy. It mainly acts as AChE inhibitor, and owing to its mild short-term CNS-stimulating potential, it acts as neuroprotectant in AD (Howe & Price, 2001). In addition to increasing cognitive function, caffeine promotes attention, vigilance, mood, and arousal. Caffeine at low to moderate doses (50–300 mg) acts as a CNS-stimulating agent, while high dose results in insomnia, restlessness, tachycardia, and even anxiety (Temple et al., 2017). It is also reported to reduce Aβ plaques and associated neurotoxicity, thereby enhancing cerebral performance in AD induced mice.

This neuroprotection efficacy of caffeine may be attributed via the caspase-3 expression reduction. In vivo animal models demonstrated its neuroprotective efficacy in AD. In the transgenic mice model, caffeine reduces the β-secretase, presenilin 1 (PS1) and Aβ plaque deposition in the hippocampus and cortex. This reduction of β-secretase is mediated via the down-regulation of cRaf-1/NF-κB (Arendash et al., 2009). It stimulates the pathway through the activation of PKA and CREB and by blocking phospho-ERK and JNK expressions in the transgenic mice mode, thereby suggesting neuronal protection in AD.

Moreover, caffeine primarily reduces tau phosphorylation and proteolytic tau fragments in mice models (THY-Tau22) (Laurent et al., 2014). Mice treated with caffeine also showed a significant improvement of spatial memory in the MWM test (Stazi et al., 2022). Caffeine also has significant antioxidant and anti-inflammatory potential (Castaldo et al., 2021), which is attributed to the phosphorylated tau protein reduction and decrease in Aβ plaques (Prasanthi et al., 2010). Thus, caffeine lowers the increased levels of oxidative stress and prevents the activation of adenosine A2A, which controls the buildup of α-β, lowers tau hyperphosphorylation, and lowers the buildup of misfolded proteins such as α-synuclein. Caffeine shows a strong antagonistic effect against adenosine A2A receptor and an agonistic effect with Nrf-2 to regulate cellular homeostasis by reducing oxidative stress (Ikram et al., 2020; Kolahdouzan & Hamadeh, 2017). These reports and working mechanisms have led to caffeine entering the clinical trial phase for AD therapy (Londzin et al., 2021).

6.4.3 Miscellaneous Phytochemicals

The production of specific omega-3 polyunsaturated fatty acids (n-3 PUFAs) consequential metabolites, such as the resolvins D (RvD) and E (RvE) series, maresins (MaR), and protectins (PD), collectively known as specialised pro-resolving mediators (SPMs), which act as potent anti-inflammatory agents and can protect against inflammation. RvDs reduce gliosis in neurodegenerative illnesses and promote serotonin levels in depression. In neurodegeneration, protectins inhibited neurite and dendritic retraction and apoptosis, but maresins decreased cell death.

Every SPM lowered the levels of pro-inflammatory cytokines. Resolvens activated mTOR and MAP/ERK signalling in models of neurological disorders and neurodegeneration, whereas resolvens and maresins both increased the NF-κB pathway in models of depression (Giacobbe et al., 2020; Avallone et al., 2019; Layé et al., 2015). Curcumin hinders the development of misfolded aggregation proteins, such as Aβ and tau, inflammation, and oxidative stress, acting as a neuroprotective agent and preventing various pathways involved in neurodegeneration (Lo Cascio et al., 2021). Additionally, resveratrol promotes mitochondrial biogenesis, which has been demonstrated to be AMPK dependent. Resveratrol may impact neuronal energy balance and contribute to the neuroprotective effects of resveratrol by activating AMPK in neurons. Resveratrol enhances the release of neurotrophy and neurite outgrowth and activates SIRT1 and vitagenes to block the harmful effects brought on by oxidative stress (Sun et al., 2010; Yadav et al., 2022). Sulforaphane act via ERK1/2, p38, JNK, and NF-κB pathway to control tau hyperphosphorylation and tangles and reduce Ab plaques by cleaving of BACE1 (Schepici et al., 2020).

6.5 THE ROLE OF ASTROCYTES IN NEURODEGENERATION OF ALZHEIMER'S DISEASE

Astrocytes are a type of glial cell found in the central nervous system (CNS). They have numerous functions, including maintaining homeostasis, regulating neurotransmitter levels, and providing support to neurons (Cai et al., 2017). However, in AD, astrocytes play a significant role in the pathology of the disease. In AD, the accumulation of Aβ plaques and tau protein tangles in the brain leads to neuronal damage and death. Astrocytes respond to this damage by becoming reactive, a process called astrogliosis (Frost & Li, 2017; Kumar et al., 2023). Reactive astrocytes release cytokines and other inflammatory molecules that can further damage neurons and exacerbate the disease. Astrocytes are also involved in the clearance of Aβ from the brain. They express various Aβ receptors, including LRP1 and RAGE, and can internalize and degrade Aβ. However, in AD, astrocytes' ability to clear Aβ is impaired, which leads to the accumulation of plaques in the brain (Hulshof et al., 2022).

Additionally, astrocytes can modulate the synaptic activity of neurons through the release of glutamate and other neurotransmitters. In AD, reactive astrocytes can release excessive amounts of glutamate, leading to excitotoxicity and further neuronal damage (Di Benedetto et al., 2022).

Astrocytes are also involved in the regulation of cerebral blood flow. They can dilate or constrict blood vessels in response to changes in neuronal activity, a process called neurovascular coupling. In AD, impaired neurovascular coupling leads to decreased blood flow to the brain, which exacerbates the disease's pathology (Katsouri et al., 2020).

Moreover, recent studies have shown that astrocytes can also contribute to the spread of tau pathology in the brain. Astrocytes can accept tau from neurons and transfer it to other neurons, a process called transcellular propagation. This propagation of tau pathology is thought to contribute to the spread of the disease in the brain (Vogels et al., 2020).

6.6 CONCLUSION

There is currently no consensus on the main mechanism(s) causing the neuronal degeneration in AD; however, a number of possible pathogenic pathways have surfaced. The role of APP and Aβ in the neurodegenerative process and their connection to the neurofibrillary disease brought on by tau are important issues. Aβ is produced in neurons and then released into the extracellular environment, where it can either be destroyed or removed by astrocytes and microglia. The accumulation is caused by an increase or decrease in Aβ production and clearance. Tau is primarily expressed in neurons and is heavily regulated by several PTMs. Through many processes, abnormal PTMs, LLPS, and pathogenic tau seeds can lead to tau aggregation and accumulation. Glial cells are crucial in the processes of seeding and dispersion, where tau pathology can spread as the disease progresses. Genetic factors may influence or induce the pathophysiology of AD. Late-onset AD is mostly linked to a collection of genes concentrated in glial cells, such as APOE and TREM2, which are crucial for Aβ clearance and glial function. Early-onset AD is primarily caused by mutations in APP and PS1/2, which are involved in Aβ generation.

Phytochemicals are plant-derived compounds that have potential neuroprotective effects against AD. Several types of phytochemicals, such as flavonoids, curcumin, resveratrol, and catechins, have anti-inflammatory, antioxidant, and anti-amyloidogenic properties that can prevent or slow down neurodegenerative processes. Medicinal plants may serve as significant lead molecules in the effort to create a new class of brain communication-enhancing medications due to the increasing interest in the creation of drugs that can enhance brain function.

Another potential therapeutic approach may involve specifically targeting astroglial homeostatic cascades. It was discovered that a neuroprotective medication increased the expression of astroglial glutamate uptake transporters. Such a medication would be a potent weapon for slowing the progression of neurodegenerative illnesses. Further research is required to optimize the dosage of phytochemicals for neuroprotection. Overall, phytochemicals offer a promising avenue for the prevention and treatment of AD by reducing inflammation, oxidative stress, and accumulation of toxic proteins in the brain.

ABBREVIATIONS

AD- Alzheimer's disease; **Aβ-** Amyloid Beta; **BDNF-** Brain-Derived Neurotrophic Factor; **ACh-** Acetylcholine; **SIRT1-** Sirtuin 1; **Nrf2-** Erythroid 2-related Factor 2; **PON2-** Paraoxonase-2; **JNK-** c-Jun N-terminal Kinase; **TNF-α-** Tumour Necrosis Factor alpha; **PGC-1α-** Proliferator-Activated Receptor Gamma Coactivator 1-alpha; **MAPKs-** Mitogen-Activated Protein Kinases; **CREB-** Cyclic AMP Response Element Binding Protein; **PI3K/Akt-** Phosphoinositide 3- Kinase; **Ach-E-** Acetylcholinesterase; **NMDA-** N-Methyl-D-Aspartate; **7nAChRs-** α7 Nicotinic Acetylcholine Receptor; **NF- κB-** Nuclear Factor Kappa B; **PGE2-** Prostaglandin E2; **IL-** Interleukin; **TrkB-** Tyrosine Receptor Kinase B; **GSK-3-** Glycogen Synthase Kinase-3; **LTP-** Long-term Potentiation; **LTD-** Long-term depression; **NO-** Nitric Oxide; **NGF-** Nerve Growth Factor; **TLR-** Toll-Like Receptor; **LTB4-** Leukotriene B4; **APP-** Amyloid Precursor Protein; **BACE-1-** Beta-site Amyloid Precursor Protein Cleaving Enzyme 1; **ADAM 10-** A Disintegrin and Metalloproteinase

Domain-containing Protein 10; **$GABA_A$**_γ- Aminobutyric acid type A; **LPS-** Lipopolysaccharides; **GPx-** Glutathione Peroxidase; **CAT-** Catalase; **SOD-** Superoxide Dismutase; **iNOS-** inducible Nitric Oxide Synthase; **TAU-** Tubulin Associated Unit; **GSH-** Glutathione; **BuChE-** Butyrylcholinesterase; **FDA-** Food and Drug Administration; **sAPPβ-** Smaller β-APP fragments; **CTFβ-** C-terminal β fragment; **AICD-** APP intracellular domain; **ApoE-** Apolipoprotein E; **IDE-** Insulin degrading enzyme; **NFTs-** Neurofibrillary tangles; **MTs-** Microtubules; **P-tau-** Phosphorylated tau; **MAP-** Microtubule associated protein; **PHFs-** Paired helical filaments; **SFs-** Straight filaments; **iNFTs-** Intraneuronal NFTs; **eNFTs-** Extra neuronal NFTs; **CAA-** Cerebral amyloid angiopathy; **DPA-** Deep perforator arteriopathy; **BBB-** Blood—brain barrier; **GVD-** Granulovacuolar degeneration; **LOAD-** Late onset AD; **EOAD-** Early onset AD; **MTLs-** Medial temporal lobes; **PKA-** Protein kinase A; **CREB-** phospho-cAMP response element-binding protein; **MWM-** Morris water maze

ENDNOTE

* Corresponding author

REFERENCES

Abdullah, A., & Ravanan, P. (2018). Kaempferol mitigates endoplasmic reticulum stress induced cell death by targeting caspase 3/7. *Scientific Reports*, *8*(1), 2189. https://doi.org/10.1038/s41598-018-20499-7

Alghamdi, A., Birch, D. J. S., Vyshemirsky, V., & Rolinski, O. J. (2022). Impact of the flavonoid quercetin on beta-amyloid aggregation revealed by intrinsic fluorescence. *The Journal of Physical Chemistry B*, *126*(38), 7229–7237. https://doi.org/10.1021/acs.jpcb.2c02763

Alquezar, C., Arya, S., & Kao, A. W. (2021). Tau post-translational modifications: Dynamic transformers of tau function, degradation, and aggregation. *Frontiers in Neurology*, *11*. https://doi.org/10.3389/FNEUR.2020.595532

Altmann, A., Ng, B., Landau, S. M., Jagust, W. J., & Greicius, M. D. (2015). Regional brain hypometabolism is unrelated to regional amyloid plaque burden. *Brain : A Journal of Neurology*, *138*(Pt 12), 3734–3746. https://doi.org/10.1093/BRAIN/AWV278

Alzheimer, A. (1907). About a peculiar disease of the cerebral cortex. *Centralblatt für Nervenheilkunde Psychiatrie*, *30*(1), 177–179.

Amor, S., Peferoen, L. A., Vogel, D. Y., Breur, M., van der Valk, P., Baker, D., & van Noort, J. M. (2014). Inflammation in neurodegenerative diseases—an update. *Immunology*, *142*(2), 151–166. https://doi.org/10.1111/imm.12233

Anusha, C., Sumathi, T., & Joseph, L. D. (2017). Protective role of apigenin on rotenone induced rat model of Parkinson's disease: Suppression of neuroinflammation and oxidative stress mediated apoptosis. *Chemico-Biological Interactions*, *269*, 67–79. https://doi.org/10.1016/j.cbi.2017.03.016

Arendash, G. W., Mori, T., Cao, C., Mamcarz, M., Runfeldt, M., Dickson, A., . . . & Potter, H. (2009). Caffeine reverses cognitive impairment and decreases brain amyloid-β levels in aged Alzheimer's disease mice. *Journal of Alzheimer's Disease*, *17*(3), 661–680.

Arndt, J. W., Qian, F., Smith, B. A., Quan, C., Kilambi, K. P., Bush, M. W., Walz, T., . . . & Weinreb, P. H. (2018). Structural and kinetic basis for the selectivity of aducanumab for aggregated forms of amyloid-β. *Scientific Reports*, *8*(1), 1–16. https://doi.org/10.1038/s41598-018-24501-0

Asai, M., Iwata, N., Yoshikawa, A., Aizaki, Y., Ishiura, S., Saido, T. C., & Maruyama, K. (2007). Berberine alters the processing of Alzheimer's amyloid precursor protein to decrease Aβ secretion. *Biochemical and Biophysical Research Communications*, *352*(2), 498–502. https://doi.org/10.1016/j.bbrc.2006.11.043

Attems, J., Jellinger, K., Thal, D. R., & Van Nostrand, W. (2011). Review: Sporadic cerebral amyloid angiopathy. *Neuropathology and Applied Neurobiology*, *37*(1), 75–93. https://doi.org/10.1111/J.1365-2990.2010.01137.X

Avallone, R., Vitale, G., & Bertolotti, M. (2019). Omega-3 fatty acids and neurodegenerative diseases: New evidence in clinical trials. *International Journal of Molecular Sciences*, *20*(17). https://doi.org/10.3390/ijms20174256

Bhatti, G. K., Reddy, A. P., Reddy, P. H., & Bhatti, J. S. (2020). Lifestyle modifications and nutritional interventions in aging-associated cognitive decline and Alzheimer's disease. *Frontiers in Aging Neuroscience*, *11*, 369. https://doi.org/10.3389/fnagi.2019.00369

Borras, C., Gambini, J., Gomez-Cabrera, M. C., Sastre, J., Pallardo, F. V., Mann, G. E., & Vina, J. (2006). Genistein, a soy isoflavone, up-regulates expression of antioxidant genes: Involvement of estrogen receptors, ERK1/2, and NFkappaB. *The FASEB Journal*, *20*(12), 2136–2138. https://doi.org/10.1096/fj.05-5522fje

Bourassa, P., Tremblay, C., Schneider, J. A., Bennett, D. A., & Calon, F. (2020). Brain mural cell loss in the parietal cortex in Alzheimer's disease correlates with cognitive decline and TDP-43 pathology. *Neuropathology and Applied Neurobiology*, *46*(5), 458–477. https://doi.org/10.1111/NAN.12599

Bredesen, D. E. (2009). Neurodegeneration in Alzheimer's disease: Caspases and synaptic element interdependence. *Molecular Neurodegeneration*, *4*, 1–10.

Bredesen, D. E., John, V., & Galvan, V. (2010). Importance of the caspase cleavage site in amyloid-β protein precursor. *Journal of Alzheimer's Disease*, *22*(1), 57–63.

Bredesen, D. E., Mehlen, P., & Rabizadeh, S. (2004). Apoptosis and dependence receptors: A molecular basis for cellular addiction. *Physiological Reviews*, *84*(2), 411–430.

Busche, M. A., & Hyman, B. T. (2020). Synergy between amyloid-β and tau in Alzheimer's disease. *Nature Neuroscience*, *23*(10), 1183–1193. https://doi.org/10.1038/s41593-020-0687-6

Cai, W., Wu, J., Wang, X., Huang, J., Shen, Z., & Chen, X. (2019). Epimedium flavonoids mitigate proteotoxicity and extend healthspan via DAF-16 in C. elegans. *Traditional Medicine and Modern Medicine*, *2*(1), 19–25. https://doi.org/10.1142/s2575900019500046

Cai, Z., Wan, C.-Q., & Liu, Z. (2017). Astrocyte and Alzheimer's disease. *Journal of Neurology*, *264*, 2068–2074.

Calderaro, A., Patane, G. T., Tellone, E., Barreca, D., Ficarra, S., Misiti, F., & Lagana, G. (2022). The neuroprotective potentiality of flavonoids on Alzheimer's disease. *International Journal of Molecular Sciences*, *23*(23). https://doi.org/10.3390/ijms232314835

Calfio, C., Gonzalez, A., Singh, S. K., Rojo, L. E., & Maccioni, R. B. (2020). The emerging role of nutraceuticals and phytochemicals in the prevention and treatment of Alzheimer's disease. *Journal of Alzheimer's Disease*, *77*(1), 33–51.

Campanella, C., Pace, A., Caruso Bavisotto, C., Marzullo, P., Marino Gammazza, A., Buscemi, S., & Palumbo Piccionello, A. (2018). Heat shock proteins in Alzheimer's disease: Role and targeting. *International Journal of Molecular Sciences*, *19*(9). https://doi.org/10.3390/ijms19092603

Castaldo, L., Toriello, M., Sessa, R., Izzo, L., Lombardi, S., Narváez, A., & Grosso, M. (2021). Antioxidant and anti-inflammatory activity of coffee brew evaluated after simulated gastrointestinal digestion. *Nutrients*, *13*(12), 4368. https://doi.org/10.3390/nu13124368

Chang, W. P., Koelsch, G., Wong, S., Downs, D., Da, H., Weerasena, V., . . . & Tang, J. (2004). In vivo inhibition of Abeta production by memapsin 2 (beta-secretase) inhibitors. *Journal of Neurochemistry*, *89*(6), 1409–1416. https://doi.org/10.1111/J.1471-4159.2004.02452.X

Chen, H. Q., Jin, Z. Y., & Li, G. H. (2007). Biochanin a protects dopaminergic neurons against lipopolysaccharide-induced damage through inhibition of microglia activation and proinflammatory factors generation. *Neuroscience Letters*, *417*(2), 112–117. https://doi.org/10.1016/j.neulet.2006.11.045

Chen, X., Chen, M., Schafer, N. P., & Wolynes, P. G. (2020). Exploring the interplay between fibrillization and amorphous aggregation channels on the energy landscapes of tau repeat isoforms. *Proceedings of the National Academy of Sciences of the United States of America*, *117*(8), 4125–4130. https://doi.org/10.1073/PNAS.1921702117/-/DCSUPPLEMENTAL

Cheng, Y., He, G., Mu, X., Zhang, T., Li, X., Hu, J., . . . Du, G. (2008). Neuroprotective effect of baicalein against MPTP neurotoxicity: Behavioral, biochemical and immunohistochemical profile. *Neuroscience Letters*, *441*(1), 16–20. https://doi.org/10.1016/j.neulet.2008.05.116

Cheng, Z., Kang, C., Che, S., Su, J., Sun, Q., Ge, T., . . . Cui, R. (2022). Berberine: A promising treatment for neurodegenerative diseases. *Frontiers in Pharmacology*, *13*, 845591. https://doi.org/10.3389/fphar.2022.845591

Chtourou, Y., Fetoui, H., & Gdoura, R. (2014). Protective effects of naringenin on iron-overload-induced cerebral cortex neurotoxicity correlated with oxidative stress. *Biological Trace Element Research*, *158*(3), 376–383. https://doi.org/10.1007/s12011-014-9948-0

Coronel, R., Bernabeu-Zornoza, A., Palmer, C., Muñiz-Moreno, M., Zambrano, A., Cano, E., & Liste, I. (2018). Role of amyloid precursor protein (APP) and its derivatives in the biology and cell fate specification of neural stem cells. *Molecular Neurobiology*, *55*(9), 7107–7117. https://doi.org/10.1007/S12035-018-0914-2

Costa, L. G., Garrick, J. M., Roque, P. J., & Pellacani, C. (2016). Mechanisms of neuroprotection by quercetin: Counteracting oxidative stress and more. *Oxidative Medicine and Cellular Longevity*, *2016*, 2986796. https://doi.org/10.1155/2016/2986796

Crary, J. F., Trojanowski, J. Q., Schneider, J. A., Abisambra, J. F., Abner, E. L., Alafuzoff, I., . . . & Nelson, P. T. (2014). Primary age-related tauopathy (PART): A common pathology associated with human aging. *Acta Neuropathologica*, *128*(6), 755–766. https://doi.org/10.1007/S00401-014-1349-0

Cui, Y., Wu, J., Jung, S. C., Park, D. B., Maeng, Y. H., Hong, J. Y., . . . & Eun, S. Y. (2010). Anti-neuroinflammatory activity of nobiletin on suppression of microglial activation. *Biological and Pharmaceutical Bulletin*, *33*(11), 1814–1821. https://doi.org/10.1248/bpb.33.1814

Cummings, J., Aisen, P., Lemere, C., Atri, A., Sabbagh, M., & Salloway, S. (2021). Aducanumab produced a clinically meaningful benefit in association with amyloid lowering. *Alzheimer's Research and Therapy*, *13*(1), 1–3. https://doi.org/10.1186/S13195-021-00838-Z/METRICS

Darvesh, A. S., Carroll, R. T., Bishayee, A., Novotny, N. A., Geldenhuys, W. J., & Van der Schyf, C. J. (2012). Curcumin and neurodegenerative diseases: A perspective. *Expert Opinion on Investigational Drugs*, *21*(8), 1123–1140. https://doi.org/10.1517/13543784.2012.693479

Devi, S., Kumar, V., Singh, S. K., Dubey, A. K., & Kim, J. J. (2021). Flavonoids: Potential candidates for the treatment of neurodegenerative disorders. *Biomedicines*, *9*(2). https://doi.org/10.3390/biomedicines9020099

Di Benedetto, G., Burgaletto, C., Bellanca, C. M., Munafò, A., Bernardini, R., & Cantarella, G. (2022). Role of microglia and astrocytes in Alzheimer's disease: From neuroinflammation to Ca2+ homeostasis dysregulation. *Cells*, *11*(17), 2728.

Dickson, T. C., & Vickers, J. C. (2001). The morphological phenotype of beta-amyloid plaques and associated neuritic changes in Alzheimer's disease. *Neuroscience*, *105*(1), 99–107. https://doi.org/10.1016/S0306-4522(01)00169-5

Dong, Y., Bi, W., Zheng, K., Zhu, E., Wang, S., Xiong, Y., . . . & Cheng, Y. (2020). Nicotine prevents oxidative stress-induced hippocampal neuronal injury through alpha7-nAChR/Erk1/2 signaling pathway. *Frontiers in Molecular Neuroscience*, *13*, 557647. https://doi.org/10.3389/fnmol.2020.557647

D'Onofrio, G., Panza, F., Frisardi, V., Solfrizzi, V., Imbimbo, B. P., Paroni, G., . . . & Pilotto, A. (2012). Advances in the identification of γ-secretase inhibitors for the treatment of Alzheimer's disease. *Expert Opinion on Drug Discovery*, *7*(1), 19–37. https://doi.org/10.1517/17460441.2012.645534

D'Onofrio, G., Sancarlo, D., Ruan, Q., Yu, Z., Panza, F., Daniele, A., . . . Seripa, D. (2017). Phytochemicals in the treatment of Alzheimer's disease: A systematic review. *Current Drug Targets*, *18*(13), 1487–1498. https://doi.org/10.2174/1389450117666161102121553

Doody, R. S., Raman, R., Farlow, M., Iwatsubo, T., Vellas, B., Joffe, S., . . . & Mohs, R. (2013). A phase 3 trial of semagacestat for treatment of Alzheimer's disease. *The New England Journal of Medicine*, *369*(4), 341–350. https://doi.org/10.1056/NEJMOA1210951

Dourado, N. S., Souza, C. D. S., de Almeida, M. M. A., Bispo da Silva, A., Dos Santos, B. L., Silva, V. D. A., . . . & Costa, S. L. (2020). Neuroimmunomodulatory and neuroprotective effects of the flavonoid apigenin in in vitro models of neuroinflammation associated with Alzheimer's disease. *Frontiers in Aging Neuroscience*, *12*, 119. https://doi.org/10.3389/fnagi.2020.00119

Edison, P., Archer, H. A., Hinz, R., Hammers, A., Pavese, N., Tai, Y. F., . . . & Brooks, D. J. (2007). Amyloid, hypometabolism, and cognition in Alzheimer disease. *Neurology*, *68*(7), 501–508. https://doi.org/10.1212/01.WNL.0000244749.20056.D4

Epperly, T. D., Dunay, M. A., & Boice, J. L. (2017). Alzheimer disease: Pharmacologic and nonpharmacologic therapies for cognitive and functional symptoms. *American Family Physician*, *95*(12), 771–778.

Fan, X., Xia, L., Zhou, Z., Qiu, Y., Zhao, C., Yin, X., & Qian, W. (2022). Tau acts in concert with kinase/phosphatase underlying synaptic dysfunction. *Front Aging Neurosci*, *14*, 908881. https://doi.org/10.3389/fnagi.2022.908881

Fernandez, J. W., Rezai-Zadeh, K., Obregon, D., & Tan, J. (2010). EGCG functions through estrogen receptor-mediated activation of ADAM10 in the promotion of non-amyloidogenic processing of APP. *FEBS Letters*, *584*(19), 4259–4267. https://doi.org/10.1016/j.febslet.2010.09.022

Ferrea, S., & Winterer, G. (2009). Neuroprotective and neurotoxic effects of nicotine. *Pharmacopsychiatry*, *42*(6), 255–265. https://doi.org/10.1055/s-0029-1224138

Ferreira-Vieira, T. H., Guimaraes, I. M., Silva, F. R., & Ribeiro, F. M. (2016). Alzheimer's disease: Targeting the cholinergic system. *Current Neuropharmacology*, *14*(1), 101. https://doi.org/10.2174/1570159X13666150716165726

Friedli, M. J., & Inestrosa, N. C. (2021). Huperzine a and its neuroprotective molecular signaling in Alzheimer's disease. *Molecules*, *26*(21). https://doi.org/10.3390/molecules26216531

Frost, G. R., & Li, Y.-M. (2017). The role of astrocytes in amyloid production and Alzheimer's disease. *Open biology*, *7*(12), 170228.

Furgerson, M., Clark, J. K., Crystal, J. D., Wagner, J. J., Fechheimer, M., & Furukawa, R. (2014). Hirano body expression impairs spatial working memory in a novel mouse model. *Acta Neuropathologica Communications*, *2*(1). https://doi.org/10.1186/S40478-014-0131-9

Gachowska, M., Szlasa, W., Saczko, J., & Kulbacka, J. (2021). Neuroregulatory role of ginkgolides. *Molecular Biology Reports*, *48*(7), 5689–5697. https://doi.org/10.1007/s11033-021-06535-2

Galasso, C., Orefice, I., Pellone, P., Cirino, P., Miele, R., Ianora, A., . . . & Sansone, C. (2018). On the Neuroprotective role of astaxanthin: New perspectives? *Marine Drugs*, *16*(8). https://doi.org/10.3390/md16080247

Gao, J., Wang, W. Y., Mao, Y. W., Graff, J., Guan, J. S., Pan, L., . . . & Tsai, L. H. (2010). A novel pathway regulates memory and plasticity via SIRT1 and miR-134. *Nature*, *466*(7310), 1105–1109. https://doi.org/10.1038/nature09271

Gauthier, S., Feldman, H. H., Schneider, L. S., Wilcock, G. K., Frisoni, G. B., Hardlund, J. H., . . . & Wischik, C. M. (2016). Efficacy and safety of tau-aggregation inhibitor therapy in patients with mild or moderate Alzheimer's disease: A randomised, controlled, double-blind, parallel-arm, phase 3 trial. *Lancet (London, England)*, *388*(10062), 2873–2884. https://doi.org/10.1016/S0140-6736(16)31275-2

Giacobbe, J., Benoiton, B., Zunszain, P., Pariante, C. M., & Borsini, A. (2020). The anti-inflammatory role of omega-3 polyunsaturated fatty acids metabolites in pre-clinical models of psychiatric, neurodegenerative, and neurological disorders. *Front Psychiatry*, *11*, 122. https://doi.org/10.3389/fpsyt.2020.00122

Gibson, P. H., & Tomlinson, B. E. (1977). Numbers of Hirano bodies in the hippocampus of normal and demented people with Alzheimer's disease. *Journal of the Neurological Sciences*, *33*(1–2), 199–206. https://doi.org/10.1016/0022-510X(77)90193-9

Glenner, G. G., & Wong, C. W. (1984). Alzheimer's disease and down's syndrome: Sharing of a unique cerebrovascular amyloid fibril protein. *Biochemical and Biophysical Research Communications*, *122*(3), 1131–1135. https://doi.org/10.1016/0006-291X(84)91209-9

Goel, P., Chakrabarti, S., Goel, K., Bhutani, K., Chopra, T., & Bali, S. (2022). Neuronal cell death mechanisms in Alzheimer's disease: An insight. *Frontiers in Molecular Neuroscience*, *15*, 409. https://doi.org/10.3389/FNMOL.2022.937133/BIBTEX

González-Reyes, R. E., Nava-Mesa, M. O., Vargas-Sánchez, K., Ariza-Salamanca, D., & Mora-Muñoz, L. (2017). Involvement of astrocytes in Alzheimer's disease from a neuroinflammatory and oxidative stress perspective. *Frontiers in Molecular Neuroscience*, *10*, 427. https://doi.org/10.3389/FNMOL.2017.00427/BIBTEX

Greenberg, S. M., Bacskai, B. J., Hernandez-Guillamon, M., Pruzin, J., Sperling, R., & van Veluw, S. J. (2019). Cerebral amyloid angiopathy and Alzheimer disease—one peptide, two pathways. *Nature Reviews Neurology*, *16*(1), 30–42. https://doi.org/10.1038/s41582-019-0281-2

Grewal, A. K., Singh, T. G., Sharma, D., Sharma, V., Singh, M., Rahman, M. H., . . . & Abdel-Daim, M. M. (2021). Mechanistic insights and perspectives involved in neuroprotective action of quercetin. *Biomed Pharmacother*, *140*, 111729. https://doi.org/10.1016/j.biopha.2021.111729

Grimmig, B., Kim, S. H., Nash, K., Bickford, P. C., & Douglas Shytle, R. (2017). Neuroprotective mechanisms of astaxanthin: A potential therapeutic role in preserving cognitive function in age and neurodegeneration. *Geroscience*, *39*(1), 19–32. https://doi.org/10.1007/s11357-017-9958-x

Grosso, C., Valentão, P., Ferreres, F., & Andrade, P. B. (2013). The use of flavonoids in central nervous system disorders. *Current Medicinal Chemistry*, *20*(37), 4694–4719. https://doi.org/10.2174/09298673113209990155.

Grundke-Iqbal, I., Iqbal, K., Tung, Y. C., Quinlan, M., Wisniewski, H. M., & Binder, L. I. (1986). Abnormal phosphorylation of the microtubule-associated protein tau (tau) in Alzheimer cytoskeletal pathology. *Proceedings of the National Academy of Sciences of the United States of America*, *83*(13), 4913–4917. https://doi.org/10.1073/PNAS.83.13.4913

Gu, X. H., Xu, L. J., Liu, Z. Q., Wei, B., Yang, Y. J., Xu, G. G., . . . & Wang, W. (2016). The flavonoid baicalein rescues synaptic plasticity and memory deficits in a mouse model of Alzheimer's disease. *Behavioural Brain Research*, *311*, 309–321. https://doi.org/10.1016/j.bbr.2016.05.052

Guan, Y.-H., Zhang, L.-J., Wang, S.-Y., Deng, Y.-D., Zhou, H.-S., Chen, D.-Q., . . . & Lan-Chun Zhang, C. (2022). The role of microglia in Alzheimer's disease and progress of treatment. *Ibrain*, *8*(1), 37–47. https://doi.org/10.1002/IBRA.12023

Guo, T., Noble, W., & Hanger, D. P. (2017). Roles of tau protein in health and disease. *Acta Neuropathologica*, *133*(5), 665–704. https://doi.org/10.1007/S00401-017-1707-9

Guo, T., Zhang, D., Zeng, Y., Huang, T. Y., Xu, H., & Zhao, Y. (2020). Molecular and cellular mechanisms underlying the pathogenesis of Alzheimer's disease. *Molecular Neurodegeneration*, *15*(1), 1–37.

Haass, C., Kaether, C., Thinakaran, G., & Sisodia, S. (2012). Trafficking and proteolytic processing of APP. *Cold Spring Harbor Perspectives in Medicine*, *2*(5). https://doi.org/10.1101/CSHPERSPECT.A006270

Han, B. H., Cofell, B., Everhart, E., Humpal, C., Kang, S. S., Lee, S. K., & Kim-Han, J. S. (2022). Amentoflavone promotes cellular uptake and degradation of amyloid-beta in neuronal cells. *International Journal of Molecular Sciences*, *23*(11). https://doi.org/10.3390/ijms23115885

Hansen, D. V., Hanson, J. E., & Sheng, M. (2018). Microglia in Alzheimer's disease. *The Journal of Cell Biology*, *217*(2), 459. https://doi.org/10.1083/JCB.201709069

Hardy, J. A., & Higgins, G. A. (1992). Alzheimer's disease: The amyloid cascade hypothesis. *Science* (New York, NY), *256*(5054), 184–185. https://doi.org/10.1126/SCIENCE.1566067

Hecht, M., Krämer, L. M., von Arnim, C. A. F., Otto, M., & Thal, D. R. (2018). Capillary cerebral amyloid angiopathy in Alzheimer's disease: Association with allocortical/hippocampal microinfarcts and cognitive decline. *Acta Neuropathologica*, *135*(5), 681–694. https://doi.org/10.1007/S00401-018-1834-Y

Hemonnot, A. L., Hua, J., Ulmann, L., & Hirbec, H. (2019). Microglia in Alzheimer disease: Well-known targets and new opportunities. *Frontiers in Cellular and Infection Microbiology*, *9*(July), 233. https://doi.org/10.3389/FNAGI.2019.00233/BIBTEX

Hole, K. L., Williams, R. J., & Rendeiro, C. (2021). Flavonoids as an intervention for Alzheimer's disease: Progress and hurdles towards defining a mechanism of action1. *Brain Plasticity*, *6*(2), 167–192. https://doi.org/10.3233/bpl-200098

Hondius, D. C., Koopmans, F., Leistner, C., Pita-Illobre, D., Peferoen-Baert, R. M., Marbus, F., . . . & Smit, A. B. (2021). The proteome of granulovacuolar degeneration and neurofibrillary tangles in Alzheimer's disease. *Acta Neuropathologica*, *141*(3), 341–358. https://doi.org/10.1007/S00401-020-02261-4

Hou, X., Watzlawik, J. O., Cook, C., Liu, C. C., Kang, S. S., Lin, W. L., . . . & Springer, W. (2020). Mitophagy alterations in Alzheimer's disease are associated with granulovacuolar degeneration and early tau pathology. *Alzheimer's & Dementia: The Journal of the Alzheimer's Association*, *17*(3). https://doi.org/10.1002/ALZ.12198

Howe, M. N., & Price, I. R. (2001). Effects of transdermal nicotine on learning, memory, verbal fluency, concentration, and general health in a healthy sample at risk for dementia. *International Psychogeriatrics*, *13*(4), 465–475. https://doi.org/10.1017/S1041610201007888

Hu, S., Maiti, P., Ma, Q., Zuo, X., Jones, M. R., Cole, G. M., & Frautschy, S. A. (2015). Clinical development of curcumin in neurodegenerative disease. *Expert Review of Neurotherapeutics*, *15*(6), 629–637. https://doi.org/10.1586/14737175.2015.1044981

Huang, L. K., Chao, S. P., & Hu, C. J. (2020). Clinical trials of new drugs for Alzheimer disease. *Journal of Biomedical Science*, *27*(1), 1–13. https://doi.org/10.1186/S12929-019-0609-7/TABLES/3

Hulshof, L. A., Van Nuijs, D., Hol, E. M., & Middeldorp, J. (2022). The role of astrocytes in synapse loss in Alzheimer's disease: A systematic review. *Frontiers in Cellular Neuroscience*, *16*, 899251.

Ikram, M., Park, T. J., Ali, T., & Kim, M. O. (2020). Antioxidant and neuroprotective effects of caffeine against Alzheimer's and Parkinson's disease: Insight into the role of Nrf-2 and A2AR signaling. *Antioxidants (Basel)*, *9*(9). https://doi.org/10.3390/antiox9090902

Jeremic, D., Jiménez-Díaz, L., & Navarro-López, J. D. (2021). Past, present and future of therapeutic strategies against amyloid-β peptides in Alzheimer's disease: A systematic review. *Ageing Research Reviews*, *72*, 101496. https://doi.org/10.1016/j.arr.2021.101496

Kalb, A., von Haefen, C., Sifringer, M., Tegethoff, A., Paeschke, N., Kostova, M., . . . & Spies, C. D. (2013). Acetylcholinesterase inhibitors reduce neuroinflammation and -degeneration in the cortex and hippocampus of a surgery stress rat model. *PLoS ONE*, *8*(5), e62679. https://doi.org/10.1371/journal.pone.0062679

Karuppagounder, S. S., Madathil, S. K., Pandey, M., Haobam, R., Rajamma, U., & Mohanakumar, K. P. (2013). Quercetin up-regulates mitochondrial complex-I activity to protect against programmed cell death in rotenone model of Parkinson's disease in rats. *Neuroscience*, *236*, 136–148. https://doi.org/10.1016/j.neuroscience.2013.01.032

Kataria, R., Sobarzo-Sanchez, E., & Khatkar, A. (2018). Role of morin in neurodegenerative diseases: A review. *Current Topics in Medicinal Chemistry*, *18*(11), 901–907. https://doi.org/10.2174/1568026618666180711153416

Katsouri, L., Birch, A. M., Renziehausen, A. W., Zach, C., Aman, Y., Steeds, H., . . . & Ries, M. (2020). Ablation of reactive astrocytes exacerbates disease pathology in a model of Alzheimer's disease. *Glia*, *68*(5), 1017–1030.

Kelsey, N. A., Wilkins, H. M., & Linseman, D. A. (2010). Nutraceutical antioxidants as novel neuroprotective agents. *Molecules*, *15*(11), 7792–7814. https://doi.org/10.3390/molecules15117792

Keowkase, R., Shoomarom, N., Bunargin, W., Sitthithaworn, W., & Weerapreeyakul, N. (2018). Sesamin and sesamolin reduce amyloid-beta toxicity in a transgenic Caenorhabditis elegans. *Biomed Pharmacother, 107*, 656–664. https://doi.org/10.1016/j.biopha.2018.08.037

Khan, M. M., Raza, S. S., Javed, H., Ahmad, A., Khan, A., Islam, F., . . . & Islam, F. (2012). Rutin protects dopaminergic neurons from oxidative stress in an animal model of Parkinson's disease. *Neurotoxicity Research, 22*(1), 1–15. https://doi.org/10.1007/s12640-011-9295-2

Khanna, G., Bhandari, R., Kuhad, A., & Kuhad, A. (2023). Aducanumab. *Drugs of the Future, 44*(2), 115–121. https://doi.org/10.1358/dof.2019.44.2.2895649

Kim, H., Kim, Y. S., Kim, S. Y., & Suk, K. (2001). The plant flavonoid wogonin suppresses death of activated C6 rat glial cells by inhibiting nitric oxide production. *Neuroscience Letters, 309*(1), 67–71. https://doi.org/10.1016/S0304-3940(01)02028-6

Kim, J. H., Quilantang, N. G., Kim, H. Y., Lee, S., & Cho, E. J. (2018). Attenuation of hydrogen peroxide-induced oxidative stress in SH-SY5Y cells by three flavonoids from Acer okamotoanum. *Chemical Papers, 73*(5), 1135–1144. https://doi.org/10.1007/s11696-018-0664-7

Kim, K. A., Jung, J. H., Choi, Y. S., & Kim, S. T. (2022). Wogonin inhibits tight junction disruption via suppression of inflammatory response and phosphorylation of AKT/NF-κB and ERK1/2 in rhinovirus-infected human nasal epithelial cells. *Inflammation Research, 71*(3), 357–368. https://doi.org/10.1007/s00011-022-01542-w

Kim, M. H., Kim, S. H., & Yang, W. M. (2014). Mechanisms of action of phytochemicals from medicinal herbs in the treatment of Alzheimer's disease. *Planta Medica, 80*(15), 1249–1258. https://doi.org/10.1055/s-0034-1383038

Kolahdouzan, M., & Hamadeh, M. J. (2017). The neuroprotective effects of caffeine in neurodegenerative diseases. *CNS Neuroscience & Therapeutics, 23*(4), 272–290. https://doi.org/10.1111/cns.12684

Konrath, E. L., Passos, C. D. S., Klein-Júnior, L. C., & Henriques, A. T. (2013). Alkaloids as a source of potential anticholinesterase inhibitors for the treatment of Alzheimer's disease. *Journal of Pharmacy and Pharmacology, 65*(12), 1701–1725. https://doi.org/10.1111/jphp.12090

Kumar, A., Fontana, I. C., & Nordberg, A. (2023). Reactive astrogliosis: A friend or foe in the pathogenesis of Alzheimer's disease. *Journal of Neurochemistry, 164*(3), 309–324.

Kumar, G. P., & Khanum, F. (2012). Neuroprotective potential of phytochemicals. *Pharmacognosy Reviews, 6*(12), 81–90. https://doi.org/10.4103/0973-7847.99898

Kunjathoor, V. V., Tseng, A. A., Medeiros, L. A., Khan, T., & Moore, K. J. (2004). β-Amyloid promotes accumulation of lipid peroxides by inhibiting CD36-mediated clearance of oxidized lipoproteins. *Journal of Neuroinflammation, 1*, 23. https://doi.org/10.1186/1742-2094-1-23

Lasagna-Reeves, C. A., Castillo-Carranza, D. L., Sengupta, U., Sarmiento, J., Troncoso, J., Jackson, G. R., & Kayed, R. (2012). Identification of oligomers at early stages of tau aggregation in Alzheimer's disease. *FASEB Journal: Official Publication of the Federation of American Societies for Experimental Biology, 26*(5), 1946–1959. https://doi.org/10.1096/FJ.11-199851

Lau, F. C., Bielinski, D. F., & Joseph, J. A. (2007). Inhibitory effects of blueberry extract on the production of inflammatory mediators in lipopolysaccharide-activated BV2 microglia. *Journal of Neuroscience Research, 85*(5), 1010–1017. https://doi.org/10.1002/jnr.21205

Laurent, C., Eddarkaoui, S., Derisbourg, M., Leboucher, A., Demeyer, D., Carrier, S., . . . & Blum, D. (2014). Beneficial effects of caffeine in a transgenic model of Alzheimer's disease-like tau pathology. *Neurobiology of Aging, 35*(9), 2079–2090. https://doi.org/10.1016/j.neurobiolaging.2014.03.027

Layé, S., Madore, C., St-Amour, I., Delpech, J.-C., Joffre, C., Nadjar, A., & Calon, F. (2015). N-3 polyunsaturated fatty acid and neuroinflammation in aging and Alzheimer's disease. *Nutrition and Aging, 3*(1), 33–47. https://doi.org/10.3233/nua-150049

Leng, F., & Edison, P. (2020). Neuroinflammation and microglial activation in Alzheimer disease: Where do we go from here? *Nature Reviews Neurology, 17*(3), 157–172. https://doi.org/10.1038/s41582-020-00435-y

Li, J., Sun, M., Cui, X., & Li, C. (2022). Protective effects of flavonoids against Alzheimer's disease: Pathological hypothesis, potential targets, and structure-activity relationship. *International Journal of Molecular Sciences, 23*(17). https://doi.org/10.3390/ijms231710020

Li, Q., Wang, B.-L., Sun, F.-R., Li, J.-Q., Cao, X.-P., & Tan, L. (2018). The role of UNC5C in Alzheimer's disease. *Annals of Translational Medicine, 6*(10).

Li, R., Wang, X., Qin, T., Qu, R., & Ma, S. (2016). Apigenin ameliorates chronic mild stress-induced depressive behavior by inhibiting interleukin-1β production and NLRP3 inflammasome activation in the rat brain. *Behavioural Brain Research, 296*, 318–325. https://doi.org/10.1016/j.bbr.2015.09.031

Li, S., & Pu, X. P. (2011). Neuroprotective effect of kaempferol against a 1-methyl-4-phenyl-1, 2, 3, 6-tetrahydropyridine-induced mouse model of Parkinson's disease. *Biological and Pharmaceutical Bulletin, 34*(8), 1291–1296. https://doi.org/10.1248/bpb.34.1291

Lilienfeld, S. (2002). Galantamine—a novel cholinergic drug with a unique dual mode of action for the treatment of patients with Alzheimer's disease. *CNS Drug Reviews, 8*(2), 159–176. https://doi.org/10.1111/j.1527-3458.2002.tb00221.x

Lim, H. J., Prajapati, R., Seong, S. H., Jung, H. A., & Choi, J. S. (2023). Antioxidant and antineuroinflammatory mechanisms of kaempferol-3-O-beta-d-glucuronate on lipopolysaccharide-stimulated BV2 microglial cells through the Nrf2/HO-1 signaling cascade and MAPK/NF-kappaB pathway. *ACS Omega, 8*(7), 6538–6549. https://doi.org/10.1021/acsomega.2c06916

Lin, M. W., Chen, Y. H., Yang, H. B., Lin, C. C., & Hung, S. Y. (2020). Galantamine inhibits abeta (1–42)-induced neurotoxicity by enhancing alpha7nAChR expression as a cargo carrier for LC3 binding and abeta (1–42) engulfment during autophagic degradation. *Neurotherapeutics, 17*(2), 676–689. https://doi.org/10.1007/s13311-019-00803-7

Lo Cascio, F., Marzullo, P., Kayed, R., & Palumbo Piccionello, A. (2021). Curcumin as scaffold for drug discovery against neurodegenerative diseases. *Biomedicines, 9*(2). https://doi.org/10.3390/biomedicines9020173

Londzin, P., Zamora, M., Kąkol, B., Taborek, A., & Folwarczna, J. (2021). Potential of caffeine in Alzheimer's disease—a review of experimental studies. *Nutrients, 13*(2), 537. https://doi.org/10.3390/nu13020537

Lossi, L. (2022). The concept of intrinsic versus extrinsic apoptosis. *Biochemical Journal, 479*(3), 357–384.

Mantile, F., & Prisco, A. (2020). Vaccination against β-amyloid as a strategy for the prevention of Alzheimer's disease. *Biology, 9*(12), 1–21. https://doi.org/10.3390/BIOLOGY9120425

Martucciello, S., Masullo, M., Cerulli, A., & Piacente, S. (2020). Natural products targeting ER stress, and the functional link to mitochondria. *International Journal of Molecular Sciences, 21*(6). https://doi.org/10.3390/ijms21061905

Masters, C. L., Simms, G., Weinman, N. A., Multhaup, G., McDonald, B. L., & Beyreuther, K. (1985). Amyloid plaque core protein in Alzheimer disease and down syndrome. *Proceedings of the National Academy of Sciences of the United States of America, 82*(12), 4245–4249. https://doi.org/10.1073/PNAS.82.12.4245

McConlogue, L., Buttini, M., Anderson, J. P., Brigham, E. F., Chen, K. S., Freedman, S. B., . . . & Sinha, S. (2007). Partial reduction of BACE1 has dramatic effects on Alzheimer plaque and synaptic pathology in APP transgenic mice. *The Journal of Biological Chemistry, 282*(36), 26326–26334. https://doi.org/10.1074/JBC.M611687200

Mo, J. S., Choi, D., Han, Y. R., Kim, N., & Jeong, H. S. (2019). Morin has protective potential against ER stress induced apoptosis in renal proximal tubular HK-2 cells. *Biomed Pharmacother, 112*, 108659. https://doi.org/10.1016/j.biopha.2019.108659

Mukhin, V. N., Pavlov, K. I., & Klimenko, V. M. (2017). Mechanisms of neuron loss in Alzheimer's disease. *Neuroscience and Behavioral Physiology, 47*(5), 508–516. https://doi.org/10.1007/S11055-017-0427-X

Musiek, E. S., & Holtzman, D. M. (2015). Three dimensions of the amyloid hypothesis: Time, space and "wingmen". *Nature Neuroscience, 18*(6), 800–806. https://doi.org/10.1038/NN.4018

Nakajima, A., & Ohizumi, Y. (2019). Potential benefits of nobiletin, a citrus flavonoid, against Alzheimer's disease and Parkinson's disease. *International Journal of Molecular Sciences, 20*(14). https://doi.org/10.3390/ijms20143380

Naoi, M., Shamoto-Nagai, M., & Maruyama, W. (2019). Neuroprotection of multifunctional phytochemicals as novel therapeutic strategy for neurodegenerative disorders: Antiapoptotic and antiamyloidogenic activities by modulation of cellular signal pathways. *Future Neurology, 14*(1), FNL9. https://doi.org/10.2217/fnl-2018-0028

Navarro, F., Forján, E., Vázquez, M., Montero, Z., Bermejo, E., Castaño, M. Á., . . . & Vega, J. M. (2016). Microalgae as a safe food source for animals: Nutritional characteristics of the acidophilic microalga Coccomyxa onubensis. *Food & Nutrition Research, 60*(1). https://doi.org/10.3402/FNR.V60.30472

Nelson, P. T., Alafuzoff, I., Bigio, E. H., Bouras, C., Braak, H., Cairns, N. J., . . . & Beach, T. G. (2012). Correlation of Alzheimer disease neuropathologic changes with cognitive status: A review of the literature. *Journal of Neuropathology and Experimental Neurology, 71*(5), 362–381. https://doi.org/10.1097/NEN.0B013E31825018F7

Ng, Y. P., Or, T. C. T., & Ip, N. Y. (2015). Plant alkaloids as drug leads for Alzheimer's disease. *Neurochemistry International, 89*, 260–270. https://doi.org/10.1016/j.neuint.2015.07.018

Nisbet, R. M., & Götz, J. (2018). Amyloid-β and tau in Alzheimer's disease: Novel pathomechanisms and non-pharmacological treatment strategies. *Journal of Alzheimer's Disease: JAD, 64*(s1), S517–S527. https://doi.org/10.3233/JAD-179907

Pepeu, G., & Grazia Giovannini, M. (2017). The fate of the brain cholinergic neurons in neurodegenerative diseases. *Brain Research, 1670*, 173–184. https://doi.org/10.1016/J.BRAINRES.2017.06.023

Pereira, J. B., Janelidze, S., Ossenkoppele, R., Kvartsberg, H., Brinkmalm, A., Mattsson-Carlgren, N., . . . & Hansson, O. (2021). Untangling the association of amyloid-β and tau with synaptic and axonal loss in Alzheimer's disease. *Brain: A Journal of Neurology, 144*(1), 310–324. https://doi.org/10.1093/BRAIN/AWAA395

Piao, W. H., Campagnolo, D., Dayao, C., Lukas, R. J., Wu, J., & Shi, F. D. (2009). Nicotine and inflammatory neurological disorders. *Acta Pharmacologica Sinica, 30*(6), 715–722. https://doi.org/10.1038/aps.2009.67

Prasanthi, J. R., Dasari, B., Marwarha, G., Larson, T., Chen, X., Geiger, J. D., & Ghribi, O. (2010). Caffeine protects against oxidative stress and Alzheimer's disease-like pathology in rabbit hippocampus induced by cholesterol-enriched diet. *Free Radical Biology and Medicine, 49*(7), 1212–1220. https://doi.org/10.1016/j.freeradbiomed.2010.07.007

Rabizadeh, S., & Bredesen, D. E. (2003). Ten years on: Mediation of cell death by the common neurotrophin receptor p75NTR. *Cytokine & Growth Factor Reviews, 14*(3–4), 225–239.

Raina, P., Santaguida, P., Ismaila, A., Patterson, C., Cowan, D., Levine, M., Booker, L., & Oremus, M. (2008). Effectiveness of cholinesterase inhibitors and memantine for treating dementia: Evidence review for a clinical practice guideline. *Annals of Internal Medicine, 148*(5), 379–397. https://doi.org/10.7326/0003-4819-148-5-200803040-00009

Ramesh, M., Gopinath, P., & Govindaraju, T. (2020). Role of post-translational modifications in Alzheimer's disease. *Chembiochem: A European Journal of Chemical Biology, 21*(8), 1052–1079. https://doi.org/10.1002/CBIC.201900573

Regan, P., Whitcomb, D. J., & Cho, K. (2017). Physiological and pathophysiological implications of synaptic tau. *The Neuroscientist: A Review Journal Bringing Neurobiology, Neurology and Psychiatry, 23*(2), 137–151. https://doi.org/10.1177/1073858416633439

Reinisalo, M., Karlund, A., Koskela, A., Kaarniranta, K., & Karjalainen, R. O. (2015). Polyphenol stilbenes: Molecular mechanisms of defence against oxidative stress and aging-related diseases. *Oxidative Medicine and Cellular Longevity, 2015*, 340520. https://doi.org/10.1155/2015/340520

Rhein, V., & Eckert, A. (2007). Effects of Alzheimer's amyloid-beta and tau protein on mitochondrial function—role of glucose metabolism and insulin signalling. *Archives of Physiology and Biochemistry, 113*(3), 131–141. https://doi.org/10.1080/13813450701572288

Rivière, C., Richard, T., Quentin, L., Krisa, S., Merillon, J. M., & Monti, J. P. (2007). Inhibitory activity of stilbenes on Alzheimer's beta-amyloid fibrils in vitro. *Bioorganic & Medicinal Chemistry, 15*(2), 1160–1167. https://doi.org/10.1016/j.bmc.2006.09.069

Rodda, J., & Carter, J. (2012). Cholinesterase inhibitors and memantine for symptomatic treatment of dementia. *BMJ (Clinical Research Ed.), 344*(7856). https://doi.org/10.1136/BMJ.E2986

Roussarie, J. P., Yao, V., Rodriguez-Rodriguez, P., Oughtred, R., Rust, J., Plautz, Z., . . . & Greengard, P. (2020). Selective neuronal vulnerability in Alzheimer's disease: A network-based analysis. *Neuron, 107*(5), 821–835.e12. https://doi.org/10.1016/J.NEURON.2020.06.010

Ryan, T. M., Caine, J., Mertens, H. D., Kirby, N., Nigro, J., Breheney, K., . . . & Masters, C. L. (2013). Ammonium hydroxide treatment of Aβ produces an aggregate free solution suitable for biophysical and cell culture characterization. *PeerJ, 1*, e73.

Santa-Maria, I., Diaz-Ruiz, C., Ksiezak-Reding, H., Chen, A., Ho, L., Wang, J., & Pasinetti, G. M. (2012). GSPE interferes with tau aggregation in vivo: Implication for treating tauopathy. *Neurobiol Aging, 33*(9), 2072–2081. https://doi.org/10.1016/j.neurobiolaging.2011.09.027

Schepici, G., Bramanti, P., & Mazzon, E. (2020). Efficacy of sulforaphane in neurodegenerative diseases. *International Journal of Molecular Sciences, 21*(22). https://doi.org/10.3390/ijms21228637

Schneider, L. S., Mangialasche, F., Andreasen, N., Feldman, H., Giacobini, E., Jones, R., . . . & Kivipelto, M. (2014). Clinical trials and late-stage drug development for Alzheimer's disease: An appraisal from 1984 to 2014. *Journal of Internal Medicine, 275*(3), 251–283. https://doi.org/10.1111/JOIM.12191

Scott, L. J., & Goa, K. L. (2000). Galantamine: A review of its use in Alzheimer's disease. *Drugs, 60*, 1095–1122. https://doi.org/10.2165/00003495-200060050-00008

Scott, L. J., Kiss, T., Kawabe, T. T., & Hajós, M. (2016). Neuronal network activity in the hippocampus of tau transgenic (Tg4510) mice. *Neurobiology of Aging, 37*, 66–73. https://doi.org/10.1016/J.NEUROBIOLAGING.2015.10.002

Selkoe, D. J. (2019). Alzheimer disease and aducanumab: Adjusting our approach. *Nature Reviews. Neurology, 15*(7), 365–366. https://doi.org/10.1038/S41582-019-0205-1

Selkoe, D. J. (2000). Toward a comprehensive theory for Alzheimer's disease. Hypothesis: Alzheimer's disease is caused by the cerebral accumulation and cytotoxicity of amyloid β-protein. *Annals of the New York Academy of Sciences, 924*(1), 17–25. https://doi.org/10.1111/j.1749-6632.2000.tb05554.x

Serrano-Pozo, A., Frosch, M. P., Masliah, E., & Hyman, B. T. (2011). Neuropathological alterations in Alzheimer disease. *Cold Spring Harbor Perspectives in Medicine, 1*(1). https://doi.org/10.1101/CSHPERSPECT.A006189

Sharifi-Rad, J., Rapposelli, S., Sestito, S., Herrera-Bravo, J., Arancibia-Diaz, A., Salazar, L. A., . . . & Calina, D. (2022). Multi-target mechanisms of phytochemicals in Alzheimer's disease: Effects on oxidative stress, neuroinflammation and protein aggregation. *Journal of Personalized Medicine, 12*(9), 1515. https://doi.org/10.3390/jpm12091515

Sheng, M., Sabatini, B. L., & Südhof, T. C. (2012). Synapses and Alzheimer's disease. *Cold Spring Harbor Perspectives in Biology, 4*(5), 10. https://doi.org/10.1101/CSHPERSPECT.A005777

Shim, S. B., Lee, S. H., Chae, K. R., Kim, C. K., Hwang, D. Y., Kim, B. G., . . . & Kim, Y. K. (2008). Nicotine leads to improvements in behavioral impairment and an increase in the nicotine acetylcholine receptor in transgenic mice. *Neurochemical Research, 33*, 1783–1788. https://doi.org/10.1007/s11064-008-9629-5

Shou, J. W., & Shaw, P. C. (2022). Therapeutic efficacies of berberine against neurological disorders: An update of pharmacological effects and mechanisms. *Cells, 11*(5). https://doi.org/10.3390/cells11050796

Silva Dos Santos, J., Goncalves Cirino, J. P., de Oliveira Carvalho, P., & Ortega, M. M. (2020). The pharmacological action of kaempferol in central nervous system diseases: A review. *Frontiers in Pharmacology, 11*, 565700. https://doi.org/10.3389/fphar.2020.565700

Singh, S. K., Srivastav, S., Castellani, R. J., Plascencia-Villa, G., & Perry, G. (2019). Neuroprotective and antioxidant effect of ginkgo biloba extract against AD and other neurological disorders. *Neurotherapeutics, 16*(3), 666–674. https://doi.org/10.1007/s13311-019-00767-8

Stazi, M., Lehmann, S., Sakib, M. S., Pena-Centeno, T., Büschgens, L., Fischer, A., . . . & Wirths, O. (2022). Long-term caffeine treatment of Alzheimer mouse models ameliorates behavioural deficits and neuron loss and promotes cellular and molecular markers of neurogenesis. *Cellular and Molecular Life Sciences*, 1–18. https://doi.org/10.1007/s00018-021-04062-8

Subramanian, J., Savage, J. C., & Tremblay, M. È. (2020). Synaptic loss in Alzheimer's disease: Mechanistic insights provided by two-photon in vivo imaging of transgenic mouse models. *Frontiers in Cellular Neuroscience, 14*. https://doi.org/10.3389/FNCEL.2020.592607

Sun, A. Y., Wang, Q., Simonyi, A., & Sun, G. Y. (2010). Resveratrol as a therapeutic agent for neurodegenerative diseases. *Molecular Neurobiology, 41*(2–3), 375–383. https://doi.org/10.1007/s12035-010-8111-y

Suzuki, K., Koike, H., Matsui, H., Ono, Y., Hasumi, M., Nakazato, H., . . . & Yamanaka, H. (2002). Genistein, a soy isoflavone, induces glutathione peroxidase in the human prostate cancer cell lines LNCaP and PC-3. *International Journal of Cancer, 99*(6), 846–852. https://doi.org/10.1002/ijc.10428

Takano, K., Tabata, Y., Kitao, Y., Murakami, R., Suzuki, H., Yamada, M., . . . & Hori, O. (2007). Methoxyflavones protect cells against endoplasmic reticulum stress and neurotoxin. *American Journal of Physiology-Cell Physiology, 292*(1), C353–361. https://doi.org/10.1152/ajpcell.00388.2006

Tatsumi, S., Uchihara, T., Aiba, I., Iwasaki, Y., Mimuro, M., Takahashi, R., & Yoshida, M. (2014). Ultrastructural differences in pretangles between Alzheimer disease and corticobasal degeneration revealed by comparative light and electron microscopy. *Acta Neuropathologica Communications, 2*(1), 1–11. https://doi.org/10.1186/S40478-014-0161-3/TABLES/2

Temple, J. L., Bernard, C., Lipshultz, S. E., Czachor, J. D., Westphal, J. A., & Mestre, M. A. (2017). The safety of ingested caffeine: A comprehensive review. *Frontiers in Psychiatry, 8*, 80. https://doi.org/10.3389/fpsyt.2017.00080

Thal, D. R., Ronisz, A., Tousseyn, T., Rijal Upadhaya, A., Balakrishnan, K., Vandenberghe, R., . . . & Walter, J. (2019). Different aspects of Alzheimer's disease-related amyloid β-peptide pathology and their relationship to amyloid positron emission tomography imaging and dementia. *Acta Neuropathologica Communications, 7*(1). https://doi.org/10.1186/S40478-019-0837-9

Thibert, C., & Fombonne, J. (2010). Dependence receptors: Mechanisms of an announced death. *Cell Cycle, 9*(11), 2085–2091.

Thimmulappa, R. K., Mai, K. H., Srisuma, S., Kensler, T. W., Yamamoto, M., & Biswal, S. (2002). Identification of Nrf2-regulated genes induced by the chemopreventive agent sulforaphane by oligonucleotide microarray. *Cancer Research, 62*(18), 5196–5203.

Tun, M. K., & Herzon, S. B. (2012). The pharmacology and therapeutic potential of (-)-huperzine A. *Journal of Experimental Pharmacology*, *4*, 113–123. https://doi.org/10.2147/JEP.S27084

Urnowey, S., Ansai, T., Bitko, V., Nakayama, K., Takehara, T., & Barik, S. (2006). Temporal activation of anti- and pro-apoptotic factors in human gingival fibroblasts infected with the periodontal pathogen, Porphyromonas gingivalis: Potential role of bacterial proteases in host signalling. *BMC Microbiology*, *6*. https://doi.org/10.1186/1471-2180-6-26

Vafeiadou, K., Vauzour, D., Lee, H. Y., Rodriguez-Mateos, A., Williams, R. J., & Spencer, J. P. (2009). The citrus flavanone naringenin inhibits inflammatory signalling in glial cells and protects against neuroinflammatory injury. *Archives of Biochemistry and Biophysics*, *484*(1), 100–109. https://doi.org/10.1016/j.abb.2009.01.016

Velmurugan, B. K., Rathinasamy, B., Lohanathan, B. P., Thiyagarajan, V., & Weng, C. F. (2018). Neuroprotective role of phytochemicals. *Molecules*, *23*(10), 2485. https://doi.org/10.3390/molecules23102485

Venkatesan, R., Ji, E., & Kim, S. Y. (2015). Phytochemicals that regulate neurodegenerative disease by targeting neurotrophins: A comprehensive review. *BioMed Research International*, *2015*, 814068. https://doi.org/10.1155/2015/814068

Vogels, T., Leuzy, A., Cicognola, C., Ashton, N. J., Smolek, T., Novak, M., . . . & Zilka, N. (2020). Propagation of tau pathology: Integrating insights from postmortem and in vivo studies. *Biological Psychiatry*, *87*(9), 808–818.

Wang, D., Gao, K., Li, X., Shen, X., Zhang, X., Ma, C., . . . & Zhang, L. (2012). Long-term naringin consumption reverses a glucose uptake defect and improves cognitive deficits in a mouse model of Alzheimer's disease. *Pharmacology Biochemistry and Behavior*, *102*(1), 13–20. https://doi.org/10.1016/j.pbb.2012.03.013

Wei, P. C., Lee-Chen, G. J., Chen, C. M., Chen, Y., Lo, Y. S., & Chang, K. H. (2022). Isorhamnetin attenuated the release of interleukin-6 from beta-amyloid-activated microglia and mitigated interleukin-6-mediated neurotoxicity. *Oxidative Medicine and Cellular Longevity*, *2022*, 3652402. https://doi.org/10.1155/2022/3652402

Welcome, M. O. (2020). Blood brain barrier inflammation and potential therapeutic role of phytochemicals. *PharmaNutrition*, 11, 100177. https://doi.org/10.1016/j.phanu.2020.100177

Wen, L., Shi, D., Zhou, T., Tu, J., He, M., Jiang, Y., & Yang, B. (2020). Identification of two novel prenylated flavonoids in mulberry leaf and their bioactivities. *Food Chemistry*, *315*, 126236. https://doi.org/10.1016/j.foodchem.2020.126236

Wiersma, V. I., van Ziel, A. M., Vazquez-Sanchez, S., Nölle, A., Berenjeno-Correa, E., Bonaterra-Pastra, A., . . . & Scheper, W. (2019). Granulovacuolar degeneration bodies are neuron-selective lysosomal structures induced by intracellular tau pathology. *Acta Neuropathologica*, *138*(6), 943–970. https://doi.org/10.1007/S00401-019-02046-4

Wirths, O., & Zampar, S. (2020). Neuron loss in Alzheimer's disease: Translation in transgenic mouse models. *International Journal of Molecular Sciences*, *21*(21), 1–19. https://doi.org/10.3390/IJMS21218144

Woodruff-Pak, D. S., Vogel III, R. W., & Wenk, G. L. (2001). Galantamine: Effect on nicotinic receptor binding, acetylcholinesterase inhibition, and learning. *Proceedings of the National Academy of Sciences*, *98*(4), 2089–2094. https://doi.org/10.1073/pnas.98.4.2089

Xia, M., Cheng, X., Yi, R., Gao, D., & Xiong, J. (2016). The binding receptors of Aβ: An alternative therapeutic target for Alzheimer's disease. *Molecular Neurobiology*, *53*, 455–471.

Xia, W. (2019). γ-Secretase and its modulators: Twenty years and beyond. *Neuroscience Letters*, *701*, 162–169. https://doi.org/10.1016/J.NEULET.2019.02.011

Xu, J., Zhang, H., Yang, F., & Yu, J. X. (2013). Intervention effect of berberine on expressions of TNF-alpha and receptor type I in Abeta25–35-induced inflammatory reaction in SH-SY5Y cell lines. *Zhongguo Zhong yao za zhi= Zhongguo Zhongyao Zazhi= China Journal of Chinese Materia Medica*, *38*(9), 1327–1330.

Yadav, E., Yadav, P., Khan, M. M. U., Singh, H., & Verma, A. (2022). Resveratrol: A potential therapeutic natural polyphenol for neurodegenerative diseases associated with mitochondrial dysfunction. *Frontiers in Pharmacology*, *13*, 922232. https://doi.org/10.3389/fphar.2022.922232

Yamamoto, N., Shibata, M., Ishikuro, R., Tanida, M., Taniguchi, Y., Ikeda-Matsuo, Y., & Sobue, K. (2017). Epigallocatechin gallate induces extracellular degradation of amyloid beta-protein by increasing neprilysin secretion from astrocytes through activation of ERK and PI3K pathways. *Neuroscience*, *362*, 70–78. https://doi.org/10.1016/j.neuroscience.2017.08.030

Zhang, H., Wei, W., Zhao, M., Ma, L., Jiang, X., Pei, H., . . . & Li, H. (2021). Interaction between Aβ and tau in the pathogenesis of Alzheimer's disease. *International Journal of Biological Sciences*, *17*(9), 2181. https://doi.org/10.7150/IJBS.57078

Zhang, J., Wu, N., Wang, S., Yao, Z., Xiao, F., Lu, J., & Chen, B. (2021). Neuronal loss and microgliosis are restricted to the core of Aβ deposits in mouse models of Alzheimer's disease. *Aging Cell*, *20*(6), e13380. https://doi.org/10.1111/ACEL.13380

Zhang, N., Hu, Z., Zhang, Z., Liu, G., Wang, Y., Ren, Y., . . . & Geng, F. (2017). Protective role of naringenin against Aβ25–35-caused damage via ER and PI3K/Akt-mediated pathways. *Cellular and Molecular Neurobiology*, *38*(2), 549–557. https://doi.org/10.1007/s10571-017-0519-8

Zhang, S. Q., Obregon, D., Ehrhart, J., Deng, J., Tian, J., Hou, H., . . . & Tan, J. (2013). Baicalein reduces beta-amyloid and promotes nonamyloidogenic amyloid precursor protein processing in an Alzheimer's disease transgenic mouse model. *Journal of Neuroscience Research*, *91*(9), 1239–1246. https://doi.org/10.1002/jnr.23244

Zhang, X. X., Tian, Y., Wang, Z. T., Ma, Y. H., Tan, L., & Yu, J. T. (2021). The epidemiology of Alzheimer's disease modifiable risk factors and prevention. *The Journal of Prevention of Alzheimer's Disease*, *8*, 313–321. https://doi.org/10.14283/jpad.2021.15

Zhao, J., Liu, X., Xia, W., Zhang, Y., & Wang, C. (2020). Targeting amyloidogenic processing of APP in Alzheimer's disease. *Frontiers in Molecular Neuroscience*, *13*, 137.

Zhou, R., Yang, G., Guo, X., Zhou, Q., Lei, J., & Shi, Y. (2019). Recognition of the amyloid precursor protein by human γ-secretase. *Science* (New York, NY), *363*(6428). https://doi.org/10.1126/SCIENCE.AAW0930

7 Computational Modelling of BACE-1, AChE, and Phosphodiesterase Inhibitors as Anti-Alzheimer's Agents

Vishal S. Patil, Jagadeesh Dodakallanavar, Pukar Khanal, B M Patil, Raushan Kumar Chaudhary, Amit Kumar Shrivastava, Nongmaithem Randhoni Chanu, Darasaguppe R. Harish and Uday Venkat Mateti*

7.1 INTRODUCTION

Clinical symptoms of Alzheimer's disease (AD) include cognitive decline, confusion, poorer judgment, making decisions, apraxia, and difficulties with language. Common neuropsychiatric manifestations include melancholy, disinterest, worry, irritability, illusions, and hallucinations (Tarawneh and Holtzman, 2012; Kumar et al., 2018). Hence, AD is a complex progressive neurological disorder and one of the great health care challenges of the 21st century (Serafín et al., 2020). With an incidence that rises from 1% at age 60 to a minimum of 35% at 90 years of age, dementia is quickly becoming one of the most pressing medical issues facing the elderly (Corrada et al., 2010). Around 80% of all cases of dementia can be attributed to AD, making it the most common form of the disease (Scott and Barrett, 2007). Over 6.5 million U.S. citizens age 65 and above are now coping with AD. Unless there are significant medical advancements to prevent, slow, or cure AD, that number might reach 13.8M by 2060 (Corrada et al., 2010; Monica Moore et al., 2021).

Intracellular tangles of neurofibrillary cells and extracellular amyloidal deposits of proteins that contribute to senile plaques are two basic characteristics of this neurological deterioration process (Murphy and LeVine, 2010). Throughout the last 20 years, scientists have been inspired to explore innovative medications and potential lead hits over AD by discoveries in the discipline of AD pathogenesis (Kumar and Singh, 2015). The progressive decline in cognition that is characteristic of AD is the focus of ongoing clinical research, with the hope that improved treatments will be developed in the coming years (Weller and Budson, 2018).

Memantine, an N-methyl d-aspartate receptor antagonist, has a small effect on the disease and targets its later stages, and acetylcholinesterase inhibitors (rivastigmine, galantamine, donepezil) are the only additional classes of pharmacologic therapy offered to those suffering from AD, although enzyme acetylcholinesterase (AChE) remains a very viable focus for the treatment of symptoms in AD and cholinergic deficiency is a consistent and early observation in AD (Briggs et al., 2016). ACHE and N-methyl d-aspartate receptor antagonists are the only kinds of pharmacologic therapy available for patients with AD. Even though cholinergic insufficiency is a consistent and early observation in AD, AChE remains a very promising focus for the management of symptoms in AD; it is often overlooked (Mehta et al., 2012; Marucci et al., 2021). The pathology of AD is unknown, but aberrant synthesis or gathering of amyloid-β (Aβ) peptides have been implicated in both genetic and clinical research during the past 30 years. The β-site APP cleaving enzyme 1 (BACE1) and γ-secretase cleaves amyloid precursor protein (APP) to produce Aβ. The creation of Aβ is halted in

DOI: 10.1201/9781003412069-7

BACE1-deficient mice. For this reason, blocking BACE1 with drugs is being actively researched as a potential treatment for AD (Murphy and LeVine, 2010; Das and Yan, 2019).

Another optimistic target, phosphodiesterase (PDEs), is accountable for hydrolysing "cyclic adenosine monophosphate (cAMP) and cyclic guanosine monophosphate (cGMP)", which becomes another promising target of AD (Wu et al., 2018; Sheng et al., 2022). PDEs are engaged in neurological disorders by their regulation of cAMP and/or cGMP levels. Numerous PDE subfamilies have significant expression in the human brain. Many PDE antagonists have shown notable cognitive enhancement effects in preclinical investigations, and more than 15 people are being tested in clinical trials, leading many to believe that PDEs are attractive targets to use in the management of AD (Azevedo et al., 2014).

Naturally and synthesized coumarins (or benzopyrones) show a wide range of medicinal properties, including those of antidepressant, antibacterial, antioxidant, anti-inflammatory, antinociceptive, anti-tumour, anti-asthmatic, and anti-viral natures (Sharifi-Rad et al., 2021; Fylaktakidou et al., 2004). These have demonstrated strong AChE inhibitory action (Anand et al., 2012).

In addition, new analogues capable of suppressing Aβ aggregation have been developed through the functionalization of the aromatic centre of coumarins (Soto-Ortega et al., 2011). Plant-based coumarins have been shown to protect against intracerebroventricular injection of Aβ-induced decline in memory in mice (Nisha and Devi, 2017). New analogues with enhanced AChE inhibitory properties and other pharmacological activities, such as BACE inhibition corresponding to reduced Aβ deposition, have been designed and synthesized by the identification of key structural features within the coumarin (Anand et al., 2012). Coumarins primarily interact with the peripheral anionic site of AChE, leading researchers to focus on synthesizing inhibitors of AChE that combine the Coumarin with a moiety that binds to the enzyme's catalytic site (Peitzika and Pontiki, 2023). Hence, this research is framed to identify and infer the action of coumarin derivatives for the management of AD by identifying the inhibitors of BACE1, ACHE, and PDE4D and attenuating the progression of AD through regulating multiple pathways.

7.2 MATERIALS AND METHODS

7.2.1 Mining of Coumarins and Their Pharmacokinetic Screening

Coumarin derivatives were collated from the ChEBI website (*www.ebi.ac.uk/chebi/*), and CNS permeability was predicted based on Suenderhauf et al., 2012. The criteria for CNS permeability of compounds are as follows (1) compounds with hydrogen bond acceptor <=4 with BBB value −1.0028 to 0.4049 and (2) compounds with rotational bond <=7 with BBB value >0.4049 (Lipinski, 2004).

7.2.2 Protein Target Prediction

Prioritized coumarins probable protein targets were predicted from the SwissTargetPrediciton (Daina et al., 2019) online server (*www.way2drug.com/ge/*) by submitting the SMILES and the *Homo sapiens* species was selected. As a result, Coumarins targeting BACE-1, ACHE, and PDE4D were separated for further analysis.

7.2.3 Molecular Docking

Firstly, the ligand preparation was carried out using the POAP pipeline. The 3D structures of prioritized coumarins and known inhibitors were downloaded in '.sdf' format from the PubChem small molecule database (https://pubchem.ncbi.nlm.nih.gov/). Using the conjugate gradients approach, the structures' energy was reduced by employing mmff94 forcefield (Rappé et al., 1992). By including the "gastigers charges and polar hydrogens", the ligands were converted into ".pdbqt" format.

Secondly, the RCSB (www.rcsb.org) database was utilized to retrieve the x-ray crystallographic structure of the macromolecules BACE1 (PDB: 4IVT), ACHE (PDB: 4EY7), and PDE4D (PDB: 6INM). The ligand was docked into the binding site of the target and the P2Rank server was employed to determine the binding site of each target.

Thirdly, the POAP pipeline extension was employed to perform molecular docking via AutoDock Vina (Morris et al., 2009; Patil et al., 2022; Samdani and Vetrivel, 2018). When visualizing protein-ligand interactions with DSV 2019 (Dassault Systèmes, 2019), the inter-molecular interaction compounds with the lowest binding energies (BE) were chosen.

7.2.4 Docked Complexes Stability

The MD simulation was run for 150ns to assess the molecular stability and interactions between molecules associated with the best-docked configurations from all three different targets (www.gromacs.org) using GROMACS 2021.3 (Van Der Spoel et al., 2005; Dwivedi et al., 2022). The "Amber ff99SB-ildn force field" was employed and topologies were generated by the xleap module of AmberTools. TIP3P solvent model was added in a "rectangular box with 10.0Å boundary conditions" from the edges of protein and neutralized the system with the required amount of counter ions. "Canonical (NVT and NPT) and isobaric (NPT) ensembles" were used to equilibrate the systems for 1 ns. Throughout the MD run of 150 ns for each complex, the run coordinates were captured every 2 fs. The tools included within Gromacs were used to analyse the trajectories that were generated.

7.2.5 MM-PBSA Calculation

In the current investigation, the MM-PBSA method and the 'g_mmpbsa' program were used to compute the relative binding energy and its impact on individual residue. The values utilized to infer the BE were taken from other studies (Bhandare et al., 2019; Bhandare and Ramaswamy, 2018). 50 typical snapshots were taken along the stable trajectory visible within 100 to 150ns to calculate the binding energy.

7.2.6 Analysis of PCA and DCCM

Using MD trajectories, PCA analyses the main categories of time-dependent motions of atoms (Bhandare and Ramaswamy, 2018; Amadei et al., 1993; Amadei et al., 1996). Using the "least square fit" to the reference structure, the translational and rotational mobility of the molecule is eliminated. The eigenvalue associated with each eigenvector displays the energy contribution of that eigenvector to the motion and the eigenvectors and eigenvalues of the trajectory were determined by diagonalizing the covariance matrix by employing gromacs utility g_covar and the g_anaeig tool was employed to analyse and display the eigenvectors (Aalten et al., 1995).

7.2.7 Coumarin-Regulated Target Pathway Enrichment

The prioritized Coumarins (predicted as lead hits for BACE1, ACHE, and PDE4D) targeting various host protein molecules were submitted into STRING (Szklarczyk et al., 2016; Biradar et al., 2020) *ver.* 11.0 (https://string-db.org/), and the organism was chosen as *Homo sapiens*. To identify the regulated pathways, modulated proteins were enriched using references to the KEGG (www.genome.jp/kegg/pathway.html). In addition, AD-related pathways were discovered from published research.

7.2.8 Network Pharmacology Studies

The combined network of coumarin-regulated-protein-modulated pathways was built by Cytoscape (Shannon et al., 2003; Patil et al., 2020) ver. 3.6.1 (https://cytoscape.org/). During network elucidation, duplicate nodes and edges were deleted to reduce the likelihood of accidental matches. After that, the entire network was directed, and the total number of connections in the completed network was analysed by edge count by setting the nodes to "*low values to small size*" and "*low values to bright colors.*"

7.3 RESULTS

7.3.1 Mining of Coumarins and Their Pharmacokinetic Screening

A total of 379 coumarin derivatives were retrieved from ChEBI. The compounds were screened for hydrogen bond acceptor (HBA), rotatable bonds, and BBB permeability, and 55 compounds passed HBA (<=4) and BBB (−1.0028 to 0.4049) criteria; 14 compounds passed rotatable bonds (<=7) and BBB of (>0.4049). Out of 379 compounds, 69 compounds were prioritized, which are expected to cross CNS as per Suenderhauf et al. (2012) criteria.

7.3.2 Identification of BACE-1, AChE, and Phosphodiesterase Inhibitors

The prioritized compounds' probable protein targets were predicted by SwissTargetPrediction sever. Out of 69 compounds, 66 compounds were modulated 712 protein targets, of which 11, 21, and 16 compounds were predicted to inhibit BACE1, ACHE, and PDE4D protein molecules, respectively.

7.3.3 Active Site Prediction

The binding site residues of BACE-1, AChE, and PDE4D were predicted by P2Rank online server. The binding site residues of BACE1 are 71, 72, 73, 74, 75, 91, 93, 95, 96, 98, 130, 131, 132, 133, 134, 168, 169, 171, 176, 179, 187, 189, 231, 259, 285, 287, 289, 290, 291, 292, 293, 294, 296, 390, 393, 396, and 400. The probability of the pocket is 0.96, and its SASA point is 170nm^2. The active site residues of ACHE are 103, 105, 107, 117, 118, 151, 152, 153, 155, 156, 157, 161, 164, 233, 234, 317, 318, 320, 324, 325, 326, 327, 328, 368, 369, 372, 478, and 479. The probability of the pocket is 0.96 and its SASA point is 141nm^2. The active site residues of PDE4D (PDB ID: 6INM) are 461, 462, 503, 506, 509, 510, 511, 512, 531, 532, 573, 574, 575, 576, 620, 621, 623, 624, 631, 638, 641, 642, 645, 657, 658, 659, 660, 671, 674, and 678. The probability of the pocket is 0.99 and its SASA point is 157nm.

7.3.4 Docking Studies

From the prioritized coumarin derivatives, decursin, farnesiferol A, and 2-Acetamido-6H-dibenzo[b,d]pyran-6-one scored the least BE with BACE1, ACHE, and PDE4D, respectively. The analysed BEs for BACE1-decursin, ACHE-farnesiferol A and PDE4D- 2-Acetamido-6H-dibenzo[b,d]pyran-6-one were −8.3, −12.7, and −9.6kcal/mol, respectively, whereas, the standard molcculc complexes BACE1-Lanabecestat, ACHE-donepezil, and PDE4D-(1S)-1-[2-(1H-indol-3-yl) ethyl]-6,7-dimethoxy-3,4-dihydroisoquinoline-2(1H)-carbaldehyde (AKU) scored BEs of −8.1, −11.0, and −8.9kcal/mol. The compound and amino acid interactions are presented in Figure 7.1.

Decursin formed 9 interactions with active site residues of BACE1 i.e., Ile187, Tyr259 (2), Tyr132 (3), Leu91, Ile179, Phe169; lanabecestat formed 7 with BACE1 i.e., Ile187, Tyr132, Arg189, Val130, Phe169, Gly291, Asp289. Farnesiferol A formed 17 interactions with active site residues of ACHE i.e. Tyr103, Gly152, Tyr372 (5), His478 (3), Trp317 (2), Phe369 (2), Tyr368 (3), and Trp117 (2), and donepezil formed 7 with ACHE i.e., Phe326, Trp117, Trp317(2), Phe369, Tyr372, Tyr368. Likewise, 2-Acetamido-6H-dibenzo[b,d]pyran-6-one formed 10 interactions with active site residues of PDE4D i.e. His462 (2), Met575, Ile638(2), Met659, Phe642, Phe674 (2), and Leu621, and AKU formed 8 interactions with PDE4D: Gln671 (2), His462=O Ile638, Tyr461 (2), Phe674, and Asn623. Interestingly, all the coumarin derivatives predicted to hit BACE1, ACHE, and PDE4D were found to interact with binding site residues. Figure 7.1 shows the binding mode of lead hits and standard molecules with BACE1, ACHE, and PDE4D.

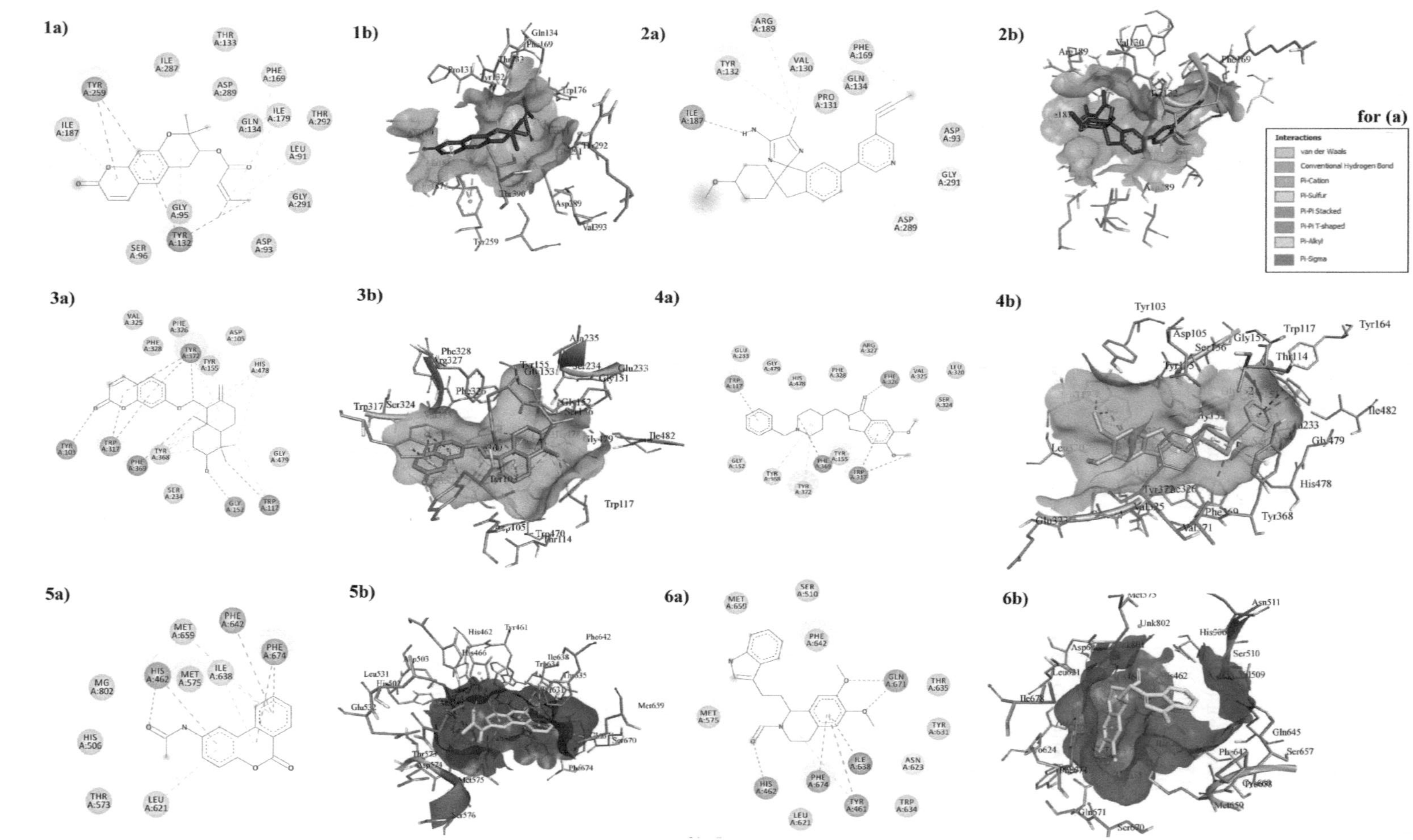

FIGURE 7.1 (a) 2D and (b) 3D interaction of (1) Decursin and (2) Lanabecestat with BACE1, (3) Farnesiferol A, and (4) Donepezil with ACHE, (5) 2-Acetamido-6H-dibenzo[b,d]pyran-6-one (6) AKU with PDE4D.

7.3.5 Stability of the Docked Complexes

7.3.5.1 Stability of Decursin and Lanabecestat with BACE1

The BACE1-lanabecestat and dercusin trajectory shows stability after an 80 ns equilibration period. The average root-mean-square deviations (RMSD) for the backbone were ~1.5 Å and ~1.9 Å, respectively, and for the complex, they were ~2.0 Å and ~2.5 Å, respectively. The residues Asp372 to Gln377 for loop showed higher residual fluctuations (RMSF) up to 4 Å and residues Val227 to *Val231* also formed maximum RMSF up to 3 Å. The C- and N-terminal regions showed residual fluctuation up to 2.5 Å. The amino acids involved in the strong and maintained nonbonded contacts (i.e., Leu91, Val130, Tyr132, Phe169, Ile179, Ile187, Arg189, Tyr259, Asp289, and Gly291) exhibited less fluctuation (~1 Å).

The structural folding and compaction of a molecular structure can be understood by its radius of gyration (Rg). The Rg indicated consistent folding and stable complex formation (Figure 7.2). In the BACE1-dercusin complex, a steady rise in Rg and SASA was seen until 80 ns and was stable throughout 150 ns. The results indicate, the opening of the binding pocket for a ligand to get buried into the pocket and the formation of a stable complex during the equilibration period of ~80ns (refer to RMSD values). After 80ns, the complex was found for higher compactness, whereas in the BACE1-lanabecestat complex, a steady increase in the Rg and SASA was seen till 150 ns; this implied binding pocket opening, which allows free movement of compound interaction within it and is further responsible for the formation of a stable complex. Decursin formed 5 *H*-bond interactions, in which 3 were consistent until 80 ns and 1 was consistent after 80 ns. This supported the RMSD, Rg, and SASA values that the complex was stable after an equilibration period of 80ns. In contrast, lanabecestat–BACE1 complex formed 4 *H*-bonds, and among them, 3 were constant until 150 ns, indicating a stable interaction throughout the simulation.

Further, Table 7.1 shows the free binding energy of complexes using MMPBSA. The estimated relative BE of dercusin and lanabecestat with BACE1 was −10.653±3.15 and −17.078±3.635 kcal/mol, respectively. In the dercusin-BACE1 complex, residues Trp258, Pro131, Glu261, Ile187, Asp192, Asp442, Ala188, Val130, and Glu257 favoured the stable complex formation by scoring contribution energy of −4.52, −4.36, −3.49, −1.75, −1.74, −1.47, −1.44, −1.25, and −1.19kJ/mol, respectively. However, Arg189, Arg255, Asn98, Asp284, Arg256, and Lys285 residues did not favour the interactions as they scored positive energy. Meanwhile, in the lanabecestat–BACE1 complex, residues Thr133, Thr390, Val393, Pro131, Ser388, Ile287, Lys136, Ser389, and Tyr259 favoured the stable complex formation by scoring contribution energy of −6.10, −4.46, −3.03, −2.20, −1.87, −1.81, −1.76, −1.70, and −1.50kJ/mol respectively. However, Tyr132, Asp93, Lys285, Arg296, and Asp289 residues did not favour the interactions as they scored positive energy. Figure 7.2 represents the structural stabilities of dercusin and lanabecestat with BACE1.

TABLE 7.1
Top Hit Complexes' Binding Free Energy and Interaction Energies by MM-PBSA Calculation

Complex	MM-PBSA (kcal/mol)				
	ΔE_{VDW}	ΔE_{ELE}	ΔG_{Sol}	ΔG_{Surf}	ΔG_{bind}
BACE1-Decursin	−26.055±2.562	−8.276±3.203	26.863±6.008	−3.185±0.331	−10.653±3.15
BACE1-Lanabecestat	−29.210±3.028	−5.66±1.266	21.254±2.933	−3.461±0.243	−17.078±3.635
PDE4D-2-Acetamido-6H-dibenzo[b,d]pyran-6-one	−25.760±3.588	1.848±2.662	12.589±4.577	−2.544±0.276	−13.865±3.393
PDE4D-AKU	−35.095±3.399	−27.749±1.588	86.437±9.663	−4.123±0.221	−19.469±10.947
ACHE-Donepezil	−42.296±2.604	−102.255±2.928	43.699±4.677	−4.941±0.244	−105.79±4.240
ACHE-Farnesiferol A	−50.251±2.75	−8.259±1.739	23.514±4.276	−4.888±0.293	−39.885±4.718

ΔE_{VDW} = van der Waals contribution; ΔE_{ELE} = electrostatic energy; ΔG_{Sol} = polar solvation free energy; ΔG_{Surf} = solvent-accessible surface area; ΔG_{bind} = Binding free energy.

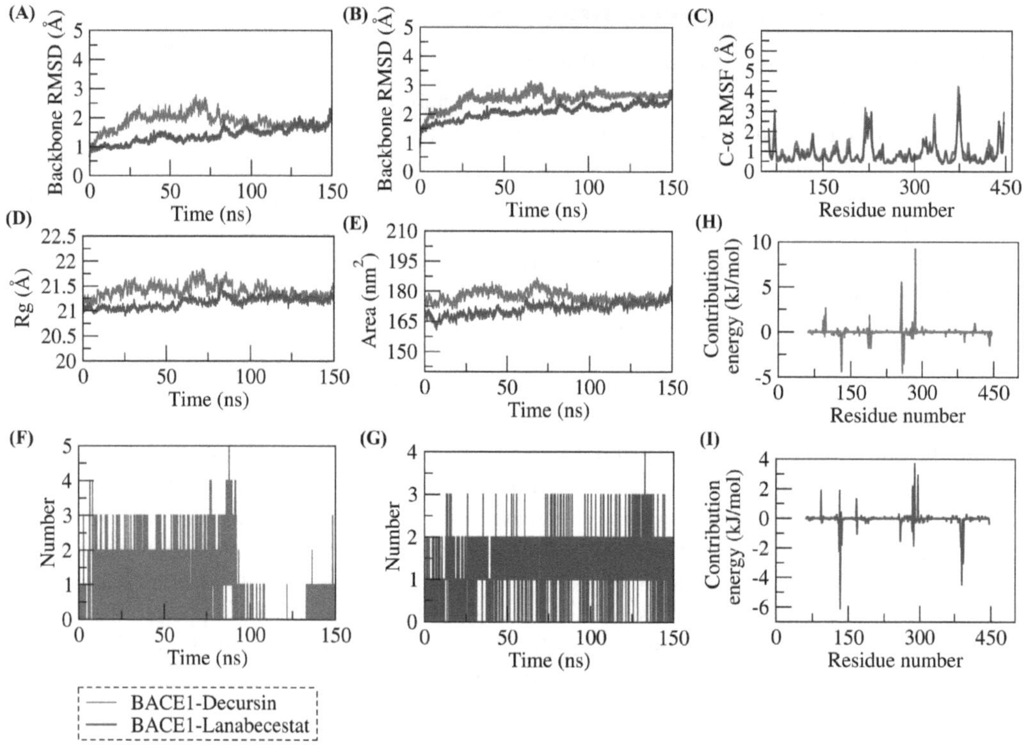

FIGURE 7.2 Parameters describing BACE1 with decursin (red) and lanabecestat (blue) complexes structural stabilities. (A) RMSD of backbone B) RMSD of complex, (C) RMSF, (D) Rg, (E) SASA, (E and F) number of H-bond interactions, and (G and H) contribution energy plot highlighting the importance of the binding pocket residues in stable complex formation

7.3.5.2 Stability of Farnesiferol A and Donepezil with ACHE

ACHE-donepezil and farnesiferol A complex shows stability after 50 ns equilibration time. The backbone average RMSD was ~1.5 Å and ~2.2 Å, respectively, and for the complex was ~2.0 Å and ~3 Å, respectively. Both N- and C-terminal showed RMSF up to ~3.8 and ~6 Å, respectively, and Tyr103, Trp117, Gly152, Trp317, Phe326, Tyr368, Phe369, Tyr372, and His478 showed lesser (~1 Å) as these are participating in ligand contact. The Rg shows very stable folding and complex generation after 50ns and the SASA was stable until 150 ns. Farnesiferol A formed 3 *H*-bonds and 2 were constant and Donepezil formed 3 *H*-bonds and 1 was consistent.

Further, the estimated relative BEs of donepezil and farnesiferol A with ACHE were −105.79±4.240 and −39.885±4.718kcal/mol, respectively. In the donepezil–ACHE complex, residues Glu316, Glu125, Trp317, Asp126, Glu216, Asp224, Asp380, Asp105, Tyr103, Phe111, Trp117, Glu323 favoured the stable complex formation by scoring contribution energy of −22.78, −14.12, −12.74, −12.42, −8.87, −8.79, −8.36, −7.04, −4.71, −4.44, −2.33, and −2.06kJ/mol, respectively. However, Lys379, Arg455, Arg250, and Arg167residue did not favour the interactions. In contrast, in the farnesiferol A–ACHE complex, residues Phe111, Tyr372, Tyr108, Trp117, Trp317, Asp105, Glu323, Asp380, Met116, Tyr103, and Glu316 favoured the stable complex formation by scoring contribution energy of −10.35, −8.21, −8.11, −8.05, −7.42, −5.97, −4.40, −4.17, −3.05, −2.62, −2.56kJ/mol respectively. However, Arg278, Arg327, and Lys379 residue did not Favor the interactions. Figure 7.3 represents the structural stabilities of Farnesiferol A and Donepezil with ACHE.

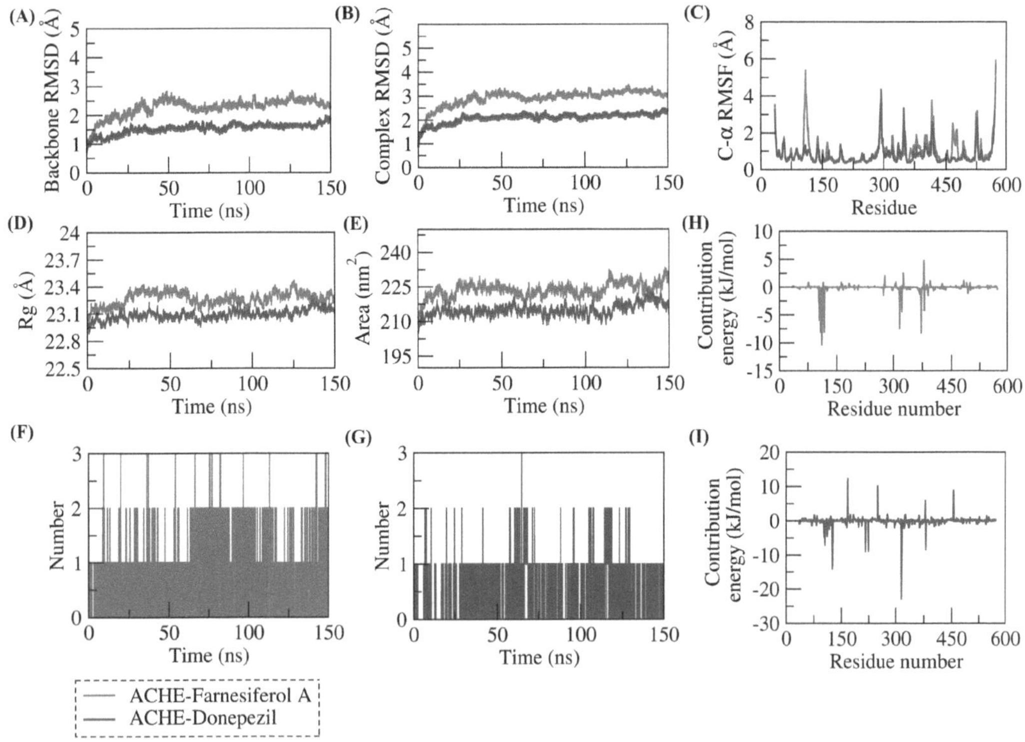

FIGURE 7.3 Parameters describing ACHE with farnesiferol A (red) and donepezil (blue) complexes structural stabilities. (A) RMSD of backbone B) RMSD of complex, (C) RMSF, (D) Rg, (E) SASA, (E and F) several H-bond interactions, and (G and H) contribution energy plot highlighting the importance of the binding pocket residues in stable complex formation.

7.3.5.3 Stability of 2-Acetamido-6H-dibenzo[b,d]pyran-6-one and AKU with PDE4D

The PDE4D with AKU and 2-Acetamido-6H-dibenzo[b,d]pyran-6-one shows stability after 30 ns equilibration time. The RMSD for the backbone was ~1.5 and 1.4 Å, respectively, and for the complex was ~2.5 Å and ~2 Å, respectively. Both N- and C-terminal RMSF was up to ~4.8 Å and residues Tyr461, His462, Met575, Leu621, Asn623, Ile638, Phe642, Met659, Gln671, and Phe674 showed lesser RMSF (~1.5 Å) as these are participating in ligand contact. The PDE4D-AKU complex had stable Rg value and stable complex formation 30ns and the PDE4D-2-Acetamido-6H-dibenzo[b,d]pyran-6-one complex showed a decrease in the Rg during equilibration time, indicates higher compactness and stable complex formation. A similar trend was observed for SSA for both the complexes, but the PDE4D-2-Acetamido-6H-dibenzo[b,d]pyran-6-one complex was more stable. 2-Acetamido-6H-dibenzo[b,d]pyran-6-one formed 2 *H*-bond interactions, in which 1 was consistent, and AKU formed 2 *H*-bonds and were not constant throughout the simulations. Further, the estimated relative BE of 2-Acetamido-6H-dibenzo[b,d] pyran-6-one and AKU with PDE4D was −13.865±3.393 and −19.469±10.947 kcal/mol, respectively. The residues Tyr677, Phe674, Ile678, Met659, Pro658, Asp676, and Gly673 favoured the stable complex formation by scoring contribution energy of −9.31, −6.55, −5.40, −4.92, −4.29, −2.53, −2.24, −2.2kJ/mol, respectively. However, Met575, Glu532, Asp574, Asp503, and Asp620 residues did not Favor the interactions. In the residues, the AKU–PDE4D complex, Asp503, Asp620, Met575, Phe674, Ile638, His502, Phe642, Pro624, Leu621, Tyr461, Glu532, and Asp574 favoured the stable complex formation by scoring contribution energy of −21.62, −9.80, −8.42, −7.63, −7.095, −5.408, −4.24, −3.90, −3.52, −3.50, −3.14, −2.62, kJ/mol respectively. However, Asn623 residue did not favour the interactions. Figure 7.4 presents the structural stabilities of 2-Acetamido-6H-dibenzo[b,d]pyran-6-one and AKU with PDE4D.

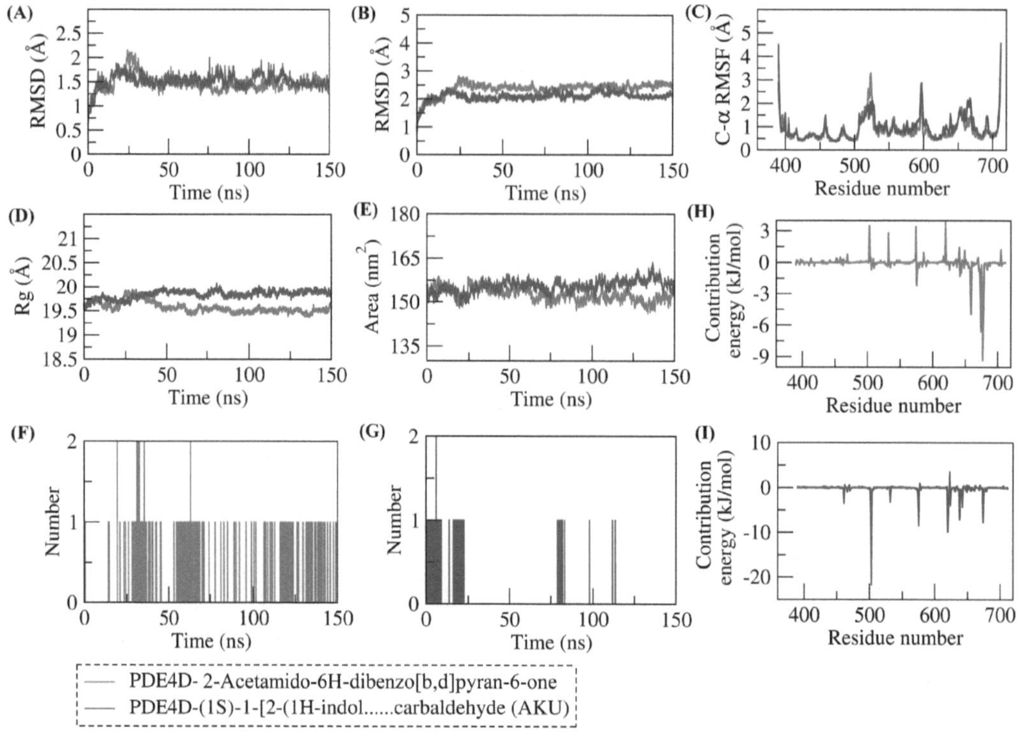

FIGURE 7.4 Parameters describing PDE4D with 2-Acetamido-6H-dibenzo[b,d]pyran-6-one(red) and AKU (blue) complexes structural stabilities. (A) RMSD of backbone B) RMSD of complex, (C) RMSF, (D) Rg, (E) SASA, (E and F) a number of H-bond interactions, and (G and H) contribution energy plot highlighting the importance of the binding pocket residues in stable complex formation.

7.3.6 Principal Component Analysis

It is observed that known inhibitors lanabecestat and donepezil with BACE1 and ACHE, respectively, show lesser conformational diversity; however, AKU with PDE4D showed larger conformational diversity. However, selected ligands—namely decursin and farnesiferol A, targeting BACE1 and ACHE, respectively—showed greater conformational diversity, and AKU with PDE4D showed less diversity. This reveals the good equilibration and stabilization of complex 2-Acetamido-6H-dibenzo[b,d]pyran-6-one with PDE4D (Figure 7.5).

7.3.7 Gene Set Pathway Enrichment and Network Analysis

A total of 39 coumarin derivatives are predicted to hit BACE1, ACHE, and PDE4D. Among them, 8 were common. 39 compounds were predicted to modulate 712 other host targets. A total of 38 therapeutic targets of AD from the TTD database were shortlisted. After matching the 712 targets with 38 targets, out of 712 targets, 18 were found to be common. In total, 39 compounds were predicted to 18 therapeutic targets of AD. The enrichment analysis of 18 targets revealed involvement in 22 molecular pathways, in which 10 pathways namely, Neuroactive ligand-receptor interaction, GABAergic synapse, Serotonergic synapse, Phenylalanine, Tyrosine, Glycine, Serine and Threonine metabolism, cGMP-PKG, and Calcium signalling pathways were directly associated with AD pathogenesis (Table 7.2 and Figure 7.6) the Along with these, Parkinson disease was also enriched. Among them, MAOA and MAOB targets were identified as hub genes.

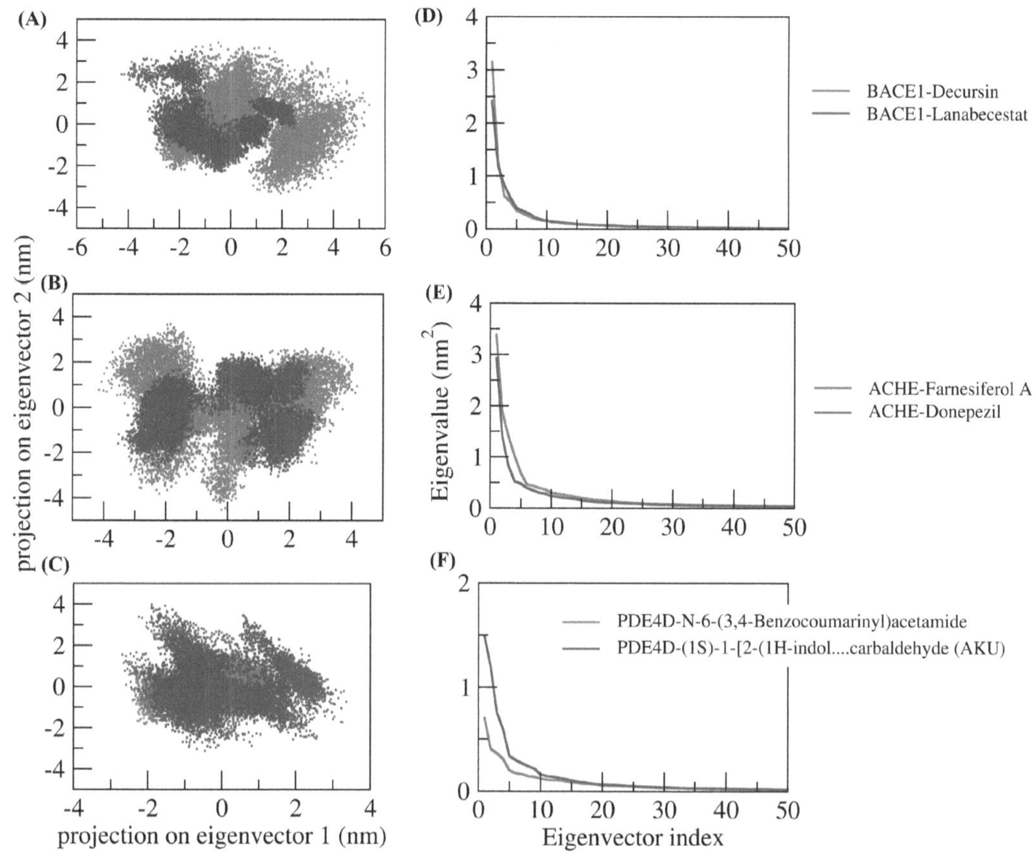

FIGURE 7.5 Principal component analysis of protein-ligand complexes: The collective motion of coumarins (red) and standard (black) with (A) BACE1, (B) ACHE, and (C) PDE4D using projections of MD trajectories on 2 eigenvectors corresponding to the first 2 principal components. The first 50 eigenvectors were plotted versus eigenvalue for coumarins (red) and standard molecule (black) with (D) BACE1, (E) ACHE, and (F) PDE4D.

TABLE 7.2

Coumarin derivatives-regulated pathways associated with AD pathogenesis.

KEGG ID	Pathway	Gene count	FDR value	Enriched gene set within the pathway
hsa04080	Neuroactive ligand-receptor interaction	9	1.79E-09	OPRD1, ADRA2A, GABRA5, ADORA2A, ADORA1, CHRM5, GABRA1, OPRM1, GABRA2
hsa04727	GABAergic synapse	3	0.005	GABRA5, GABRA1, GABRA2
hsa00360	Phenylalanine metabolism	2	0.0076	MAOA, MAOB
hsa04726	Serotonergic synapse	3	0.0076	APP, MAOA, MAOB
hsa04022	cGMP-PKG signalling pathway	3	0.0133	OPRD1, ADRA2A, ADORA1
hsa00350	Tyrosine metabolism	2	0.0146	MAOA, MAOB
hsa00260	Glycine, serine, and threonine metabolism	2	0.0158	MAOA, MAOB
hsa04020	Calcium signalling pathway	3	0.0169	EGFR, ADORA2A, CHRM5
hsa05012	Parkinson disease	3	0.025	ADORA2A, MAOA, MAOB

FDR: False discovery Rate

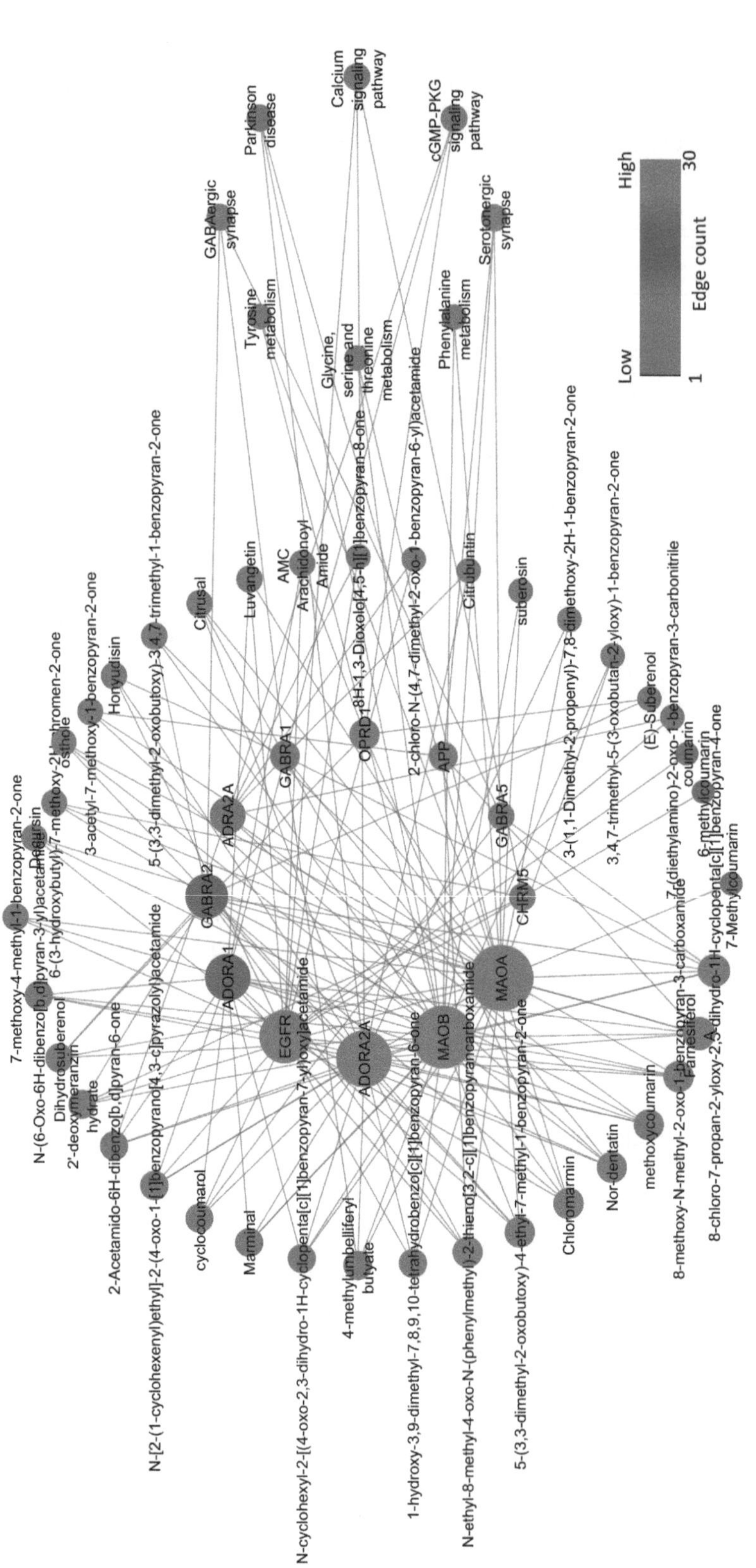

FIGURE 7.6 Network interaction of Coumarin derivatives with their targets and regulated pathways associated with concerning the KEGG database. In the network interaction, the node with a small size represents minimum edge count and vice versa.

7.4 DISCUSSION

Memory loss and difficulties retaining knowledge are classic symptoms of Alzheimer's disease, which affects the central cholinergic neuronal circuits that are essential for learning and memory (progressive cognitive impairment) (Breijyeh and Karaman, 2020). Synthetic ACHE and NMDA inhibitors are widely used in the current pharmacotherapy of AD. Nevertheless, these drugs are known to have a variety of side effects and lose their effects as the illness progresses owing to the death of brain cells and an acetylcholine deficiency (Patil et al., 2021).

On the other hand, β-secretase, an aspartyl protease, is thought to be the rate-limiting step in Aβ production by cleaving the APP in the lumina. One of the benefits of inhibiting BACE1 is that it stops Aβ from being formed in the first place while processing APP (Vassar et al., 2009; Hampel et al., 2021). The use of BACE1 inhibitors enhanced neurological function and overcame the neurological abnormalities inflicted on mice of the AD model, confirming the crucial role played by β-secretase in the development of AD (Hampel et al., 2021). In the 1970s, research on a transgenic fly with a learning defect revealed the importance of phosphodiesterases (PDEs) in cognition (Tibbo et al., 2019). PDEs can destroy cyclic nucleotides. The PDE4D gene was shown to act on cyclic adenosine monophosphate (cAMP) (Tibbo et al., 2019). There is substantial evidence linking abnormal cAMP signalling to the cognitive deficits seen in neurodegenerative illnesses such as AD.

Herbal plants are used routinely in medicine to treat a wide range of illnesses and conditions. It has been experimentally demonstrated that numerous Coumarin derivatives play a chief role in the management of AD (Pisani et al., 2022; Hamulakova et al., 2017). The present work attempted to identify the BACE1, ACHE, and PDE4D inhibitors from coumarin derivatives (polyphenolic compounds) for the management of AD. Initially, we identified 379 coumarins derivatives, in which 69 compounds were expected to cross CNS as per Suenderhauf et al., criteria (compounds with hydrogen bond acceptor <=4 with BBB value −1.0028 to 0.4049 and compounds with rotational bond <=7 with BBB value >0.4049). Out of 69 compounds, 11, 21, and 16 derivatives of coumarin were predicted to hit BACE1, ACHE, and PDE4D, respectively, from the SwissTargetPrediction server.

To evaluate the probable mode of these compounds against BACE1, ACHE, and PDE4D, we employed docking and MD. Here, all the shortlisted Coumarins scored the BE <-6.0 kcal/mol. Further, except a few compounds (Decursin and 8-methoxy-N-methyl-2-oxo-1-benzopyran-3-carboxamide with BACE1; N-[2-(1-cyclohexenyl)ethyl]-2-(4-oxo-1-[1]benzopyrano[4,3-c]pyrazolyl) acetamide and methoxycoumarin with ACHE; Nor-dentatin and AMC Arachidonoyl Amide with PDE4D), all compounds exhibited a minimum of one *H*-bond in all three targets.

Aβ42 is the only monomeric form of amyloid-β (Aβ) that aggregates into active conformation species and is likely the cause of toxicity in AD, whereas all monomeric forms of Aβ are generated by BACE1 (O'Brien and Wong, 2011). BACE1 is found in higher concentrations and at higher rates of activity in AD brains and bodily fluids, lending credence to the idea that BACE1 is important in the progression of AD (Hampel et al., 2021); in this study, BACE1 was majorly targeted by decursin (−8.3kcal/mol) via forming 9 interaction with active site residues. Interestingly, Park et al. reported the presence of decursin in the 95% ethanolic extract of *Angelica gia*s and reported as potent inhibitor of BACE1 (35.5% inhibition at 0.5 μg/mL) as well as of ACHE and MAOB with moderate inhibition of MAOA (Park et al., 2022).

One of the most important enzymes in the cholinergic nerve system is acetylcholinesterase (AChE). Several different kinds of neurons degenerate during AD, but forebrain cholinergic neurons are particularly hard hit, with acetylcholine levels falling in tandem (Chen et al., 2022). Current AD treatment relies heavily on AChE inhibitors, which boost cholinergic transmission despite an overall decline in AChE activity in the AD brain. Nevertheless, these inhibitors have very limited and temporary therapeutic benefits (Chen et al., 2022).

In this study, ACHE is chiefly targeted by farnesiferol A (−12.7kcal/mol) via forming 17 interactions with binding site residues. Also, the other 20 compounds were interacting with binding site residues in the range of 2 to 17. Among them, decursin was found third lead for ACHE. Adhami et al. reported farnesiferol A isolated from *Ferula gummosa* Boiss showed a strong ACHE antagonism effect with IC_{50} of 17 μg/mL compared to physostigmine (0.8 μg/mL) (Adhami et al., 2013). Previously, Kang et al, (2001) reported that the AChE activity in the hippocampus was considerably reduced by 34% when decursin was treated at a dosage of 1mg/kg. Decursin and its related compounds decursinol (isolated from *Angelica gigas*) also showed potent ACHE antagonism effect with IC_{50} of 2.8×10^{-4}M and 2.8×10^{-5}M (Kang et al., 2001). Likewise, in this study, PDE4D was targeted by 2-Acetamido-6H-dibenzo[b,d]pyran-6-one and other 15 compounds with BE range from −9.6 to −7.2kcal/mol and interacting with active site residues number ranging from 6 to 15.

These findings corroborate the previous findings and support the use of coumarin derivatives for management. Importantly, the prioritized coumarins against BACE1, ACHE, and PDE4D were shown to be highly equilibrated and stable throughout the simulation, having a minimal number of different conformations and occupying correspondingly less conformational space than known inhibitors, as revealed by the examination of complexes for MD simulation, MMPBSA, and PCA.

In addition, this research highlighted the significance of regulating certain pathways by a group of coumarins i.e., neuroactive ligand-receptor interaction, GABAergic synapse, serotonergic synapse, phenylalanine, tyrosine, glycine, serine and threonine metabolism, cGMP-PKG, and calcium signalling pathways that are directly linked to AD pathogenesis. These findings support the "multi-compound–multi protein" interaction theory, which suggests that many molecules (of a similar category) may be used to target the diverse AD pathogenesis. Inhibitors of BACE1, ACHE, and PDE4D were found, accounting for 9.475% of the total coumarins studied, which is a noteworthy finding. In addition, the "*right molecule-right target-right duration*" evaluation may lead to tachyphylaxis toward disease pathogenesis, which might be resolved by the "*multi compounds-multi protein*" interaction rather than the "*single compound-single target*" interaction hypothesis (Khanal et al., 2022a, 2022b).

7.5 CONCLUSION

The current study investigated the multiple coumarin derivatives to identify the probable BACE1, ACHE, and PDE4D inhibitors for the management of AD. This study reported probable leads, namely decursin, farnesiferol A, and 2-Acetamido-6H-dibenzo[b,d]pyran-6-one, against AD by inhibiting BACE1, ACHE, and PDE4D, respectively. Interestingly, all the predicted coumarin derivatives were found to interact with binding site residues of the selected proteins. In addition, the enrichment analysis revealed these compounds to regulate identified the regulation of various pathways like Neuroactive ligand-receptor interaction, GABAergic synapse, serotonergic synapse, phenylalanine, tyrosine, glycine, serine and threonine metabolism, cGMP-PKG, and calcium signalling pathways that are directly concerned with AD pathogenesis. These results revealed the potential therapeutic role of coumarin derivative-enriched plant extracts as resources against AD. We emphasize that the results reported here are only preliminary and require additional validation through wet-lab experimental techniques, as they are based on processor simulations.

ENDNOTE

* Corresponding author

REFERENCES

Aalten, D. V., Findlay, J. B. C., Amadei, A., & Berendsen, H. J. C. (1995). Essential dynamics of the cellular retinol-binding protein evidence for ligand-induced conformational changes. *Protein Engineering, Design and Selection*, *8*(11), 1129–1135.

Adhami, H. R., Scherer, U., Kaehlig, H., Hettich, T., Schlotterbeck, G., Reich, E., & Krenn, L. (2013). Combination of bioautography with HPTLC—MS/NMR: A fast identification of acetylcholinesterase inhibitors from galbanum. *Phytochemical Analysis*, *24*(4), 395–400.

Amadei, A., Linssen, A. B. M., & Berendsen, H. J. (1993). Essential dynamics of proteins. *Proteins: Structure, Function, and Bioinformatics*, *17*(4), 412–425.

Amadei, A., Linssen, A. B. M., De Groot, B. L., Van Aalten, D. M. F., & Berendsen, H. J. C. (1996). An efficient method for sampling the essential subspace of proteins. *Journal of Biomolecular Structure and Dynamics*, *13*(4), 615–625.

Anand, P., Singh, B., & Singh, N. (2012). A review on coumarins as acetylcholinesterase inhibitors for Alzheimer's disease. *Bioorganic & Medicinal Chemistry*, *20*(3), 1175–1180.

Azevedo, M. F., Faucz, F. R., Bimpaki, E., Horvath, A., Levy, I., de Alexandre, R. B., . . . & Stratakis, C. A. (2014). Clinical and molecular genetics of the phosphodiesterases (PDEs). *Endocrine Reviews*, *35*(2), 195–233.

Bhandare, V. V., Kumbhar, B. V., & Kunwar, A. (2019). Differential binding affinity of tau repeat region R2 with neuronal-specific β-tubulin isotypes. *Scientific Reports*, *9*(1), 10795.

Bhandare, V. V., & Ramaswamy, A. (2018). The proteinopathy of D169G and K263E mutants at the RNA Recognition Motif (RRM) domain of tar DNA-binding protein (tdp43) causing neurological disorders: A computational study. *Journal of Biomolecular Structure and Dynamics*, *36*(4), 1075–1093.

Biradar, P., Patil, V., Joshi, H., Khanal, P., & Mallapur, S. (2020). Experimental validation and network pharmacology evaluation to decipher the mechanism of action of Erythrina variegata L. bark against scopolamine-induced memory impairment in rats. *Advances in Traditional Medicine*, 1–14.

Breijyeh, Z., & Karaman, R. (2020). Comprehensive review on Alzheimer's disease: Causes and treatment. *Molecules*, *25*(24), 5789.

Briggs, R., Kennelly, S. P., & O'Neill, D. (2016). Drug treatments in Alzheimer's disease. *Clinical Medicine*, *16*(3), 247.

Chen, Z. R., Huang, J. B., Yang, S. L., & Hong, F. F. (2022). Role of cholinergic signaling in Alzheimer's disease. *Molecules*, *27*(6), 1816.

Corrada, M. M., Brookmeyer, R., Paganini-Hill, A., Berlau, D., & Kawas, C. H. (2010). Dementia incidence continues to increase with age in the oldest old: The 90+ study. *Annals of Neurology*, *67*(1), 114–121.

Daina, A., Michielin, O., & Zoete, V. (2019). SwissTargetPrediction: Updated data and new features for efficient prediction of protein targets of small molecules. *Nucleic Acids Research*, *47*(W1), W357–W364.

Das, B., & Yan, R. (2019). A close look at BACE1 inhibitors for Alzheimer's disease treatment. *CNS Drugs*, *33*(3), 251–263.

Dassault Systèmes Biovia. (2019). *DassaultSystèmes*. San Diego: Discovery S. Available at: https://discover.3ds.com/discovery-studio-visualizer-download.

Dwivedi, P. S., Patil, V. S., Khanal, P., Bhandare, V. V., Gurav, S., Harish, D. R., . . . & Roy, S. (2022). System biology-based investigation of Silymarin to trace hepatoprotective effect. *Computers in Biology and Medicine*, *142*, 105223.

Fylaktakidou, K. C., Hadjipavlou-Litina, D. J., Litinas, K. E., & Nicolaides, D. N. (2004). Natural and synthetic coumarin derivatives with anti-inflammatory/antioxidant activities. *Current Pharmaceutical Design*, *10*(30), 3813–3833.

Hampel, H., Vassar, R., De Strooper, B., Hardy, J., Willem, M., Singh, N., . . . & Vergallo, A. (2021). The β-secretase BACE1 in Alzheimer's disease. *Biological Psychiatry*, *89*(8), 745–756.

Hamulakova, S., Kozurkova, M., & Kuca, K. (2017). Coumarin derivatives in pharmacotherapy of Alzheimer's disease. *Current Organic Chemistry*, *21*(7), 602–612.

Kang, S. Y., Lee, K. Y., Sung, S. H., Park, M. J., & Kim, Y. C. (2001). Coumarins isolated from angelica g igas inhibit acetylcholinesterase: Structure– activity relationships. *Journal of Natural Products*, *64*(5), 683–685.

Khanal, P., Patil, V. S., Bhandare, V. V., Dwivedi, P. S., Shastry, C. S., Patil, B. M., . . . & Roy, S. (2022b). Computational investigation of benzalacetophenone derivatives against SARS-CoV-2 as potential multi-target bioactive compounds. *Computers in Biology and Medicine*, *146*, 105668.

Khanal, P., Patil, V. S., Bhandare, V. V., Patil, P. P., Patil, B. M., Dwivedi, P. S., . . . & Roy, S. (2022a). Systems and in vitro pharmacology profiling of diosgenin against breast cancer. *Frontiers in Pharmacology*, *13*.

Kumar, A., Sidhu, J., Goyal, A., & Tsao, J. W. (2018). Alzheimer disease. In: *StatPearls* [Internet]. Treasure Island (FL): StatPearls Publishing, 2023 January. PMID: 29763097.

Kumar, A., & Singh, A. (2015). A review on Alzheimer's disease pathophysiology and its management: An update. *Pharmacological Reports*, *67*(2), 195–203.

Lagunin, A., Ivanov, S., Rudik, A., Filimonov, D., & Poroikov, V. (2013). DIGEP-Pred: Web service for in silico prediction of drug-induced gene expression profiles based on structural formula. *Bioinformatics*, *29*(16), 2062–2063.

Lipinski, C. A. (2004). Lead-and drug-like compounds: The rule-of-five revolution. *Drug Discovery Today: Technologies*, *1*(4), 337–341.

Marucci, G., Buccioni, M., Dal Ben, D., Lambertucci, C., Volpini, R., & Amenta, F. (2021). Efficacy of acetylcholinesterase inhibitors in Alzheimer's disease. *Neuropharmacology*, *190*, 108352.

Mehta, M., Adem, A., & Sabbagh, M. (2012). New acetylcholinesterase inhibitors for Alzheimer's disease. *International Journal of Alzheimer's Disease, 2012.*

Monica, Moore M., Díaz-Santos, M., & Vossel, K. (2021). *Alzheimer's Association 2021 Facts and Figures Report.* Alzheimer's Association, University of California, Los Angeles.

Morris, G. M., Huey, R., Lindstrom, W., Sanner, M. F., Belew, R. K., Goodsell, D. S., & Olson, A. J. (2009). AutoDock4 and AutoDockTools4: Automated docking with selective receptor flexibility. *Journal of Computational Chemistry, 30*(16), 2785–2791.

Murphy, M., & LeVine III, H. (2010). Alzheimer's disease and the β-amyloid peptide. *Journal of Alzheimer's Disease, 19*(1), 311–323.

Nisha, S. A., & Devi, K. P. (2017). Gelidiella acerosa protects against Aβ 25–35-induced toxicity and memory impairment in Swiss Albino mice: An in vivo report. *Pharmaceutical Biology, 55*(1), 1423–1435.

O'Brien, R. J., & Wong, P. C. (2011). Amyloid precursor protein processing and Alzheimer's disease. *Annual Review of Neuroscience, 34*, 185–204.

Park, J. E., Mun, S. K., Yee, S. T., & Kim, H. (2022). Evaluation of inhibitory activities of Sophora flavescens and Angelica gigas Nakai root extracts against monoamine oxidases, cholinesterases, and β-secretase. *Processes, 10*(5), 880.

Patil, V. S., Deshpande, S. H., Harish, D. R., Patil, A. S., Virge, R., Nandy, S., & Roy, S. (2020). Gene set enrichment analysis, network pharmacology and in silico docking approach to understand the molecular mechanism of traditional medicines for the treatment of diabetes mellitus. *Journal of Proteins and Proteomics, 11*, 297–310.

Patil, V. S., Harish, D. R., Vetrivel, U., Deshpande, S. H., Khanal, P., Hegde, H. V., . . . & Jalalpure, S. S. (2022). Pharmacoinformatics analysis reveals flavonoids and diterpenoids from Andrographis paniculata and Thespesia populnea to target hepatocellular carcinoma induced by hepatitis B virus. *Applied Sciences, 12*(21), 10691.

Patil, V. S., Meena, H., & Harish, D. R. (2021). Erythrina variegata L. bark: An untapped bioactive source harbouring therapeutic properties for the treatment of Alzheimer's disease. *In Silico Pharmacology, 9*(1), 51.

Peitzika, S. C., & Pontiki, E. (2023). A review on recent approaches on molecular docking studies of novel compounds targeting acetylcholinesterase in Alzheimer disease. *Molecules, 28*(3), 1084.

Pisani, L., Catto, M., Muncipinto, G., Nicolotti, O., Carrieri, A., Rullo, M., . . . & Altomare, C. (2022). A twenty-year journey exploring coumarin-based derivatives as bioactive molecules. *Frontiers in Chemistry, 10.*

Rappé, A. K., Casewit, C. J., Colwell, K. S., Goddard III, W. A., & Skiff, W. M. (1992). UFF, a full periodic table force field for molecular mechanics and molecular dynamics simulations. *Journal of the American Chemical Society, 114*(25), 10024–10035.

Samdani, A., & Vetrivel, U. (2018). POAP: A GNU parallel based multithreaded pipeline of open babel and AutoDock suite for boosted high throughput virtual screening. *Computational Biology and Chemistry, 74*, 39–48.

Scott, K. R., & Barrett, A. M. (2007). Dementia syndromes: Evaluation and treatment. *Expert Review of Neurotherapeutics, 7*(4), 407–422.

Serafín, V., Gamella, M., Pedrero, M., Montero-Calle, A., Razzino, C. A., Yáñez-Sedeño, P., . . . & Pingarrón, J. M. (2020). Enlightening the advancements in electrochemical bioanalysis for the diagnosis of Alzheimer's disease and other neurodegenerative disorders. *Journal of Pharmaceutical and Biomedical Analysis, 189*, 113437.

Shannon, P., Markiel, A., Ozier, O., Baliga, N. S., Wang, J. T., Ramage, D., . . . & Ideker, T. (2003). Cytoscape: A software environment for integrated models of biomolecular interaction networks. *Genome Research, 13*(11), 2498–2504.

Sharifi-Rad, J., Cruz-Martins, N., López-Jornet, P., Lopez, E. P. F., Harun, N., Yeskaliyeva, B., . . . & Cho, W. C. (2021). Natural coumarins: Exploring the pharmacological complexity and underlying molecular mechanisms. *Oxidative Medicine and Cellular Longevity, 2021.*

Sheng, J., Zhang, S., Wu, L., Kumar, G., Liao, Y., Pratap, G. K., & Fan, H. (2022). Inhibition of phosphodiesterase: A novel therapeutic target for the treatment of mild cognitive impairment and Alzheimer's disease. *Frontiers in Aging Neuroscience, 14.*

Soto-Ortega, D. D., Murphy, B. P., Gonzalez-Velasquez, F. J., Wilson, K. A., Xie, F., Wang, Q., & Moss, M. A. (2011). Inhibition of amyloid-β aggregation by coumarin analogs can be manipulated by functionalization of the aromatic center. *Bioorganic & Medicinal Chemistry, 19*(8), 2596–2602.

Suenderhauf, C., Hammann, F., & Huwyler, J. (2012). Computational prediction of blood-brain barrier permeability using decision tree induction. *Molecules*, *17*(9), 10429–10445.

Szklarczyk, D., Morris, J. H., Cook, H., Kuhn, M., Wyder, S., Simonovic, M., Santos, A., Doncheva, N. T., Roth, A., Bork, P., Jensen, L. J., & Von Mering, C. (2016). The STRING database in 2017: Quality-controlled protein—protein association networks, made broadly accessible. *Nucleic Acids Research*, *45*, D362–D368.

Tarawneh, R., & Holtzman, D. M. (2012). The clinical problem of symptomatic Alzheimer disease and mild cognitive impairment. *Cold Spring Harbor Perspectives in Medicine*, *2*(5), a006148.

Tibbo, A. J., Tejeda, G. S., & Baillie, G. S. (2019). Understanding PDE4's function in Alzheimer's disease; a target for novel therapeutic approaches. *Biochemical Society Transactions*, *47*(5), 1557–1565.

Van Der Spoel, D., Lindahl, E., Hess, B., Groenhof, G., Mark, A. E., & Berendsen, H. J. (2005). GROMACS: Fast, flexible, and free. *Journal of Computational Chemistry*, *26*(16), 1701–1718.

Vassar, R., Kovacs, D. M., Yan, R., & Wong, P. C. (2009). The β-secretase enzyme BACE in health and Alzheimer's disease: Regulation, cell biology, function, and therapeutic potential. *Journal of Neuroscience*, *29*(41), 12787–12794.

Weller, J., & Budson, A. (2018). Current understanding of Alzheimer's disease diagnosis and treatment. *F1000Research*, *7*.

Wu, Y., Li, Z., Huang, Y. Y., Wu, D., & Luo, H. B. (2018). Novel phosphodiesterase inhibitors for cognitive improvement in Alzheimer's disease: Miniperspective. *Journal of Medicinal Chemistry*, *61*(13), 5467–5483.

8 Therapeutic Application of Natural Products in Alzheimer's Disease Using Computational Methods

Muhammad Aqib Shabbir, Muhammad Naveed, Maleeha Manzoor, Hafiz Muhammad Hurraira, Noor-ul-Ain and Fatima Zaidi*

8.1 INTRODUCTION

8.1.1 Background of Alzheimer's Disease

Alzheimer's disease (AD) is a neurodegenerative disorder that progresses over time and is marked by a decline in cognitive function, memory impairment, and alterations in behaviour and personality. It is a prevalent aetiology of cognitive decline in the elderly population and has a significant global impact, affecting millions of individuals. The incidence of AD is on the rise owing to the demographic shift towards an aging population. It is anticipated that the condition will impact over 100 million individuals by the year 2050 (Bis et al., 2020).

The aetiology of AD remains uncertain, although it is hypothesized to result from a multifactorial interplay of genetic, environmental, and lifestyle determinants; it is known to be intricate and to encompass a multitude of genes. The apolipoprotein E (APOE) ε4 allele is widely recognized as the primary genetic risk factor for AD, as it has been linked to an elevated susceptibility to the onset of the condition. Environmental and lifestyle variables, such as head trauma, hypertension, tobacco use, and poor dietary habits, have also been linked to the development of Alzheimer's disease, in addition to hereditary predisposition (Lloret et al., 2019). Figure 8.1 represents the pathophysiology of AD.

The primary pathological features of AD are the development of extracellular senile plaques, which comprise amyloid-β peptide (Aβ), and the formation of intracellular neurofibrillary tangles, which are composed of hyperphosphorylated tau proteins (Soria Lopez et al., 2019). The aforementioned hallmarks are believed to be implicated in the pathogenesis of the neurodegeneration and cognitive impairment that are commonly observed in AD. It is widely accepted that the buildup of Aβ initiates a series of reactions resulting in oxidative stress, inflammation, and ultimately neuronal apoptosis. It is hypothesized that the development of neurofibrillary tangles can impede the typical operation of neurons and impede the conveyance of crucial molecules and signals within the cellular environment (Lloret et al., 2019).

Notwithstanding the progress made in comprehending the fundamental pathology of Alzheimer's disease, the process of devising novel therapies has proven to be a protracted and arduous undertaking (Long & Holtzman, 2019). The demand for novel and efficacious treatments for AD is urgent, and the quest for innovative therapies is currently underway.

 DOI: 10.1201/9781003412069-8

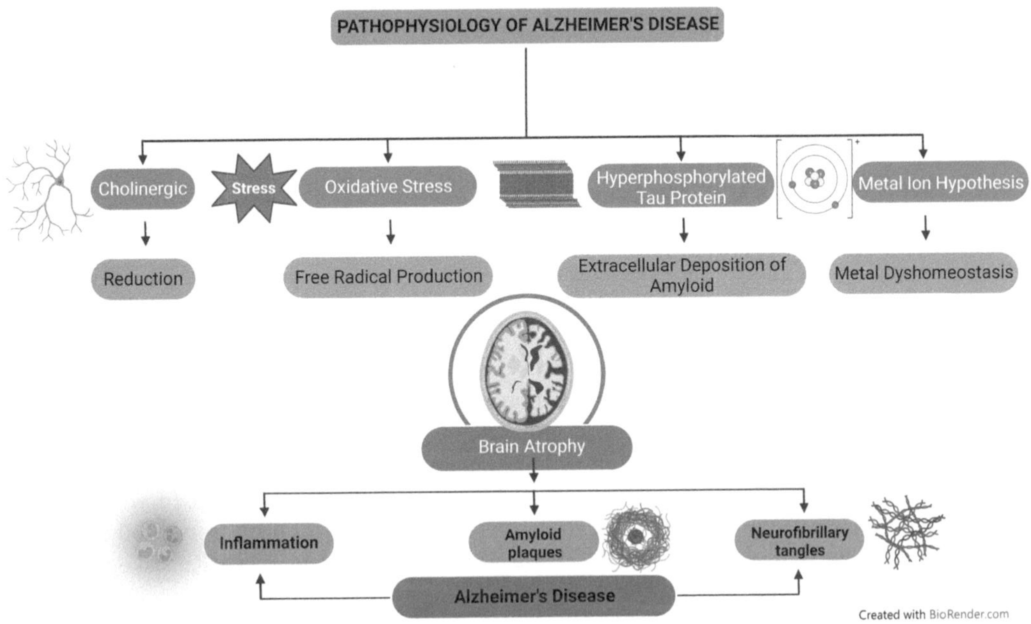

FIGURE 8.1 The pathophysiology of Alzheimer's disease (Bis et al., 2020).

8.1.2 Types of Alzheimer Disease

(i) As per the scientific literature, AD is a multifaceted and diverse ailment that is categorized into various subtypes:

(ii) **Early-onset familial Alzheimer's disease (EOFAD):** This particular manifestation of Alzheimer's disease is infrequent and predominantly impacts individuals who have not yet reached the age of 65. The aetiology of EOFAD can be attributed to genetic mutations in specific genes, namely APP, PSEN1, and PSEN2. This particular form of AD exhibits heightened aggressiveness and follows an autosomal dominant pattern of inheritance (Cummings et al., 2020).

(iii) **Late-onset sporadic Alzheimer's disease (LOSAD):** The most prevalent manifestation of AD is the variant that primarily impacts individuals aged 65 years and above. The aetiology of LOSAD is believed to be multifactorial, involving intricate interplay between genetic and environmental factors, rather than being attributed to a singular genetic mutation.

(iv) **Down syndrome-associated Alzheimer's disease (DS-AD):** The extra copy of chromosome 21 that causes Down syndrome also increases a person's risk of developing a form of AD, with the onset of the disease often occurring in their 50s or 60s (Atri, 2019).

(v) **Lewy body dementia (LBD):** Lewy bodies, abnormal protein aggregates, are a hallmark of LBD, a form of neurodegeneration. Individuals with LBD can experience cognitive impairment, such as memory loss, confusion, and changes in mood and behaviour, that are akin to those observed in AD.

(vi) **Vascular dementia:** Cerebrovascular dementia is attributed to cerebrovascular disease, which results in cerebral blood vessel damage and subsequent cerebral hypoperfusion and hypoxia, leading to cognitive impairment. Vascular dementia is known to elicit symptoms that are akin to those observed in AD, including but not limited to memory impairment and challenges with cognitive functions such as problem-solving and decision-making.

It is noteworthy that the categorization of AD is frequently complex, and a considerable number of individuals diagnosed with AD may exhibit a blend of diverse AD subtypes. A reliable diagnosis of Alzheimer's disease often requires a thorough study of the patient's health history, cognitive and neurological examinations, and imaging analysis (Scheltens et al., 2021).

8.1.3 Overview of Current Treatment Options for the Alzheimer's Disease

As per the extant scientific literature, AD lacks a cure, and the existing treatments are geared towards mitigating symptoms and decelerating the advancement of the ailment. The subsequent discourse provides a comprehensive summary of the contemporary therapeutic modalities available for the management of AD:

i. **Cholinesterase inhibitors:** Pharmaceutical agents, namely donepezil, rivastigmine, and galantamine, are designed to augment the concentrations of acetylcholine within the central nervous system. Acetylcholine, a neurotransmitter implicated in memory and learning processes, exhibits decreased levels in individuals diagnosed with AD . Cholinesterase inhibitors function by inhibiting the degradation of acetylcholine, and are frequently employed in the management of mild to moderate AD (Knapskog et al., 2021).
ii. **NMDA receptor antagonists:** The objective of this drug category, exemplified by memantine, is to regulate glutamatergic neurotransmission within the brain. An important neurotransmitter, glutamate is thought to play a role in the neurodegenerative processes behind AD as well as cognitive activities including learning and memory. In the treatment of mild to severe AD, N-methyl-D-aspartate (NMDA) receptor antagonists are commonly used because of their ability to reduce the detrimental effects of glutamate overproduction (Lei et al., 2021). They are known to exert their pharmacological effects by inhibiting the impact of excess glutamate and are commonly utilized in the treatment of moderate to severe AD (Cummings et al., 2020).
iii. **Nonpharmacological treatments:** Apart from pharmacological interventions, nonpharmacological modalities, including cognitive and physical therapy, can be employed to effectively address the symptoms of AD. The aforementioned interventions have the potential to enhance cognitive abilities, sustain bodily movement, and mitigate behavioural and psychological manifestations (Scheltens et al., 2021).

It is noteworthy that the efficacy of interventions for AD can exhibit inter-individual variability and that the selection of a particular intervention modality will hinge on the gravity of the ailment, the symptomatic manifestation of the individual, and additional factors such as their general health and medical past (Guzman-Martinez et al., 2021). A recommended approach for individuals with Alzheimer's disease often involves a multidisciplinary strategy that integrates both pharmacological and nonpharmacological interventions.

8.1.4 Importance of Natural Products in Alzheimer's Disease Treatment

The scientific literature indicates that natural products have garnered attention in the quest for remedies for AD owing to their capacity as a reservoir of novel drugs. The significance of natural products in the management of AD is attributed to several factors:

i. **Diverse chemical structures:** Natural products exhibit a diverse array of chemical structures, some of which are absent in artificially synthesized pharmaceuticals. One, the rich variety of substances available is a potentially fruitful source for the discovery of new therapeutic agents for the management of AD.

ii. **Traditional use:** Numerous natural products possess a rich historical background of employment in conventional medicine and have been utilized for the management of diverse ailments, including AD. The conventional application of these compounds serves as substantiation for their prospective therapeutic advantages (Ogbodo et al., 2022).
iii. **Bioactivity:** Numerous naturally occurring substances have exhibited bioactivity, encompassing properties such as antioxidant, anti-inflammatory, and neuroprotective effects. The aforementioned properties render them promising contenders for the management of AD, a condition typified by oxidative stress, inflammation, and neurodegeneration (Lei et al., 2021).
iv. **Safety profile:** Natural products are often derived from plant, animal, and microbial sources and have a long history of safe use. This safety profile makes them an attractive alternative to synthetic drugs, which may have side effects (Tiwari et al., 2019).
v. **Cost-effectiveness:** Natural products are frequently accessible and can be manufactured at a reduced cost, rendering them a compelling substitute for synthetic medications, which may incur high expenses (Villain & Dubois, 2019).

Although natural products have the potential to offer therapeutic benefits in the management of AD, it is crucial to acknowledge that the utilization of natural products for medicinal intentions is not devoid of obstacles. The validation of natural products as potential treatments for AD requires rigorous scientific studies to establish their efficacy and safety (Licinio & Wong, 2023). Furthermore, the variability of natural products' quality and the potential for contamination and lack of standardization may raise concerns. Figure 8.2 represents the natural therapeutics aid in the treatment of Alzheimer's disease (Bhat et al., 2022).

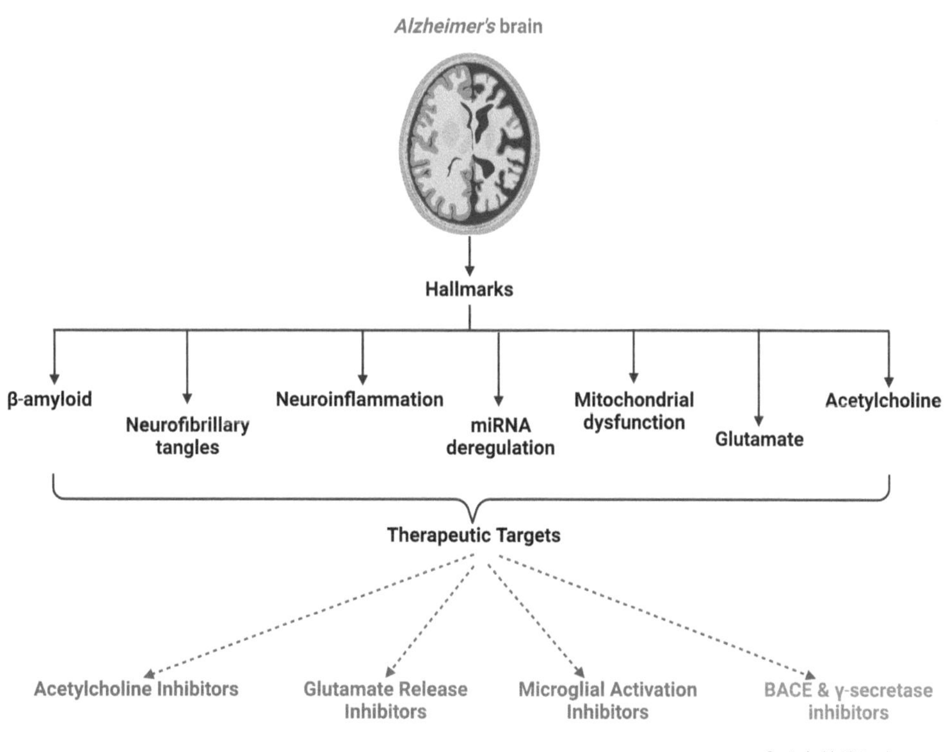

FIGURE 8.2 The natural therapeutics aid in the treatment of Alzheimer's disease (Bhat et al., 2022).

8.1.5 Computational Methods in Natural Product Research

The employment of computational techniques has gained significant prominence in the field of natural product research, particularly in the quest for novel therapeutic interventions for Alzheimer's disease, as evidenced by scholarly literature (Guzman-Martinez et al., 2021). The term "computational methods" pertains to the utilization of computer-based algorithms and mathematical models for the examination of natural products and their viability as therapeutic agents. The utilization of computational methods in natural product research is deemed valuable due to several reasons:

i. **Screening large databases:** The utilization of computational techniques enables the expeditious examination of extensive collections of natural products, encompassing substances derived from botanical, zoological, and microbial origins. This approach may facilitate the identification of compounds with promising therapeutic potential for AD.
ii. **Predictive models:** Implementing computational tools enables the quick examination of enormous arrays of natural products derived from plants, animals, and microbes. This technique offers the potential to accelerate the identification of molecules with significant therapeutic potential for Alzheimer's disease (AD).
iii. **Virtual screening:** The usage of computer simulations to forecast the binding of natural products to particular targets, such as proteins or enzymes, is known as virtual screening, a computational technique. This approach may facilitate the identification of natural compounds with the capacity to impede the molecular targets implicated in the pathogenesis of Alzheimer's disease.
iv. **Structure-activity relationships:** The use of computer approaches allows for the rapid study of large collections of natural products derived from botanical, zoological, and microbiological sources. This method is useful in unraveling structure-activity correlations, which speeds up the identification of molecules with promising treatment potential for Alzheimer's disease (AD).
v. **High-throughput methods:** Using high-throughput computational approaches, researchers may quickly explore vast reservoirs of natural products generated from botanical, zoological, and microbiological sources. This high-throughput technique accelerates the identification of molecules with promising Alzheimer's disease (AD) treatment potential e (Van Dyck et al., 2023).

To conclude, computational techniques have a significant contribution in the field of natural product research and possess the capability to considerably enhance the exploration of novel therapies for AD. It is noteworthy that computational techniques constitute a solitary facet of the diverse array of tools employed in natural product research, and their outcomes necessitate empirical validation.

8.2 NATURAL PRODUCTS AND ALZHEIMER'S DISEASE

8.2.1 Types of Natural Products That Use for the Treatment of Alzheimer's Disease

The following natural products have been investigated for their prospective therapeutic benefits in the context of Alzheimer's disease. Notwithstanding, the precise mechanisms through which they could enhance cognitive function and/or impede disease progression remain incompletely comprehended, necessitating further investigation (Lei et al., 2021).

i. **Curcumin:** Turmeric is a commonly used spice in Indian cuisine that contains curcumin as its active compound. The literature indicates that it exhibits anti-inflammatory, antioxidant, and anti-amyloid characteristics. The formation and breakdown of amyloid plaques are significant features of Alzheimer's disease, and research has demonstrated the inhibitory effects of curcumin on these processes. Moreover, studies have demonstrated that curcumin can augment neurogenesis, a phenomenon characterized by the generation of fresh neurons within the brain (Kunkle et al., 2019).
ii. **Resveratrol:** Resveratrol is a type of polyphenol that can be found in various food sources, including grapes and red wine. Research has demonstrated the presence of anti-inflammatory and antioxidant characteristics. Studies have demonstrated that Resveratrol possesses the ability to stimulate the SIRT1 gene, a genetic component implicated in the aging process and associated with neuroprotective effects (Khan et al., 2020). In addition, resveratrol has been shown in trials to reduce the production of beta-amyloid, a protein that creates plaques in the brains of Alzheimer's patients.
iii. **Quercetin:** Quercetin, a flavonoid, is present in a variety of fruits and vegetables, including apples, onions, and broccoli. Research has demonstrated its potential to exhibit anti-inflammatory and antioxidant characteristics. Studies have demonstrated the potential of Quercetin to mitigate the generation of beta-amyloid and tau proteins, which are known to contribute to the formation of neurofibrillary tangles in the brains of individuals afflicted with Alzheimer's disease (Emrani et al., 2020).
iv. **EGCG (Epigallocatechin gallate):** EGCG is a type of catechin that is commonly present in green tea. Research has demonstrated its potential to exhibit antioxidant and anti-inflammatory characteristics. The efficacy of EGCG in mitigating the production of beta-amyloid and tau has been demonstrated. Furthermore, animal studies have demonstrated that EGCG can improve cognitive function and memory.
v. ***Ginkgo biloba* extract:** The extract of *Ginkgo biloba* is obtained from the foliage of the Ginkgo tree. Research has demonstrated its possession of antioxidant and anti-inflammatory characteristics. The administration of *Ginkgo biloba* extract has demonstrated efficacy in enhancing cognitive function and memory among individuals diagnosed with Alzheimer's disease. This effect may be attributed to the extract's potential to augment cerebral blood flow (Revi, 2020).
vi. **Huperzine A:** Huperzine A is a botanical alkaloid that is extracted from the *Huperzia serrata* plant, commonly known as Chinese club moss. Research has demonstrated that it possesses the ability to impede the degradation of acetylcholine, a crucial neurotransmitter that plays a role in cognitive functions such as memory retention and acquisition. Research has demonstrated that Huperzine A can enhance cognitive performance in individuals diagnosed with Alzheimer's disease.
vii. ***Bacopa monnieri* :** *Bacopa monnieri* is a botanical species frequently employed in traditional Ayurvedic medicinal practices. Research has demonstrated that it possesses antioxidant and anti-inflammatory characteristics. Studies conducted on both animals and humans have demonstrated that *Bacopa monnieri* possesses the ability to enhance cognitive function and memory (Lei et al., 2021).
viii. **Ashwagandha:** Ashwagandha is a botanical species that is frequently employed in the practice of Ayurveda. Research has demonstrated the presence of anti-inflammatory and antioxidant characteristics. Studies have demonstrated the potential of Ashwagandha to decrease the synthesis of beta-amyloid and tau. Furthermore, research has demonstrated that Ashwagandha can enhance cognitive performance and memory in both animal models and human subjects (Association, 2019).
ix. **Ginseng:** The root of ginseng is frequently employed in the practice of traditional Chinese medicine. The literature indicates that it exhibits anti-inflammatory and antioxidant

characteristics. Research has demonstrated the potential of ginseng to enhance cognitive abilities and memory in both animal models and human subjects (Athar et al., 2021).

x. **Omega-3 fatty acids:** Omega-3 fatty acids, being polyunsaturated fatty acids, are deemed indispensable for the maintenance of optimal brain function. According to existing research, the ingestion of omega-3 fatty acids may potentially mitigate inflammation, augment cerebral blood circulation, and optimize cognitive performance. Several studies have indicated that the consumption of omega-3 fatty acids may impede the process of cognitive decline in individuals afflicted with Alzheimer's disease (Atri, 2019).

xi. **Vitamin E:** Vitamin E functions as an antioxidant, safeguarding cells against the deleterious effects of free radicals. According to research, vitamin E has been found to potentially mitigate oxidative stress and inflammation in the brain, which are considered to be underlying factors in the onset and advancement of AD. Clinical research on the effectiveness of vitamin E as a therapy for AD have shown results that are inconsistent (Breijyeh & Karaman, 2020).

xii. **Vitamin C:** Antioxidant vitamin C aids in defending cells against harm from free radicals. According to studies, vitamin C may lessen the oxidative stress and inflammation that are thought to be responsible for the onset and progression of Alzheimer's disease in the brain. However, there have been conflicting findings from clinical trials testing the effectiveness of vitamin C in treating AD (Athar et al., 2021).

xiii. **Melatonin:** A hormone called melatonin controls the sleep–wake cycle. According to certain research, melatonin may have neuroprotective properties and may lessen brain inflammation and oxidative stress. Melatonin may help Alzheimer's patients sleep better and think more clearly, according to some research (Button et al., 2019).

xiv. **Berberine:** An alkaloid found in plants called berberine has long been utilized in traditional medicine. According to certain research, berberine may have neuroprotective properties and could lessen brain inflammation and oxidative stress. There is some evidence that berberine may help Alzheimer's patients with their cognitive function.

xv. **Apigenin:** A flavonoid called apigenin can be discovered in many different fruits and vegetables. According to certain research, apigenin may have neuroprotective properties and could lessen brain inflammation and oxidative stress. Apigenin may help Alzheimer's patients' cognitive abilities, according to some research (Cummings et al., 2020).

8.2.2 Overview of Natural Products with Potential Therapeutic Effects in Alzheimer's Disease

Alzheimer's disease (AD) is a neurological condition marked by memory loss and increasing cognitive decline. It is a complicated illness with numerous underlying pathophysiological pathways, making the development of efficient treatments difficult. Natural products' varied chemical compositions and pharmacological actions have led to an increased interest in them as potential therapeutics for AD

(Emrani et al., 2020).

Numerous natural compounds with anti-inflammatory, antioxidant, neuroprotective, and anti-amyloid characteristics have been shown to have potential therapeutic effects for AD. In addition to omega-3 fatty acids, these natural compounds also contain terpenes like ginkgolides and huperzine A, alkaloids like berberine and galantamine, polyphenols like resveratrol and curcumin, flavonoids like quercetin and apigenin, and flavonoids like apigenin. It has been demonstrated that omega-3 fatty acids, which are present in fatty fish and some plants, have neuroprotective properties and may aid in reducing brain inflammation, which is thought to contribute to the advancement of AD. Vitamin E, which is included in nuts and seeds, has been demonstrated to possess antioxidant capabilities and may aid in reducing oxidative brain damage (Guzman-Martinez et al., 2021). Vitamin C, which is present in fruits and vegetables, has anti-inflammatory and antioxidant properties. Melatonin, a hormone made by the pineal

> gland, has been demonstrated to have neuroprotective effects and may aid in regulating individuals with AD's disordered sleep-wake cycles. It has been demonstrated that the plant chemical berberine, which is present in many plants including goldenseal and Oregon grape, has anti-inflammatory and anti-amyloid capabilities. It has been demonstrated that the flavonoid apigenin, which is present in parsley, celery, and chamomile tea, has anti-inflammatory and antioxidant characteristics
>
> (Knapskog et al., 2021).

Although natural compounds have the potential to be effective AD treatments, further study is required to completely understand their mechanisms of action and assess their safety and efficacy in human clinical trials.

8.3 EXAMPLES OF NATURAL PRODUCTS STUDIED USING COMPUTATIONAL METHODS

To investigate the possible therapeutic effects of natural compounds in AD, computational approaches have been widely explored. Natural goods that have undergone in-depth computer analysis include some of the following:

i. **Curcumin:** The turmeric plant contains a polyphenolic substance called curcumin. Anti-inflammatory and antioxidant activities have been demonstrated for it. Computational studies show that curcumin interacts with multiple targets important in the pathogenesis of AD, including beta-amyloid and tau proteins.

ii. **Resveratrol:** A polyphenolic substance called resveratrol can be found in nuts, berries, and grapes. It has been demonstrated to have anti-inflammatory and antioxidant effects. Resveratrol can interact with a variety of Alzheimer's disease targets, including beta-amyloid and tau proteins, according to computational research.

iii. **Epigallocatechin gallate (EGCG):** Green tea contains a catechin called EGCG. It has been demonstrated to have anti-inflammatory and antioxidant effects. Computer simulations have demonstrated that EGCG can interact with a variety of Alzheimer's disease targets, including beta-amyloid and tau proteins.

iv. **Ginkgo biloba:** A traditional Chinese plant called *Ginkgo biloba* has been used for generations to cure a variety of conditions, including memory loss. Ginkgo biloba can interact with numerous Alzheimer's disease targets, including the proteins acetylcholinesterase, beta-amyloid, and tau, according to computational research.

v. **Huperzine A:** An alkaloid called huperzine A is present in Chinese club moss plants. It has been demonstrated to possess acetylcholinesterase inhibitory action, which can enhance cognition and memory. Huperzine A can interact with a variety of Alzheimer's disease targets, including beta-amyloid and tau proteins, according to computational research.

8.4 CURRENT STATE OF RESEARCH ON NATURAL PRODUCTS IN ALZHEIMER'S DISEASE

Although research on natural treatments for Alzheimer's disease is still in its early stages, interest in the potential therapeutic effects of several substances originating from plants, animals, and other natural sources is growing. Computational techniques have been used to investigate a number of natural compounds, including Omega-3 fatty acids, Vitamins E and C, melatonin, berberine, and apigenin (Lei et al., 2021). Preclinical research on these natural compounds have demonstrated potential for lowering inflammation, preventing oxidative stress, and enhancing cognitive function.

However, despite the fact that computational approaches might offer insightful information about the potential efficacy and safety of natural compounds, they cannot replace experimental validation.

Extensive preclinical and clinical research is needed to evaluate the therapeutic potential and safety of herbal remedies for the treatment of Alzheimer's illness (Licinio & Wong, 2023).

In general, the utilization of natural products holds great potential for the advancement of novel treatments for Alzheimer's disease. Moreover, the application of computational techniques can facilitate the identification and enhancement of these chemical entities. Additional investigation is required to thoroughly examine the potential of natural products in addressing this debilitating ailment (Licinio & Wong, 2023).

8.5 COMPUTATIONAL METHODS IN NATURAL PRODUCT RESEARCH

8.5.1 Molecular Docking

8.5.1.1 Methods of Molecular Docking

The computational methodology of molecular docking is employed to simulate the interactions between a target protein and a small molecule, such as a natural product. The methodology entails situating the naturally occurring substance within a simulated setting and evaluating the energetic properties of the natural product's interactions with the protein of interest (Tiwari et al., 2019). The objective is to forecast the binding affinity of the natural compound towards the target protein, which serves as an indicator of the degree of binding strength between the natural product and the target.

The determination of the binding affinity of the natural product to the target protein involves the computation of the energy differential between the complex and the combined energies of the natural product and the target protein when they are separate entities (Volicer, 2020).

8.5.1.2 Applications of Molecular Docking in Alzheimer's Disease Research

In order to find new treatment strategies for this terrible condition, molecular docking is widely used in the field of Alzheimer's disease (AD) research. Several ways in which molecular docking can be used to gain insight into Alzheimer's disease have been uncovered:

a) The objective is to identify natural products that exhibit the potential to inhibit target proteins implicated in the pathogenesis of Alzheimer's disease. Molecular docking is a viable approach for the identification of natural products that possess the ability to inhibit the beta-amyloid peptide, a crucial factor in the pathogenesis of Alzheimer's disease (Licinio & Wong, 2023).
b) The optimization of drug design utilizing natural products. The utilization of molecular docking techniques can facilitate the alteration of the chemical composition of naturally occurring substances, thereby enhancing their capacity to bind, exhibit specificity, and demonstrate efficacy. This approach may facilitate the optimization of drug design that relies on natural products (Lloret et al., 2019).
c) Enhancing the efficacy of pharmaceutical exploration endeavours. The integration of molecular docking with high-throughput techniques, such as automated liquid handling systems and robotics, can facilitate the swift screening of vast quantities of natural products. This approach can enable the identification of promising candidates for treating AD. The implementation of this approach has the potential to greatly enhance the efficacy of endeavours related to the discovery of pharmaceuticals (Lei et al., 2021).

To conclude, molecular docking is a significant computational technique that is extensively employed in the field of natural product research, particularly in the quest for novel therapies for Alzheimer's disease. This methodology facilitates the expeditious evaluation of extensive repositories of natural compounds and can furnish significant insights into the molecular interplay between natural compounds and target proteins, along with the binding potency of natural compounds towards target proteins (Revi, 2020). The aforementioned observations may be utilized to establish a hierarchy of compounds for subsequent investigation and to direct the development of novel pharmaceuticals for Alzheimer's disease.

8.5.2 Molecular Dynamics Simulations

8.5.2.1 Method in MD Simulations

Molecular dynamics simulations (MDS) are a class of computational techniques employed to investigate the temporal evolution of atoms and molecules. Molecular Dynamics Simulation (MDS) entails the computation of the conventional equations of motion for every atom present within a given system. This is achieved through the utilization of a force field, which serves to describe the interactions that occur between the atoms (Swanson et al., 2021). The outcome is a system's trajectory over a period of time, which can offer significant insights into the system's dynamics and energetics.

Molecular dynamics simulation (MDS) is a frequently employed technique in the field of natural product research to investigate the interplay between natural products and target proteins. Molecular dynamics simulations (MDS) can yield comprehensive insights into the binding interactions that occur between a natural product and a protein, including the stability and flexibility of the resulting complex.

8.6 APPLICATIONS IN ALZHEIMER'S DISEASE RESEARCH

Molecular Dynamic Studies has several applications in AD research, including:

i. Anticipating the binding conformation of natural compounds to specific protein targets. Molecular dynamics simulations (MDS) can yield comprehensive insights into the structural and energetic characteristics of the protein-ligand complex, thereby enabling the identification of the most stable and energetically favourable binding conformation of the natural product-target protein complex.
ii. Investigating the kinetics of the interaction between the natural product and its corresponding target protein. Multidimensional scaling (MDS) has the potential to offer valuable insights into the temporal dynamics of the natural product-target protein complex, specifically with regard to its flexibility and stability. The aforementioned data has the potential to enhance the efficacy of drug design utilizing natural products and to formulate novel therapeutic approaches for Alzheimer's disease (Mercerón-Martínez et al., 2021)
iii. The investigation of the impact of mutations on the protein complex targeted by natural products. Molecular Dynamics Simulation (MDS) can be employed as a tool to investigate the impact of mutations in the target protein on the binding affinity of the natural product. The aforementioned data can be utilized to discern natural substances that exhibit efficacy against diverse iterations of the specified protein, and to direct the development of novel pharmaceuticals for Alzheimer's disease (Serrano-Pozo et al., 2021).

To conclude, Molecular Dynamic Simulation (MDS) is a beneficial instrument in the field of natural product investigation, particularly in the pursuit of novel therapies for Alzheimer's Disease (AD). It is a technique that enables the investigation of the energetic and dynamic aspects of the interplay between natural compounds and target proteins. This approach offers significant advantages in terms of understanding the binding affinity, stability, and flexibility of the resulting complex, thereby providing valuable insights into the underlying molecular mechanisms. The aforementioned data can be utilized to enhance the design of pharmaceuticals derived from natural sources and to formulate novel therapeutic approaches for Alzheimer's disease.

8.7 IN SILICO PREDICTION OF SIDE EFFECTS AND TOXICITY

The in silico approach is a computational technique employed to anticipate the plausible side effects and toxicity of natural products. The present approach entails the utilization of computational tools and algorithms to scrutinize the chemical composition of natural products and forecast their plausible impacts on the human physiology (Graff-Radford et al., 2021).

The utilization of quantitative structure-activity relationship (QSAR) models is a prominent method for in silico forecasting of side effects and toxicity. Quantitative Structure-Activity Relationship (QSAR) models employ statistical and computational techniques to establish a correlation between the chemical structure of natural products and their corresponding biological activity and toxicity.

8.7.1 Importance in Natural Product Research

In silico prediction of side effects and toxicity is an important tool in natural product research for several reasons:

a) The timely detection of potential adverse effects and toxicity. The utilization of in silico prediction enables the timely detection of potential adverse effects and toxicity of natural products, prior to their experimental evaluation in animal or human subjects. The implementation of this approach has the potential to mitigate the duration and expenses associated with pharmaceutical development, while concurrently enhancing the safety profile of drugs derived from natural products (Mary et al., 2023).
b) The process of choosing natural products that exhibit reduced side effects and toxicity. The utilization of in silico prediction is a viable approach for the preliminary screening of a vast array of natural products, with the aim of identifying those that exhibit reduced potential for adverse side effects and toxicity. These selected candidates can then be subjected to further testing and development.
c) The optimization of drugs based on natural products. The utilization of in silico prediction has the potential to optimize the chemical composition of drugs based on natural products, thereby mitigating their potential for adverse effects and toxicity, while simultaneously preserving their therapeutic efficacy (Lloret et al., 2019).
d) Regulatory compliance refers to the adherence of an organization to the laws, regulations, and guidelines set forth by governing bodies and regulatory agencies. The utilization of in silico prediction holds potential for evaluating the safety of drugs derived from natural products and guaranteeing their adherence to regulatory standards.

In conclusion, the utilization of in silico prediction methods for the assessment of side effects and toxicity is a crucial aspect of natural product research, particularly in the pursuit of novel therapeutic interventions for Alzheimer's disease. The utilization of in silico prediction enables the timely detection of potential adverse effects and toxicity, the choice of natural products with reduced side effects and toxicity, the enhancement of natural product-derived medications, and adherence to regulatory standards. The utilization of this approach has the potential to enhance the safety and effectiveness of drugs derived from natural products, while also decreasing the duration and expenses associated with drug development (Tolar et al., 2020).

8.8 ADVANTAGES AND DISADVANTAGES OF USING COMPUTATIONAL METHODS IN NATURAL PRODUCT RESEARCH

The utilization of computational methods has emerged as a pivotal instrument in the realm of natural product research, encompassing the exploration for novel remedies for Alzheimer's disease. Nevertheless, these approaches possess both benefits and drawbacks.

8.9 ADVANTAGES OF USING COMPUTATIONAL METHODS IN NATURAL PRODUCT RESEARCH

a) **Time and Cost Savings:** Computational methods offer a notable benefit in natural product research by reducing both time and cost expenditures. The utilization of in silico screening can facilitate the identification of potential candidates for further investigation. Additionally, molecular modelling and simulation techniques can be employed to enhance the optimization of drug candidates prior to their synthesis and subsequent evaluation in animal or human subjects (Serrano-Pozo et al., 2021).
b) **Predictive Power:** The utilization of computational techniques can offer a degree of prognostic capability that is arduous to attain solely through empirical approaches. Molecular modelling and simulation techniques have the capability to forecast the binding affinity of a natural product towards a protein target, as well as the impact of a structural alteration on the biological activity of the natural product (Knapskog et al., 2021).
c) **Large-Scale Analysis:** The utilization of computational techniques is viable for the examination of extensive datasets of natural product compounds, enabling the identification of intricate patterns and trends that may prove challenging to discern through manual analysis (Athar et al., 2021).

8.10 CHALLENGES AND LIMITATIONS

8.10.1 Limitations of Computational Methods

The utilization of computational methods has become a crucial instrument in the field of natural product research, enabling the effective evaluation and enhancement of potential therapeutic agents. Notwithstanding, it is imperative to take into account various constraints that are linked to the utilization of computational techniques (Long & Holtzman, 2019).

i. A significant constraint pertains to the precision of the computational models employed. The efficacy of computational methods is contingent upon the utilization of mathematical models that depict intricate biological processes. However, these models may occasionally oversimplify or inaccurately portray reality. Inaccurate predictions of compound activity, toxicity, and pharmacokinetics may arise, potentially leading to the development of ineffective or unsafe therapies (Athar et al., 2021; Mary et al., 2023).
ii. A further constraint pertains to the accessibility and calibre of data employed in computational models. The efficacy of computational techniques is contingent upon the utilization of substantial datasets for the purpose of model training and validation. The precision of the prognostications generated is heavily influenced by the calibre and comprehensiveness of the data employed. The utilization of incomplete or biased data sets may result in imprecise prognostications and impede the advancement of efficacious therapeutic interventions (Mercerón-Martínez et al., 2021).
iii. Moreover, the utilization of computational resources for modelling purposes can be substantial, encompassing both hardware and software aspects. Large-scale modelling and simulation studies often necessitate high-performance computing resources, thereby

impeding the accessibility and affordability of these techniques for smaller research groups or institutions (Association, 2019).

iv. Lastly, the comprehension of computational outcomes may present difficulties, especially for individuals lacking expertise in the field. The utilization of computational techniques has the potential to produce copious amounts of intricate data, necessitating the application of sophisticated statistical and computational proficiencies to proficiently scrutinize and construe the information.

Despite these constraints, computational techniques persist as a valuable instrument in the field of natural product investigation, providing a potent strategy for the detection and enhancement of prospective medicinal agents. With the ongoing advancements and enhancements in computational techniques, it is possible to alleviate these constraints, thereby facilitating the discovery and development of drugs with enhanced precision and efficacy (Atri, 2019).

8.11 THE NEED FOR EXPERIMENTAL VALIDATION

Although computational methods can be efficacious instruments for natural product investigation, it is crucial to bear in mind that they constitute merely a fraction of the drug discovery procedure. Experimental validation is necessary to confirm the accuracy of computational predictions and evaluate the effectiveness and safety of potential therapeutic interventions.

i. Experimental validation holds significant importance in natural product research owing to the intricate composition of natural products, which often comprise a multitude of compounds exhibiting varied biological activities. The intricate interplay between these chemical compounds and the relevant biological system may pose a challenge in accurately forecasting outcomes solely through computational approaches (Long & Holtzman, 2019). Furthermore, the chemical composition, level of purity, and overall quality of natural products can exhibit significant variation, thereby influencing their biological efficacy and potential toxicity.
ii. Experimental validation is a useful approach to overcome the constraints of computational techniques, including deficiencies in modelling and insufficient data sets. Through the process of comparing computational predictions with experimental findings, researchers can enhance the precision of their models and augment their efficacy. Additionally, this approach enables them to pinpoint domains where additional data is required to enhance the accuracy of predictions.
iii. In addition, regulatory bodies such as the FDA mandate experimental verification to establish the safety and effectiveness of prospective treatments prior to granting approval for human usage. Insufficient experimental validation may result in the creation of therapies that are hazardous or ineffective, thereby endangering patients (Mary et al., 2023).

To summarize, although computational techniques are a valuable asset in the field of natural product research, their reliability must be verified through experimental validation to evaluate the effectiveness and safety of potential treatments. The integration of computational and experimental methodologies can enhance the precision of prognostications and facilitate the creation of secure and efficacious natural product remedies for Alzheimer's disease and other ailments (Graff-Radford et al., 2021).

8.12 ETHICAL CONSIDERATIONS

The investigation of natural products, related to all medical research, presents significant ethical considerations that necessitate meticulous attention. The aforementioned factors encompass the ethical treatment of human participants in clinical trials, the acquisition and utilization of natural resources, and the plausible ramifications of natural product research on the surrounding ecosystem and indigenous populations.

i. An ethical aspect that warrants attention in the realm of natural product research pertains to the acquisition and utilization of natural products. The utilization of natural products sourced from rare or endangered species may result in detrimental effects on both ecosystems and local communities. It is imperative for researchers to ensure that they are procuring natural products in an ethical and sustainable manner, and that their research endeavours do not exacerbate the overexploitation of natural resources (Ogbodo et al., 2022).
ii. Another ethical aspect to be taken into account in the research of natural products is the well-being of human participants involved in clinical trials. Although clinical trials are an essential part of the process of developing new drugs, they do come with certain inherent dangers for the human subjects who participate in them. These concerns include the possibility of unfavourable results as well as the potential for harm from unproven therapeutic procedures (Athar et al., 2021). It is imperative for researchers to adhere to established guidelines and ethical principles when conducting clinical trials, in order to safeguard the well-being of human participants.
iii. Ultimately, the exploration of natural products can yield societal benefits that extend beyond the mere creation of novel treatments. The commercialization of natural products has the potential to result in the exploitation of indigenous communities and the unauthorized use of their traditional knowledge. It is imperative for researchers to conduct their research in a manner that upholds the rights and interests of all stakeholders, which encompasses local communities and indigenous populations.

To summarize, the study of natural products in research brings forth significant ethical considerations that necessitate meticulous attention. It is imperative for researchers to conduct their research in an ethical manner that safeguards the welfare of human subjects, upholds the rights and interests of all stakeholders, and mitigates any adverse effects on the environment and local communities. Through the consideration of ethical implications, researchers can devise natural product therapies that are both safe and effective for treating Alzheimer's disease and other ailments, while concurrently advancing sustainable and socially responsible research practices (Soria Lopez et al., 2019).

The creation of reliable and efficient natural product treatments for Alzheimer's disease presents a formidable task, however, the utilization of computational techniques and experimental verification may serve to mitigate some of the associated difficulties. The incorporation of ethical considerations is a crucial aspect of natural product research, necessitating that researchers uphold sustainable and socially responsible research practices throughout their investigations.

8.13 CONCLUSION

8.14 SUMMARY OF KEY POINTS

The neurodegenerative disorder known as Alzheimer's disease is characterized by a steady deterioration in cognitive ability and memory function. Despite the considerable amount of research conducted, a cure for AD remains elusive, and the treatments currently available exhibit restricted effectiveness. There is promising evidence indicating that natural products obtained from various sources such as plants, animals, and other natural entities possess therapeutic potential for the management of AD (Association, 2019). The properties of these molecules can be studied, and their efficacy and safety can be evaluated, using computational methods like molecular docking and molecular dynamics simulations. Various natural compounds, such as omega-3 fatty acids, vitamin E, vitamin C, melatonin, berberine, and apigenin, have been the subject of computational investigations and have demonstrated potential in mitigating inflammation, safeguarding against oxidative stress, and enhancing cognitive performance in preclinical research (Cummings et al., 2020). Despite the utility of computational methods in natural product research, their employment is circumscribed by certain constraints, necessitating experimental validation to comprehensively assess the therapeutic viability and safety of these agents. The ethical implications associated with researching natural products must also be duly considered. In general, natural products exhibit potential as a viable pathway for the advancement of novel treatments for AD (Guzman-Martinez et al., 2021). Additional investigation is required to thoroughly examine the capabilities of these compounds and to formulate efficacious therapies for this debilitating ailment.

8.15 FUTURE DIRECTIONS FOR NATURAL PRODUCT RESEARCH IN ALZHEIMER'S DISEASE

The study of natural products for the treatment of AD is a field that is currently being actively researched. There exist various potential avenues for future research that could result in the creation of novel and efficacious therapies. Finding and studying naturally occurring compounds that show activity against several pathways involved in the development of AD is a promising approach. Certain natural compounds have showed combined anti-inflammatory and antioxidant characteristics, which could potentially be useful in alleviating the negative effects of oxidative stress and chronic inflammation inside the brain (Mercerón-Martínez et al., 2021).

An additional avenue for further investigation pertains to the examination of combinations of natural products, which could potentially yield synergistic outcomes in relation to cognitive function and other facets of AD. In animal models of AD, the combination of omega-3 fatty acids and vitamin E has been shown to have a beneficial effect on cognitive function. Furthermore, the advancement of novel computational methodologies and technologies has the potential to significantly augment the investigation of natural products as a therapeutic intercession for Alzheimer's disease (Guzman-Martinez et al., 2021).

Artificial intelligence and machine learning have the potential to enhance the screening and identification process of natural products with potential therapeutic effects, thereby increasing efficiency. There is a pressing requirement for additional clinical trials to assess the safety and effectiveness of natural products in treating Alzheimer's disease. Although preclinical investigations have exhibited potential, it is crucial to convert these discoveries into clinical application to ultimately confer advantages to individuals afflicted with Alzheimer's disease (Soria Lopez et al., 2019). In general, the outlook for natural product research in the context of Alzheimer's disease appears to be encouraging. Further exploration of these compounds has the potential to yield novel and efficacious treatments for this debilitating condition.

8.16 IMPLICATIONS FOR PATIENTS AND HEALTH CARE PROVIDERS

The ramifications of the research on natural products conducted in the setting of Alzheimer's disease are substantial for both patients and health care practitioners. The utilization of natural products as a potential alternative or supplementary treatment for Alzheimer's disease presents a promising avenue, as it may offer greater efficacy and fewer adverse effects compared to conventional pharmaceutical interventions. It is noteworthy that natural products are subject to different regulations compared to pharmaceuticals, necessitating a meticulous assessment of their safety and effectiveness.

Medical professionals should be aware of the benefits and risks associated with using natural goods to treat Alzheimer's disease, and should inform their patients of these considerations. It is imperative to educate patients regarding the significance of seeking advice from a health care professional prior to consuming any natural products, as certain products may potentially interact with other medications or elicit unfavourable effects (Streit et al., 2021). In general, the exploration of natural products in the context of Alzheimer's disease holds promise for the development of novel therapeutic interventions that may enhance the quality of life for affected individuals and their loved ones. Additional investigation is required to comprehensively comprehend the mechanisms of operation and safety of natural substances. Therefore, health care professionals should meticulously evaluate the evidence prior to advising patients on their utilization.

8.17 FINAL RECOMMENDATIONS

Although the healing potential of natural remedies in the treatment of Alzheimer's disease seems evident, it is advisable to further investigate this area in order to identify novel natural products that exhibit enhanced efficacy and reduced toxicity (Villain & Dubois, 2019). The utilization of computational techniques can facilitate this investigation by forecasting the associations between natural substances and the targets implicated in the pathogenesis of AD. It is imperative to corroborate the computational findings through empirical investigations.

In addition, it is imperative to consider ethical implications in the realm of natural product research, particularly with regards to the procurement and ecological viability of such resources (Tiwari et al., 2019). Natural products may provide a supplementary therapeutic avenue for AD management for both health care providers and patients. Prior consultation with a health care provider is crucial when considering the integration of natural products into a treatment regimen, as certain natural products may potentially interact with other pharmaceuticals. In general, further investigation into natural goods for the treatment of Alzheimer's disease has the potential to yield novel, efficacious, and safe therapeutic interventions for this debilitating ailment (Atri, 2019).

ENDNOTE

* Corresponding author

REFERENCES

Association, A. S. (2019). 2019 Alzheimer's disease facts and figures. *Alzheimers Dement*, *15*(3), 321–387.

Athar, T., Al Balushi, K., & Khan, S. A. (2021). Recent advances on drug development and emerging therapeutic agents for Alzheimer's disease. *Mol Biol Rep*, *48*(7), 5629–5645. http://doi.org/10.1007/s11033-021-06512-9

Atri, A. (2019). The Alzheimer's disease clinical spectrum: Diagnosis and management. *Med Clin N Am*, *103*(2), 263–293. http://doi.org/10.1016/j.mcna.2018.10.009

Bhat, B. A., Almilaibary, A., Mir, R. A., Aljarallah, B. M., Mir, W. R., Ahmad, F., & Mir, M. A. (2022). Natural therapeutics in aid of treating Alzheimer's disease: A green gateway toward ending quest for treating neurological disorders. *Appl Plant Sec Metab Pain Neur, II.*

Bis, J. C., Jian, X., Kunkle, B. W., Chen, Y., Hamilton-Nelson, K. L., Bush, W. S., . . . & Renton, A. E. (2020). Whole exome sequencing study identifies novel rare and common Alzheimer's-Associated variants involved in immune response and transcriptional regulation. *Mol Psychiatry*, *25*(8), 1859–1875.

Breijyeh, Z., & Karaman, R. (2020). Comprehensive review on Alzheimer's disease: Causes and treatment. *Molecules*, *25*(24). http://doi.org/10.3390/molecules25245789

Button, E. B., Robert, J., Caffrey, T. M., Fan, J., Zhao, W., & Wellington, C. L. (2019). HDL from an Alzheimer's disease perspective. *Curr Opin Lipidol*, *30*(3), 224–234. http://doi.org/10.1097/mol.0000000000000604

Cummings, J., Lee, G., Ritter, A., Sabbagh, M., & Zhong, K. (2020). Alzheimer's disease drug development pipeline: 2020. *Alzheimers Dement Translat Res Clin Interv*, *6*(1), e12050.

Emrani, S., Lamar, M., Price, C. C., Wasserman, V., Matusz, E., Au, R., . . . & Libon, D. J. (2020). Alzheimer's/vascular spectrum dementia: Classification in addition to diagnosis. *J Alzheimers Disease*, *73*(1), 63–71. http://doi.org/10.3233/jad-190654

Graff-Radford, J., Yong, K. X. X., Apostolova, L. G., Bouwman, F. H., Carrillo, M., Dickerson, B. C., . . . & Murray, M. E. (2021). New insights into atypical Alzheimer's disease in the era of biomarkers. *Lancet Neurol*, *20*(3), 222–234. http://doi.org/10.1016/s1474-4422(20)30440-3

Guzman-Martinez, L., Calfío, C., Farias, G. A., Vilches, C., Prieto, R., & Maccioni, R. B. (2021). New frontiers in the prevention, diagnosis, and treatment of Alzheimer's disease. *J Alzheimers Disease*, *82*(s1), S51–S63. http://doi.org/10.3233/jad-201059

Khan, S., Barve, K. H., & Kumar, M. S. (2020). Recent advancements in pathogenesis, diagnostics and treatment of Alzheimer's disease. *Curr Neuropharmacol*, *18*(11), 1106–1125. http://doi.org/10.2174/1570159x18666200528142429

Knapskog, A. B., Engedal, K., Selbæk, G., & Øksengård, A. R. (2021). Alzheimer's disease—diagnosis and treatment. *Tidsskr Nor Laegeforen*, *141*(7). http://doi.org/10.4045/tidsskr.20.0919

Kunkle, B. W., Grenier-Boley, B., Sims, R., Bis, J. C., Damotte, V., Naj, A. C., . . . & Amlie-Wolf, A. (2019). Genetic meta-analysis of diagnosed Alzheimer's disease identifies new risk loci and implicates Aβ, tau, immunity and lipid processing. *Nat Genet*, *51*(3), 414–430.

Lei, P., Ayton, S., & Bush, A. I. (2021). The essential elements of Alzheimer's disease. *J Biol Chem*, *296*, 100105. http://doi.org/10.1074/jbc.REV120.008207

Licinio, J., & Wong, M.-L. (2023). Advances in Molecular Psychiatry—March 2023: Mitochondrial function, stress, neuroinflammation—bipolar disorder, psychosis, and Alzheimer's disease. *Mol Psychiatry*, *28*, 968–971. Nature Publishing Group.

Lloret, A., Esteve, D., Lloret, M. A., Cervera-Ferri, A., Lopez, B., Nepomuceno, M., & Monllor, P. (2019). When does Alzheimer's disease really start? The role of biomarkers. *Int J Mol Sci*, *20*(22). http://doi.org/10.3390/ijms20225536

Long, J. M., & Holtzman, D. M. (2019). Alzheimer disease: An update on pathobiology and treatment strategies. *Cell*, *179*(2), 312–339.

Mary, A., Eysert, F., Checler, F., & Chami, M. (2023). Mitophagy in Alzheimer's disease: Molecular defects and therapeutic approaches. *Mol Psychiatry*, *28*(1), 202–216.

Mercerón-Martínez, D., Ibaceta-González, C., Salazar, C., Almaguer-Melian, W., Bergado-Rosado, J. A., & Palacios, A. G. (2021). Alzheimer's disease, neural plasticity, and functional recovery. *J Alzheimers Dis*, *82*(s1), S37–S50. http://doi.org/10.3233/jad-201178

Ogbodo, J. O., Agbo, C. P., Njoku, U. O., Ogugofor, M. O., Egba, S. I., Ihim, S. A., . . . & Upasani, C. D. (2022). Alzheimer's disease: Pathogenesis and therapeutic interventions. *Curr Aging Sci*, *15*(1), 2–25. http://doi.org/10.2174/1874609814666210302085232

Revi, M. (2020). Alzheimer's disease therapeutic approaches. *Adv Exp Med Biol*, *1195*, 105–116. http://doi.org/10.1007/978-3-030-32633-3_15

Scheltens, P., De Strooper, B., Kivipelto, M., Holstege, H., Chételat, G., Teunissen, C. E., . . . & van der Flier, W. M. (2021). Alzheimer's disease. *Lancet*, *397*(10284), 1577–1590. http://doi.org/10.1016/s0140-6736(20)32205-4

Serrano-Pozo, A., Das, S., & Hyman, B. T. (2021). APOE and Alzheimer's disease: Advances in genetics, pathophysiology, and therapeutic approaches. *Lancet Neurol*, *20*(1), 68–80. http://doi.org/10.1016/s1474-4422(20)30412-9

Soria Lopez, J. A., González, H. M., & Léger, G. C. (2019). Alzheimer's disease. *Handb Clin Neurol*, *167*, 231–255. http://doi.org/10.1016/b978-0-12-804766-8.00013-3

Streit, W. J., Khoshbouei, H., & Bechmann, I. (2021). The role of microglia in sporadic Alzheimer's disease. *J Alzheimers Dis*, *79*(3), 961–968. http://doi.org/10.3233/jad-201248

Swanson, C. J., Zhang, Y., Dhadda, S., Wang, J., Kaplow, J., Lai, R. Y. K., . . . & Cummings, J. L. (2021). A randomized, double-blind, phase 2b proof-of-concept clinical trial in early Alzheimer's disease with lecanemab, an anti-Aβ protofibril antibody. *Alzheimers Res Ther*, *13*(1), 80. http://doi.org/10.1186/s13195-021-00813-8

Tiwari, S., Atluri, V., Kaushik, A., Yndart, A., & Nair, M. (2019). Alzheimer's disease: Pathogenesis, diagnostics, and therapeutics. *Int J Nanomedicine*, *14*, 5541–5554. http://doi.org/10.2147/ijn.S200490

Tolar, M., Abushakra, S., & Sabbagh, M. (2020). The path forward in Alzheimer's disease therapeutics: Reevaluating the amyloid cascade hypothesis. *Alzheimers Dement*, *16*(11), 1553–1560. http://doi.org/10.1016/j.jalz.2019.09.075

Van Dyck, C. H., Swanson, C. J., Aisen, P., Bateman, R. J., Chen, C., Gee, M., . . . & Cohen, S. (2023). Lecanemab in early Alzheimer's disease. *N Engl J Med*, *388*(1), 9–21.

Villain, N., & Dubois, B. (2019). Alzheimer's disease including focal presentations. *Semin Neurol*, *39*(2), 213–226. http://doi.org/10.1055/s-0039-1681041

Volicer, L. (2020). Physiological and pathological functions of beta-amyloid in the brain and Alzheimer's disease: A review. *Chin J Physiol*, *63*(3), 95.

9 Analysis of Potential Alzheimer's Disease Inhibitors Based on Computational Modelling with Multiple Targets

Ankita Dubey, Vrishti M Thakur, Raja Chakraborty, G.N. Darwhekar, Rupesh Kushwah and Rakhi Khabiya**

9.1 INTRODUCTION

Globally 47.5 million people have been given an Alzheimer's disease (AD) diagnosis, and 20.8 million people suffer from dementia. These figures are only anticipated to rise as the aging population grows (Meyer & Schellack, 2016). AD has become a significant public health issue, putting a tremendous strain on the resources of the person, family, society, and health care system (Weintraub et al., 2012).

After the age of 60, the incidence of dementia generally doubles every 10 years but varies by country. More than 4 million individuals in Africa are currently affected by dementia, which is predicted to afflict 818,106 people annually (Kelley & Petersen, 2007). The growing prevalence is thought to be mostly caused by the aging population. There is no correlation between a particular gender and the incidence or prevalence of AD or dementia. However, more women than men experience the condition in terms of numbers, especially after the age of 85. The longer life expectancy of women explains this (Defina et al., 2013; Durazzo et al., 2014).

Dementia is a common condition, with Alzheimer's disease being the most common type. The German psychiatrist Alois Alzheimer gave this condition its name. Neuritic plaques and neurofibrillary tangles are features of the neurodegenerative illness known as Alzheimer's disease, which progresses slowly (Breijyeh & Karaman, 2020). The medial temporal lobe and neocortical structures, which make up the majority of the injured region of the brain, gradually assemble amyloid-beta peptide (A) (Figure 9.1) (Bohush et al., 2019).

Alois Alzheimer, a scientist, investigated the brain of a person who had memory loss and a personality change before passing away. He discovered the existence of beta-amyloid plaques along with a sizable loss of neurons (J. G. Li et al., 2020). He described the condition as a horrible cerebral cortex disease. Emil Kraepelin initially used the term "Alzheimer's disease" to describe this condition in the 8th edition of his psychiatry textbook (Sims & Williams, 2016a). Brain disorders such as AD or other conditions such as infections, anomalies in the lungs and blood vessels that lower the quantity of oxygen given to the brain, intoxications, nutritional deficiencies, tumours, vitamin B12 deficiencies, and other conditions can cause cognitive abilities to gradually deteriorate (Bondi et al., 2017).

Robert Katzman's compilation of information demonstrating the histological similarity of senile and presenile AD and his subsequent proposal that AD was the fourth most prevalent cause of mortality in the elderly based on epidemiological data marked a turning point in our understanding of dementia (Knibb et al., 2006). At that time, formal research diagnostic criteria for AD were created, and the diagnostic criteria for dementia were enhanced (Dickson, 2001). These standards

DOI: 10.1201/9781003412069-9

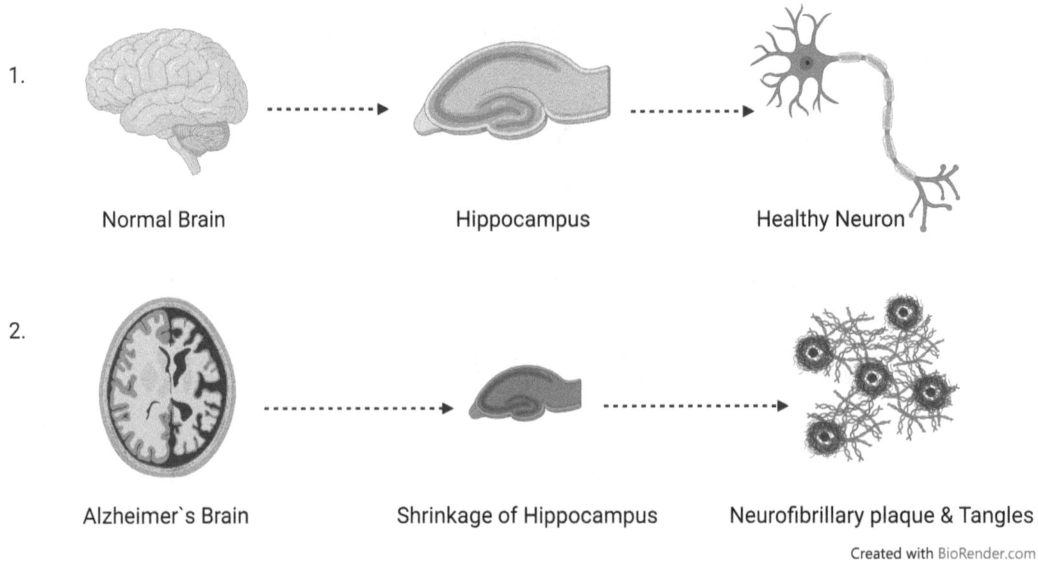

FIGURE 9.1 Differentiation between normal brain and Alzheimer's brain.

are included in the DSM-III (American Psychiatric Association, 1980) and the 10th version of the International Statistical Classification of Diseases and Related Health Problems-ICD (World Health Organisation, 1992; Burns & Iliffe, 2009)].

9.1.1 Effect on Memory

In the vast majority of instances, memory loss is the first symptom and is universal. Memory loss frequently manifests gradually, making it easy to mistake it for normal aging and to identify it as the early Alzheimer's disease phases only in hindsight (Karantzoulis & Galvin, 2011). It is required to first determine the extent to which memory loss is an issue and for whom the problem is a barrier because it is a common and serious problem in dementia (Khan et al., 2016; Abubakar et al., 2022).

Even modest changes to an individual's appearance are part of normal aging, as are some changes to the higher mental abilities or "cognitive" functions (McKhann et al., 2011). A decline in memory function characterizes dementia. However, it can also signal further dysfunctions that should always be taken into account in any assessment (Jack et al., 2011).

Long- and short-term memory are both markers of memory performance (Chen et al., 2012; Roman Laskowski et al., 1983). The technique that allows one to remember a new phone number while dialling it is short-term memory that is later developed into working memory (Toledo et al., 2017). Long-term memory enables the retention of a recognizable phone number from day to day and year to year (Y. Xie et al., 2020).

9.1.2 Types of Memories

9.1.2.1 Episodic and Semantic Memory

It appears that different forms of knowledge are stored in different ways. It has been accepted to draw a distinction between procedural and declarative memories instead of the initial separation between episodic and semantic memories (Justice et al., 2015). While semantic memories are information without a context, episodic memories are for specific events. An example of episodic memory is recalling what I had for breakfast; semantic memory is recalling that the word "breakfast" refers to a morning meal (Ray et al., 2011).

9.1.2.2 Declarative and Procedural Memory

This category may include episodic and semantic memories (which represent the memory for facts). Procedure memory is used for routines and skills and may incorporate certain sensory stimuli (Zhang et al., 2014). For instance, knowing how to drive a car is a declarative memory, whereas knowing how the engine operates is a procedural memory. In general, older people can learn just as much as younger people, but it will take them longer to reach the same level of learning since they process and "digest" knowledge more slowly (Barthélemy et al., 2020).

9.1.3 Signalling Pathway of Alzheimer's Disease

AlzPathway1, a database that aggregates signalling pathways connected to AD from various research, was developed by a study team. Apoptosis, acetylcholine synthesis, the Wnt signalling route, ubiquitin-mediated proteolysis, the cleavage and degradation of Aβ, NFT buildup and the apolipoprotein E (ApoE)-cholesterol pathway, aberrant ceramide aggregation, reactive oxidation, and the neurotrophins signalling pathway (Xiang et al., 2016).

9.1.4 Physiopathology of Alzheimer's Illness

The Aβ-peptide accumulation in the brain that results from an imbalance between the production and elimination of this protein is believed to be the fundamental cause of the illness, according to the amyloid cascade theory (one of the most generally accepted theories of AD pathogenesis, though it is still up for discussion). This process results in the production of NFT, among other occurrences that have been noted (Sommerlad et al., 2018).

Amyloid precursor protein (APP), a naturally occurring protein crucial for maintaining brain homeostasis, is converted into the 36- to 43-amino-acid-long Aβ peptide. APP gene locus duplication and Down syndrome, an uncommon form of early onset of familial origin, are associated with greater odds of early-onset AD because of the APP gene's placement on chromosome 21. The increase in cerebral Aβ-peptide caused by overexpression of APP is thought to lead to its deposition (Giri et al., 2016).

The two primary pathways for processing APP are currently understood to be amyloidogenic and nonamyloidogenic and secretase-mediated and γ and β secretase-mediated. As the enzyme α secretase breaks down APP, a soluble molecule known as sAPPα is created. This molecule is anticipated to have significant roles in neuronal survival and plasticity as well as excitotoxicity defence. It is also thought to have neuroprotective characteristics (Tapiola et al., n.d.).

By Aβ-secretase cleaving APP, the Aβ peptide is generated (mainly BACE1 enzyme). In this process, β-secretase breaks down APP to create the carboxy-terminal complex, which is linked to cell membranes, and the soluble APP fragment (sAPP), a mediator of neuronal death (McKeith et al., 2005). The latter is broken down into the Aβ peptide by a complex called Aβ-secretase made up of four proteins: presenilin 1 or 2, APH-1 (formerly known as pharynx-defective-1), nicastrin, and PEN-2 (presenilin enhancer-2). The 40 amino acid type (A 40) predominates among the Aβ peptides, which are produced in sizes ranging from 38 to 43 amino acids (Fagan et al., 2006).

Under healthy conditions, the amyloidogenic and nonamyloidogenic pathways coexist in balance, with the latter being more prevalent (Sims & Williams, 2016b). Compared with Aβ-40, the Aβ-42 peptide aggregates more frequently. According to immunohistochemical studies, Aβ42 is initially deposited and found in greater amounts in the amyloid plaques seen in AD patients (Sweeney et al., 2019). On the other hand, there is currently no proof that cerebral amyloidosis and serum Aβ-42 levels are related. The CSF fluid of AD patients shows a drop in Aβ42 levels, which can be partially attributed to greater a-amyloid plaque deposition (Huang et al., 2009). Further evidence of the link between the Aβ-42 peptide and the pathophysiology of AD is provided by the observation that early-onset familial AD types are caused by mutations in the APP and presenilin genes, resulting in a relative rise in Aβ-42 levels (Skoog & Gustafson, 2006).

9.1.5 Stages of Alzheimer's Disease

There are total 6 stages of AD:

Stage 1 and 2: Preclinical AD
Stage 3: AD with MCI
Stage 4: AD with mild dementia.
Stage 5: AD with moderate dementia
Stage 6: AD with severe dementia

(Brenowitz et al., 2021, Gomes Neves et al., n.d.).

Preclinical AD, a stage of asymptomatic AD with biomarker evidence of AD, advances through changes in moderate cognitive impairment (MCI) and/or mild behavioural impairment (MBI) and ultimately results in AD dementia (Galvin, 2015). Preclinical AD, the first stage on the AD continuum, is typified by a protracted asymptomatic phase during which people with AD pathology do not show any indications of cognitive or functional impairment or a change in how they normally function (Falcon et al., 2018).

Age, sex, and apolipoprotein E (ApoE) status are only a few of the variables that can affect whether preclinical AD develops into MCI owing to AD (with or without MBI) (Toukan & Rahman, n.d.). A person with underlying AD pathology has a higher chance of developing MCI or AD dementia than someone without it. In six longitudinal cohorts that were observed for an average of 3.8 years, 20% of preclinical AD patients went on to acquire MCI as a result of AD (Evans, 2014). A second research by Cho et al. found that 29.1% of people with AD in preclinical stages went on to acquire MCI as a result of AD, with a 4-year average follow-up rate (Galvin, 2015).

9.1.6 Risk Factor of AD

There are 5 novel early disease indicators or risk factors: sleep deprivation, hypoxia, dietary habits, gut microbiome, and hearing loss (Figure 9.2) (N. M. Li et al., 2019).

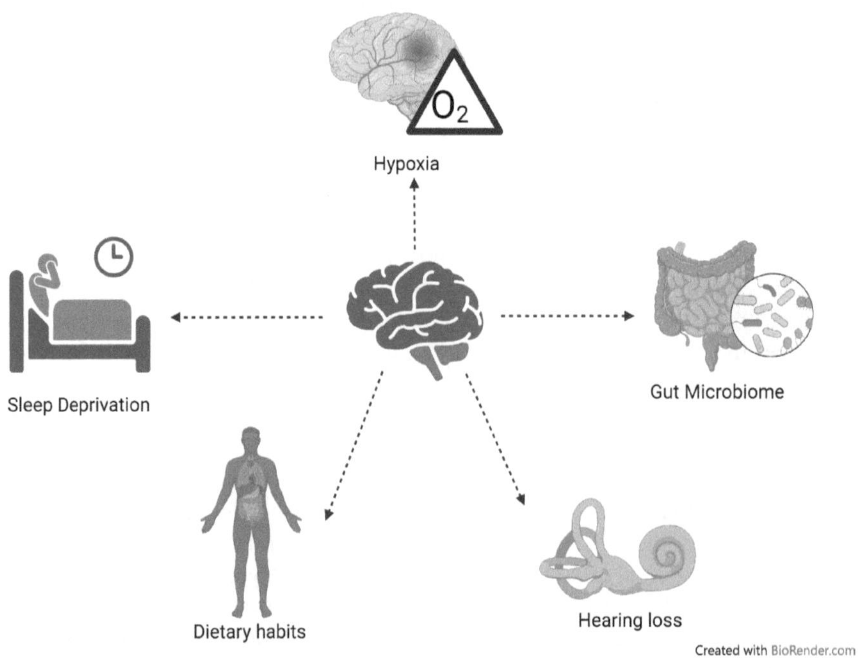

FIGURE 9.2 Risk Factors in Alzheimer's disease.

9.2 COMPUTATIONAL MODELLING

Computational modelling plays a potential key role in the expansion of science and technology through visualization capabilities and enhanced exploration (Katchalski-Katzirtt et al., 1992). Using computer software for virtual screening, structure- or ligand-based drug designing and discovery are possible (Zvěřová et al., 2013). There is a range of computer software accessible for structural affinity by selective structure-based design and lead optimization of some physicochemical properties for the advancement of drugs. Another perspective for drug design and discovery includes combinatorial chemistry, biology, high-throughput screening, and drug metabolism (ADME) groups that are joined together (Hart & Read, 1992).

Computational tools are necessary for computer-aided drug design because these processes are a cost-effective, reliable, and time-saving technique. The two novel drug discoveries and designs that are widely used are structure-based drug design and rational drug design (Abagyan et al., 1993) Structure-based drug design means that we use three-dimensional structures (protein, enzymes, and receptors) to design drugs/new drugs (ligand), and finding novel drugs based on information about a biological target is called rational drug design (Jones et al., 1995). Both techniques work with the help of computers and software by checking receptor-drug binding affinity. Around 189 computational tools are used for drug design and discovery, and a few of them are shown in Table 9.1.

TABLE 9.1
Different Computational Tools and Their Information Regarding Drug Discovery and Designing

Software Name	Year of availability	Organization name	Country name
DOCK	1982	UCSF (University of California, San Francisco)	USA
ChemOffice/ChemDraw	1985	PerkinElmer	US
Winmostar software packages	1989	X-Ability Co., Ltd.	Japan
AutoDock	1990	Scripps Research	USA
SOFT Docking	1991	University of California	USA
DockVision	1992	University of Alberta	Canada
PROCHECK	1993	Department of Biochemistry and Molecular Biology, University College	UK
ICM-Dock	1994	Molsoft L.L.C.	USA
FLOG	1994	Merck Research Laboratories	USA
DIVALI	1995	University of California-San Francisco	USA
RasMol	1995	GlaxoWellcome Medicines Research Centre	UK
Gromacs	1995	The University of Groningen	Netherlan ds
Hammerhead	1996	Arris Pharmaceutical Corporation	USA
LIGIN	1996	Weizmann Institute of Science	Israel
GASP	1996	National Human Genome Research Institute	Maryland
VMD	1996	University of Illinois	USA
Open Eye Scientific Software tools	1997	Open Eye Scientific	USA
FTDOCK	1997	BiomolecularModelling Laboratory	UK
QXP	1997	Novartis Pharmaceuticals Corporation	Switzerland
SANDOCK	1998	The University of Edinburgh	UK
BOSS	1998	Yale University Department of Chemistry	USA
MCDOCK	1999	Georgetown University Medical Center	USA
CATS	1999	F. Hoffmann-La Roche Ltd	Switzerland
FlexX	2000	BioSolvIT	Germany
Combinatorial library design	2000	CombiChem, Inc.	USA
FTrees	2000	BioSolvIT	Germany

(Continued)

TABLE 9.1
(Continued)

Software Name	Year of availability	Organization name	Country name
Ascalaph	2000	University of Stockholm	Sweden
MOLARIS-XG	2000	University of Southern California	USA
SEED	2001	University of Zürich	Switzerland
PocketMol	2001	Wright State University	USA
QMOL	2001	Cornell University	USA
YASARA	2002	Center for Molecular and Biomolecular Informatics	Netherlan ds
Pymol	2002	Schrodinger	USA
HADDOCK	2003	Utrecht University	Netherlands
LigandFitBioVia	2003	BIOVIA	USA
GEMDOCK	2004	NCTU (National Chiao Tung University)	Taiwan
Glide Schrödinger	2004	Schrodinger	USA
UCSF Chimera	2004	University of California	USA
SuperPose	2004	University of Alberta	Canada
SRS 3D	2004	LION bioscience AG	Germany
MEDock	2005	National Taiwan University	Taiwan
FoldX	2005	UniversiteitBrussel	Belgium
Visualizer DS	2005	BIOVIA	USA
AURAmol	2005	The University of York	UK (England)
PowerMV	2005	National Institute of Statistical Sciences	USA
PLANTS	2006	University of Konstanz	Germany
PSI-DOCK	2006	Peking University	China
Molegro Virtual Docker	2006	MolegroApS	Denmark
BDT	2006	UniversitatRovira i Virgili	Spain
SiMMap	2006	University of Copenhagen	Denmark
ParDOCK	2007	Indian Institute of Technology	India
EADock	2007	Swiss Institute of Bioinformatics	Switzerland
FLIPDock	2007	The Scripps Research Institute	United States
pyDock/pyDockSER	2007	University of Cambridge	United Kingdom
WinDock	2007	Howard University College of Medicine	USA
Spartan	2007	Wavefunction, Inc.	USA
HyperChem	2007	Institute of Molecular Function	Japan
Bioclipse	2007	Uppsala University	Sweden
MS-DOCK	2008	INSERM, University Paris Descartes	France
HomDock	2008	Chil^2—Molecular Design for Science and Technology	Germany
ISE-Dock	2008	Hebrew University of Jerusalem	Israel
Q-Dock	2008	Georgia Institute of Technology	USA
PharmaGist	2008	Tel Aviv University	Israel
DOVIS	2008	Biotechnology High Performance Computing Software Applications Institute (BHSAI)	USA
Align-it™	2008	Silicos NV	Belgium
DOCK Blaster	2009	University of California San Francisco	USA
Docking Server	2009	Virtua Drug Ltd	Hungary
Catalyst	2009	BIOVIA	USA
FINDSITE-LHM	2009	Stanford University	USA
AutoDockVina	2010	Scripps Research	United States

Software Name	Year of availability	Organization name	Country name
VLifeMDS	2010	VLife Sciences Technologies Pvt. Ltd.	India
ParaDockS	2010	Martin-Luther-University Halle-Wittenberg	Germany
GriDock	2010	UniversitadegliStudi di Milano	Italy
DockoMatic	2010	Boise State University	USA
BioDrugScreen	2010	Indiana University School of Medicin	USA
PharmMapper	2010	Chinese Academy of Sciences	China
SwissDock	2011	Swiss Institute of Bioinformatics	Switzerland
Pose & Rank	2011	University of California	USA
Pharmer/ZINC Pharmer	2011	University of Pittsburgh	USA
GalaxyDock	2012	Seoul National University	South Korea
idTarget	2012	National Taiwan University	Taiwan
1-Click Docking Mcule	2012	Mcule, Inc.	United States
Vina MPI	2013	University of Tennessee	USA
FINDSITEcomb	2013	Center for the Study of Systems Biology, School of Biology, Georgia Institute of Technology	USA
Sirius	2013	Friedrich-Schiller-Universität Jena	Germany
rDock	2014	Vernalis Research and University of York	UK
MAPS	2014	Scienomics SARL	France
PyRx	2015	The Scripps Research Institute	USA
Aggregator Advisor	2015	University of California	USA
MOLS 2.0	2016	University of Madras	India
ADMET Predictor	2016	Simulations Plus, Inc.	USA
MOLGEN	2016	University of Bayreuth, Bayreuth,	Germany
MOE	2018	Chemical Computing Group	Canada
Tinker	2018	Washington University	USA
LigPrep	2018	Schrödinger, LLC	USA

(*Source:* Abagyan et al., 1993; Kramer et al., n.d.; Miller et al., 1994; Sticht et al., 1995; Taylor & Burnett, 2000).

9.3 COMPUTATIONAL PROTEIN MODEL OF ALZHEIMER'S DISEASE

All of the computed protein models for Alzheimer's disease are provided below, using protein model data from the RCSB protein data repository.

9.3.1 1AML (Structure of Amyloid A-4-(1-40)-Peptide of Alzheimer's Disease

The solution structure of the full-length A4-(I –40)-peptide was determined via limited molecular dynamics simulations and two-dimensional (NMR) nuclear magnetic resonance spectroscopy. A4-(1–40) human substitute during several days, 40% by vol. trifluoroethanol/water, kept the peptide soluble and prevented it from aggregating. The 203 consecutive and in the nuclear Overhauser enhancement spectroscopy (NOESY), the medium-range cross-peaks were found, which allowed for the obvious assignment of all spin systems. Long-range NOESY cross-peaks, which would have shown the peptide's tertiary structure, were not present. Sequential chemical-shift analysis, calculations of restricted molecular dynamics using the NOE, medium-range NOESY data, and the identification of two helices—Ile31-Met35 and Glnl5-Asp23—as the major secondary structure constituents. The remaining portion of the conformation of the peptide is a random coil (Barnham et al., 2003).

9.3.2 5O3L (Tau Filament Cryo-EM Structure)

An Alzheimer's patient's brain was used to create this structure, which has atomic models and maps with 3.4–3.5 resolution using cryo-electron microscopy (cryo-EM). Tau aggregation begins with the formation of a coupled cross-helix from the two identical protofilaments that make up filament cores (Fitzpatrick et al., 2017). Tau protein residues 306–378 make up the protofilaments in question. These are ultrastructural polymorphs because pairs of straight and helical filaments have variable inter-proto filament packing. The patient-derived material may be used in cryo-EM to characterize amyloid filaments at the atomic level, opening the door for research into a number of neurodegenerative illnesses (Crescenzi et al., 2002).

9.3.3 1OWT (Amyloid Precursor Protein {APP} for Alzheimer's Disease Copper Domain Structure)

Copper is a major source of free radical generation in the brain. Cells have developed intricate metal transport networks to protect against oxidative damage caused by metals. A key factor regulating the homeostasis of copper in neurons is the protein (APP). APP knockout mice had more copper in their cerebral cortices than did APP-overexpressing transgenic mice; significantly reducing in vitro amyloid production is possible with copper binding to APP. A unique copper-binding site in the APP copper-binding domain (CuBD) structure encourages Cu(I) coordination. The CuBD's role as a neuronal metallo transporter is supported by the CuBD's surface location, structural resemblance to copper chaperones, and the importance of APP in maintaining neuronal copper homeostasis (Otzen et al., 2000).

9.3.4 1CQN (Protein Tetramer and Hydrophobic Alzheimer Homology Zipped Together)

A molecular understanding of the mechanisms behind deposition and protein fibrillation is significantly hampered by the precipitation of particular amyloidogenic sequences, such as the Alzheimer peptide (b-AP), and their inability to dissolve in solution. A soluble scaffold from a b-AP homology is built using the monomeric protein S6. The tetramer-formed S6 construct with the greatest b-AP homology crystallizes as an intermolecular antiparallel b-sheet, which connects the molecules. Additionally, following refolding, this construct exhibits enhanced coil aggregation, and fibrils are formed by a 14-mer peptide containing the modified region. The overall hydrophobicity of the sticky segment is related to intermolecular interaction, which suggests that the wild-type protein contains "structural gatekeepers." To stop aggregation, these charged side chains sever long stretches of hydrophobic residues in the primary structure (S. Zhang et al., 2000).

9.3.5 1IYT (Dementia Amyloid β-peptide Solution Structure (1–42) in an Apolar Microenvironment)

The structural investigation of these peptides is difficult in aqueous solutions because of their propensity to aggregate. Only lately have the NMR structures of Ab-(1–40) and Ab-(1–42) in SDS micelles or in aqueous trifluoroethanol been determined. The findings of these investigations suggest that two helical parts are joined by a flexible kink, albeit it was challenging to precisely quantify the placement and length of the helical stretches and, more importantly, to determine whether the kink area has a preferred form. In the search for a solvent that would permit a more precise structure determination, a thorough solvent scan was conducted. The findings revealed that Ab-(1–42) exhibited a high inclination to assume helical conformations in aqueous solutions of fluorinated alcohols. Two helical areas comprising residues 8–25 and 28–38 are shown in the 3D NMR structure of Ab (1-42). These sections are joined by a typical type I b-turn. Unexpected similarities between this structure and the C-terminal moiety of influenza hemagglutinin's fusion domain point to a direct mechanism of neurotoxicity (Lührs et al., 2005).

9.3.6 1AAP (Alzheimer's Disease Protein Precursor: The Protease Inhibitor Domain's X-ray Structure)

The domain of the Kunitz protease inhibitor (APPI) found in the precursor Alzheimer's amyloid protein may have a role in the proteolytic processes resulting in the amyloid buildup in the brain. The APPI crystal structure has been constructed and enhanced to make it simpler to pinpoint the inhibitor's physiological target. The molecular properties of the APPI structure will be reflected in the sequences at the inhibitor-protease interface of the relevant protease target. Even though the tertiary folds of APPI and the Kunitz inhibitor BPTI are quite similar overall, one of the two loops that bind to proteases' backbone conformations goes through a significant change.

Similar loop sequences are present in a variety of Kunitz inhibitors, indicating that the structural change is conserved and may play a significant role in determining the specificity of the inhibitor. In contrast to Val-34 and Arg-17 in BPTI, Phe-34 and Met-17 in APPI produce an exposed hydrophobic surface in a different area of the protease binding loops. It is evident how limited this shift in target sequences for serine proteases is when the APPI structure is superimposed on the BPTI complex. Here, a complementary pair of nonpolar side chains on kallikrein A competes with polar side chains on trypsin on the hydrophobic surface of APPI (Kong et al., 2007).

9.3.7 1HZ3 (In Water, Peptide A Assumes a Collapsed Coil Shape)

The soluble peptide Aβ's self-assembly into the amyloid associated with Alzheimer's disease is thought to include a conformational shift. Consequently, the solution conformation of Aβ is very important. In water, where it has no -helical or -sheet structure, A is compressed into a tiny cluster of loops, strands, and twists, in contrast to research in other solvents. Electrostatic and Van der Waals forces are principally responsible for conformational stability. 25% of the surface is covered by a sizable, noticeable, unbroken hydrophobic area. Given that the tight coil shape appears to be metastable and fibrillization produces an intermolecular β-sheet secondary structure, a general alteration in conformation is extremely likely. The development of lower-energy amyloid fibrils in the soluble precursor during the development of amyloid deposits and changes in solvation thermodynamics during this process are the two key driving forces suggested by a molecular hypothesis of amyloidosis (Farlow et al., 2012).

9.3.8 6HRE (Tau Filaments from Many Cases of Sporadic and Inherited Alzheimer's Disease Take on a Cold)

In 17 cases of AD (15 sporadic and 2 inherited) and 2 cases of atypical AD, the frontal cortex was studied using electron microscopy with immuno-gold negative stain and electron cryo-microscopy (posterior cortical atrophy) to look for straight filaments (SFs) and paired helical filaments (PHFs). In 3 instances of AD, 1 hereditary and 2 sporadic, Cryo-EM was utilized to pinpoint the high-resolution structures of SFs and PHFs from the frontal brain. Moreover, immuno-EM was used to analyse the PHFs and SFs of several cortical and subcortical brain areas. Every AD scenario has more PHFs than SFs.

Utilizing cryo-EM, it was found that SFs and PHFs are made up of two protofilaments that are C-shaped and have an integrated cross-/helix structure, as was previously mentioned for one case of AD. Two more amino acids were identified at the ends of the protofilament in these structures, which were seen here in higher resolution. The results of immuno-EM supported the findings of cryo-EM, showing that tau repetitions 3 and 4 were present in the filament cores of all instances of AD but not the N-terminal portions of repeats 1 and 2. These results demonstrate that tau filament structures do not significantly differ between AD patients. The development of new diagnostic procedures and therapeutic approaches will depend heavily on this knowledge in order to comprehend the mechanisms underlying tau filament generation (Spiess et al., 1990).

9.3.9 2BEG (Alzheimer's Amyloid-(1–42) Fibrils' 3D Structure)

The most severe form of neurodegenerative condition, amyloid-(A) amyloidogenesis, appears to be what causes Alzheimer's disease. The 3D structure of A's constituent fibrils (1–42). Quenched hydrogen deuterium-exchange NMR, side-chain packing constraints from paired mutagenesis studies, and hydrogen-bonding constraints from earlier solid-state NMR research were used to generate this structure. Residues 18–42, but not residues 1–17, constitute the strand-turn-strand motif, which is composed of two intermolecular, parallel, in-register sheets made up of residues 18–26 and 31–42. For the protofilament's repeating structure to develop, at least 2 molecules of A (1–42) are needed. The (nl) th molecule's even-numbered strand 2 and the (nl) th molecule's odd-numbered strand 1 residues combine to produce intermolecular side-chain interactions. Cooperativity, sequence selectivity, and apparent unidirectionality of A fibril formation can all be attributed to the interaction pattern that results inβ-strands at the extremities of the fibrils being partly unpaired. Moreover, it offers a structural foundation for fibrillation inhibitors (Leng et al., 2016).

9.3.10 2FK1 (The Copper-binding Domain of the Alzheimer's Amyloid Precursor Protein Is Structurally Studied to Understand How It Binds Copper Ions)

Dementia is mostly brought on by AD, which is primarily brought on by the amyloid peptide (A), which is produced by the proteolytic cleavage of the amyloid precursor protein. Through the extracellular Cu-binding domain of the APP, it may regulate copper (Cu) transport (CuBD) as a metalloprotein. Levels are decreased by adherence of Cu to the CuBD, indicating that a Cu mimic may have therapeutic promise. In spite of having distinct crystal lattices, the apo CuBD atomic structures from 3 different crystal forms are described here. Tyr168, His147, His151, and 2 water molecules are the metal ligands in Cu2+-bound CuBD, and they are organized in a square pyramidal shape. Studies of the site's fine structure utilizing electron paramagnetic resonance and prolonged X-ray absorption support the idea that it is akin to a Type 2 nonblue Cu core. Despite the likelihood that Met170 is a ligand, prior studies revealed that it serves as a crucial electron donor in CuBD's capacity to lower Cu ions. The design of a Cu+-bound CuBD and Cu2+-bound CuBD is essentially the same because one of the water ligands was lost. Due to the site's structure being unfavorable for Cu+, CuBD might easily transfer Cu ions to other proteins by this method (Hassan et al., 2018).

9.4 INHIBITORS OF ALZHEIMER'S DISEASE BASED ON COMPUTATIONAL MODELLING

9.4.1 AZD3293

Hassan et al. discovered the molecule AZD3293 (Figure 9.3) which is being tested to investigate AD. Docking was done between AZD3293 and BACE1(Xie et al., 2013).

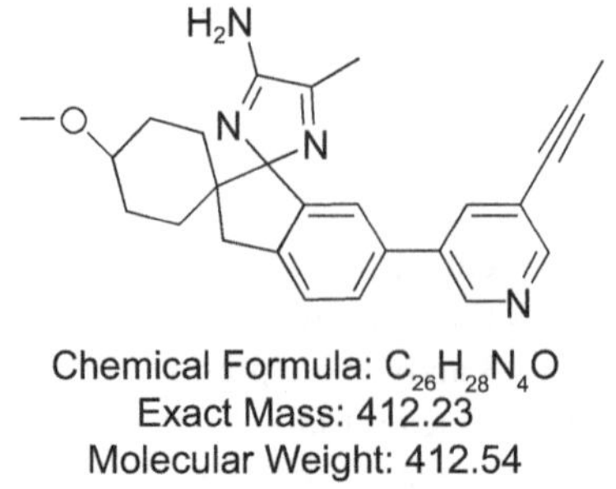

Chemical Formula: $C_{26}H_{28}N_4O$
Exact Mass: 412.23
Molecular Weight: 412.54

FIGURE 9.3 Chemical Structure of AZD 3293.

At Asp32 and Lys107, with distances of 2.95 and 2.68, respectively, 2 hydrogen bonds were found between the amino acids in AZD3293 and BACE1. Our findings indicate that AZD3293 has good drug-like behaviour and may be a promising therapeutic agent. Also, their good lead-like behaviour and therapeutic potential were supported by their pharmacokinetic features, such as ADMET properties (Xie et al., 2013). Except for little lethality in metabolism and toxicity, AZD3293 showed positive findings in all metrics, including absorption, distribution, and excretion. The hepatotoxicity behaviour of AZD3293 is shown, and this behaviour may have catastrophic effects on the body. The RMSD of AZD3293 was 0.2 nm. Peptide-based transition state analogues known as BACE1 inhibitors are created using the amyloid precursor protein (APP) residues that remain after β-secretase cleaves the APP. This drug is in phase II clinical trial (Ashraf et al., 2016).

9.4.2 Tacrine Hybrids

AChE and 5e (Figure 9.4) were docked. The ARG 289, TRP 279, HIS 440, TRP 84, and PHE 330 were the binding bonds for 5e. 5e demonstrated strong dual anti-AChE/BChE activity with IC50 values below the nanomolar range (Cheng et al., 2019). The optimal distance for a diamino linker between IAA and tacrine according to SAR analysis in the AChE bioassay was 6 carbon chains. Moreover, 5e has far stronger antidementia and nootropic potentials than tacrine when it comes to altering neuronal network activity (Cheng et al., 2019). The tacrine hybrids IAA may offer a novel model for the creation of multifunctional anti-AD medications, according to the biological evaluation and computational analyses in the current study (Cheng et al., 2019).

9.4.3 Donepezil and Its Analog

The peripheral binding site, a narrow and deep canyon known as catalytic site and the peripheral anionic site (PAS), which is composed of three amino acids: histidine (H447), serine (S203), and glutamic acid, make up the active area of the AChE enzyme (E334). Tryptophan with a possible "π" stacking and the donepezil benzyl ring (W86) (Makarian et al., 2022). Moreover, we observed potential interactions with the AChE catalytic triad amino acids H447 and S203. The carbonyl group of donepezil then formed a hydrogen bond with the amino acid phenylalanine (F295), which functioned as the hydrogen bond donor. The piperidine ring, which is located in the middle of donepezil and interacts with D74 and Y337, and the indanone ring, which is adjacent to F338, F295, Y341, and W286, both provide strong hydrophobic interactions (SvedružiĆ et al., 2021).

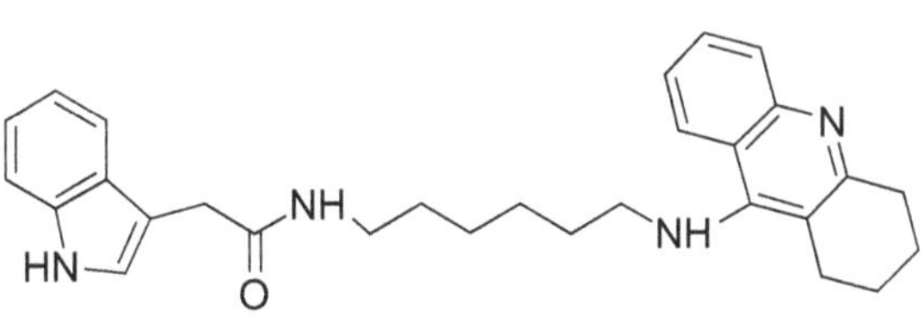

5e

Chemical Formula: $C_{29}H_{34}N_4O$
Exact Mass: 454.27
Molecular Weight: 454.62

FIGURE 9.4 Chemical Structure of Tacrine Hybrids.

The interactions between the piperidine ring and donepezil are quite similar and showed the same hydrogen bonding with phenylalanine (F295). It displayed a comparable docking rating with that of donepezil. Donepezil as approved by the FDA has an IC_{50} within its range. According to our kinetic investigations, this inhibitor forms a mixed-type bond with the enzyme (Svedružić et al., 2021).

The peripheral anionic and catalytic active sites are comparable with donepezil. Without steric or other interferences, the pyridylmethyl moiety of donepezil can successfully replace the benzyl ring. According to molecular docking studies, also identified this analogue's most likely binding location and interactions in the human AChE binding pocket (Svedružić et al., 2021) (Figure 9.5).

9.4.4 Semagacestat

The β-secretase semagacestat (Figure 9.6) joins to its substrate in a complex protein (PDB ID:6IYC). According to molecular docking studies, the presence of the substrate allows for the simultaneous binding of up to 4 semagacestat molecules to β-secretase. There are 2 drug molecules tucked away between presenilin and the Aph1 subunit, and 1 is attached to the active site tunnel's ends (Svedružić et al., 2021). The medications bind to presenilin sites on the membrane's interior and exterior. The substrate, the active site's Asp 385 and Asp 257, and the nearby PAL motif are all hidden beneath the protein surface (Ala 434, Leu 433, Pro 435). This medication's phase 3 clinical trial is currently in progress (Doody et al., 2013).

9.4.5 DYRK1A Inhibitors

The most promising ligands are 11352, 15938, 21830, and 21878 (Figure 9.7). A conserved area made up of the amino acid residues, the ATP binding pocket in the DYRK1A kinase is made up of Phe238 (gatekeeper; gk), Phe170, Ser242, Glu239 (gk+1), Lys188 (catalytic), Leu241 (gk+3), Asn292, and Asp307 and is required for the binding of inhibitors (Neagoie et al., 2012; Shahroz et al., 2022). Leu241's (gk+3) hydrogen bond interactions were crucial selection factors. A scaffold made of bis-indole indigo was one of the discovered ligands and was known to be a DYRK1A inhibitor. These inhibitors are in phase II clinical trials (Shahroz et al., 2022).

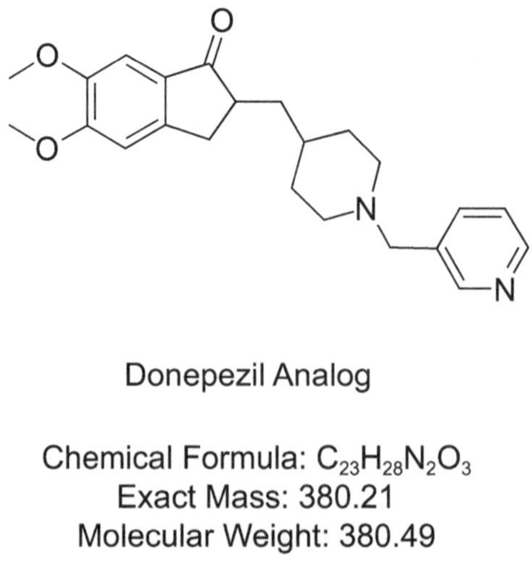

FIGURE 9.5 Chemical structure of the donepezil analogue.

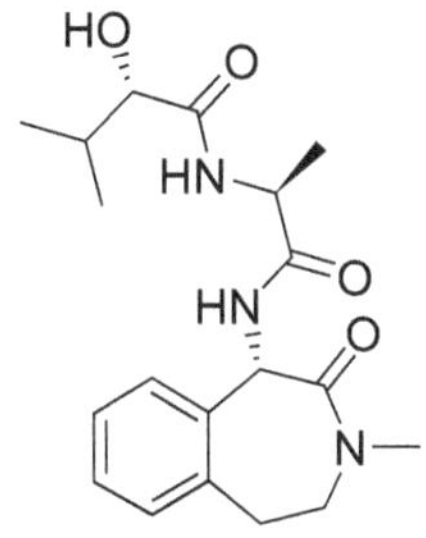

Semagacestat

Chemical Formula: $C_{19}H_{27}N_3O_4$
Exact Mass: 361.20
Molecular Weight: 361.44

FIGURE 9.6 Chemical structure of semagacestat.

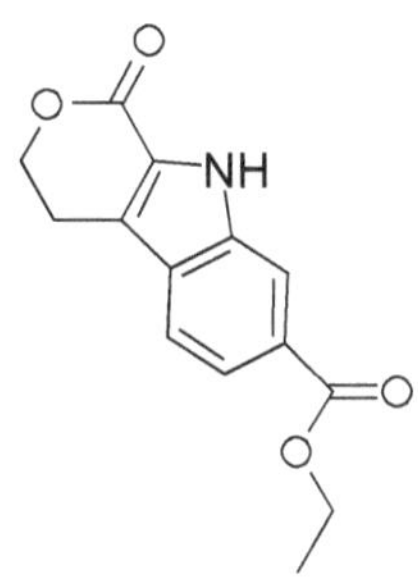

21878

Chemical Formula: $C_{14}H_{13}NO_4$
Exact Mass: 259.08
Molecular Weight: 259.26

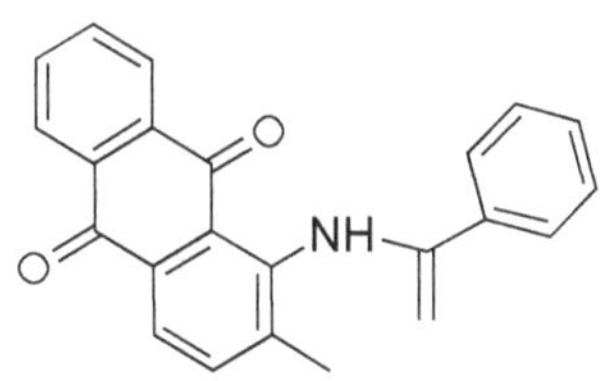

15938

Chemical Formula: $C_{23}H_{17}NO_2$
Exact Mass: 329.13
Molecular Weight: 339.39

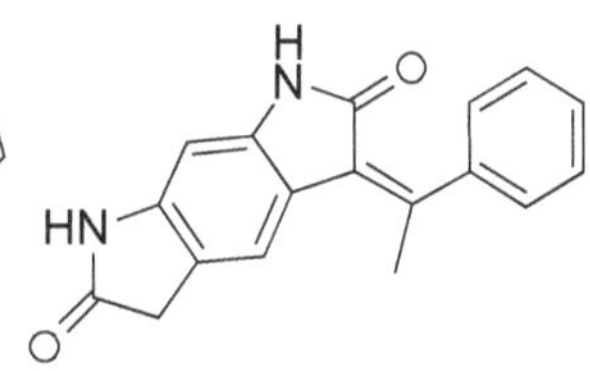

11352

Chemical Formula: $C_{18}H_{14}N_2O_2$
Exact Mass: 290.11
Molecular Weight: 290.32

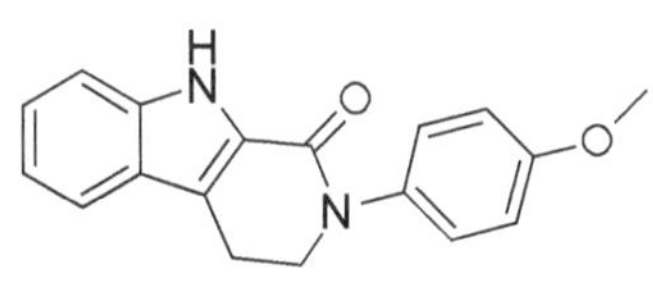

21830

Chemical Formula: $C_{18}H_{16}N_2O_2$
Exact Mass: 292.12
Molecular Weight: 292.34

FIGURE 9.7 Chemical structure of DYRK1A inhibitors.

9.5 CONCLUSION

Alzheimer's disease is radically altering people's regular routines every day, and many medications have been developed and sold worldwide to tackle this condition. Today, novel molecules are being explored using computer models. Computational modelling is a proven method for the discovery of various agonists and antagonists because the use of software tools to discover new molecles saves time and is economical. The tabulated form of information about worldwide used computation tools will help in selecting the desired software for drug discovery and designing. Certain protein analogues available in the protein data bank were also discussed. Inhibitors discovered with the help of computational methods are also given in this chapter; for example, donepezil, tacrine hybrids, etc. proved to be effective agents in clinical trials.

ENDNOTE

* Corresponding author

REFERENCES

Abagyan, R., Totrov, M., & Kuznetsov, D. (1993). ICM-A new method for protein modelling and design: Applications to docking and structure prediction from the distorted native conformation docking and global energy optimization. *Journal of Computational Chemistry*, *15*(5), 994.

Abubakar, M. B., Sanusi, K. O., Ugusman, A., Mohamed, W., Kamal, H., Ibrahim, N. H., Khoo, C. S., & Kumar, J. (2022). Alzheimer's disease: An update and insights into pathophysiology. *Frontiers in Aging Neuroscience*, *14*. Frontiers Media S.A. https://doi.org/10.3389/fnagi.2022.742408

Ashraf, Z., Alamgeer, Rasool, R., Hassan, M., Ahsan, H., Afzal, S., Afzal, K., Cho, H., & Kim, S. J. (2016). Synthesis, bioevaluation and molecular dynamic simulation studies of dexibuprofen-antioxidant mutual prodrugs. *International Journal of Molecular Sciences*, *17*(12). https://doi.org/10.3390/ijms17122151

Barnham, K. J., McKinstry, W. J., Multhaup, G., Galatis, D., Morton, C. J., Curtain, C. C., Williamson, N. A., White, A. R., Hinds, M. G., Norton, R. S., Beyreuther, K., Masters, C. L., Parker, M. W., & Cappai, R. (2003). Structure of the Alzheimer's disease amyloid precursor protein copper binding domain. A regulator of neuronal copper homeostasis. *Journal of Biological Chemistry*, *278*(19), 17401–17407. https://doi.org/10.1074/jbc.M300629200

Barthélemy, N. R., Horie, K., Sato, C., & Bateman, R. J. (2020). Blood plasma phosphorylated-tau isoforms track CNS change in Alzheimer's disease. *Journal of Experimental Medicine*, *217*(11). https://doi.org/10.1084/JEM.20200861

Bohush, A., Bieganowski, P., & Filipek, A. (2019). Hsp90 and its co-chaperones in neurodegenerative diseases. *International Journal of Molecular Sciences*, *20*(20). MDPI AG. https://doi.org/10.3390/ijms20204976

Bondi, M. W., Edmonds, E. C., & Salmon, D. P. (2017). Alzheimer's disease: Past, present, and future. *Journal of the International Neuropsychological Society*, *23*(9–10) Special Issue, 818–831. Cambridge University Press. https://doi.org/10.1017/S135561771700100X

Breijyeh, Z., & Karaman, R. (2020). Comprehensive review on Alzheimer's disease: Causes and treatment. *Molecules*, *25*(24). https://doi.org/10.3390/MOLECULES25245789

Brenowitz, W. D., Xiang, Y., McEvoy, C. T., Yang, C., Yaffe, K., Le, W. D., & Leng, Y. (2021). Current Alzheimer disease research highlights: Evidence for novel risk factors. *Chinese Medical Journal*, *134*(18), 2150–2159. Lippincott Williams and Wilkins. https://doi.org/10.1097/CM9.0000000000001706

Burns, A., & Iliffe, S. (2009). Alzheimer's disease. *BMJ (Online)*, 338(7692), 467–471. BMJ Publishing Group. https://doi.org/10.1136/bmj.b158

Chen, C. T., Peng, H. P., Jian, J. W., Tsai, K. C., Chang, J. Y., Yang, E. W., Chen, J. B., Ho, S. Y., Hsu, W. L., & Yang, A. S. (2012). Protein-protein interaction site predictions with three-dimensional probability distributions of interacting atoms on protein surfaces. *PLoS ONE*, *7*(6). https://doi.org/10.1371/journal.pone.0037706

Cheng, Z. Q., Zhu, K. K., Zhang, J., Song, J. L., Muehlmann, L. A., Jiang, C. S., Liu, C. L., & Zhang, H. (2019). Molecular-docking-guided design and synthesis of new IAA-tacrine hybrids as multifunctional AChE/BChE inhibitors. *Bioorganic Chemistry*, *83*, 277–288. https://doi.org/10.1016/j.bioorg.2018.10.057

Crescenzi, O., Tomaselli, S., Guerrini, R., Salvadori, S., D'Ursi, A. M., Temussi, P. A., & Picone, D. (2002). Solution structure of the Alzheimer amyloid β-peptide (1–42) in an apolar microenvironment: Similarity with a virus fusion domain. *European Journal of Biochemistry*, *269*(22), 5642–5648. https://doi.org/10.1046/j.1432-1033.2002.03271.x

Defina, P. A., Moser, R. S., Glenn, M., Lichtenstein, J. D., & Fellus, J. (2013). Alzheimer's disease clinical and research update for health care practitioners. *Journal of Aging Research*, *2013*. https://doi.org/10.1155/2013/207178

Dickson, D. W. (2001). Neuropathology of Alzheimer's disease and other dementias. *Clinics in Geriatric Medicine*, *17*(2), 209–228.

Doody, R. S., Raman, R., Farlow, M., Iwatsubo, T., Vellas, B., Joffe, S., Kieburtz, K., He, F., Sun, X., Thomas, R. G., Aisen, P. S., Siemers, E., Sethuraman, G., & Mohs, R. (2013). A phase 3 trial of semagacestat for treatment of Alzheimer's disease. *New England Journal of Medicine*, *369*(4), 341–350. https://doi.org/10.1056/nejmoa1210951

Durazzo, T. C., Mattsson, N., & Weiner, M. W. (2014). Smoking and increased Alzheimer's disease risk: A review of potential mechanisms. *Alzheimer's and Dementia*, *10*(3 Suppl.). https://doi.org/10.1016/j.jalz.2014.04.009

Evans, D. A. (2014). History of the Harvard ChemDraw Project. *Angewandte Chemie—International Edition*, *53*(42), 11140–11145). Wiley-VCH Verlag. https://doi.org/10.1002/anie.201405820

Fagan, A. M., Mintun, M. A., Mach, R. H., Lee, S. Y., Dence, C. S., Shah, A. R., LaRossa, G. N., Spinner, M. L., Klunk, W. E., Mathis, C. A., DeKosky, S. T., Morris, J. C., & Holtzman, D. M. (2006). Inverse relation between in vivo amyloid imaging load and cerebrospinal fluid Abeta;42 in humans. *Annals of Neurology*, *59*(3), 512–519. https://doi.org/10.1002/ana.20730

Falcon, B., Zhang, W., Schweighauser, M., Murzin, A. G., Vidal, R., Garringer, H. J., Ghetti, B., Scheres, S. H. W., & Goedert, M. (2018). Tau filaments from multiple cases of sporadic and inherited Alzheimer's disease adopt a common fold. *Acta Neuropathologica*, *136*(5), 699–708. https://doi.org/10.1007/s00401-018-1914-z

Farlow, M., Arnold, S. E., Van Dyck, C. H., Aisen, P. S., Snider, B. J., Porsteinsson, A. P., Friedrich, S., Dean, R. A., Gonzales, C., Sethuraman, G., Demattos, R. B., Mohs, R., Paul, S. M., & Siemers, E. R. (2012). Safety and biomarker effects of solanezumab in patients with Alzheimer's disease. *Alzheimer's and Dementia*, *8*(4), 261–271. https://doi.org/10.1016/j.jalz.2011.09.224

Fitzpatrick, A. W. P., Falcon, B., He, S., Murzin, A. G., Murshudov, G., Garringer, H. J., Crowther, R. A., Ghetti, B., Goedert, M., & Scheres, S. H. W. (2017). Cryo-EM structures of tau filaments from Alzheimer's disease. *Nature*, *547*(7662), 185–190. https://doi.org/10.1038/nature23002

Galvin, J. E. (2015). The Quick Dementia Rating System (QDRS): A rapid dementia stagingtool. *Alzheimer's and Dementia: Diagnosis, Assessment and Disease Monitoring*, *1*(2), 249–259. https://doi.org/10.1016/j.dadm.2015.03.003

Giri, M., Zhang, M., & Lü, Y. (2016). Genes associated with Alzheimer's disease: An overview and current status. *Clinical Interventions in Aging*, *11*, 665–681. Dove Medical Press Ltd. https://doi.org/10.2147/CIA.S105769

Gomes Neves, R., Carvalho Silva, J., & Duarte Teodoro, V. (n.d.). *Computational Modelling in Science, Technology, Engineering and Mathematics Education*. http://modellus.fct.unl.pt

Hart, T. N., & Read, R. J. (1992). A multiple-start monte carlo docking method. *Proteins Structure, Function, and Genetics*, *13*.

Hassan, M., Shahzadi, S., Seo, S. Y., Alashwal, H., Zaki, N., & Moustafa, A. A. (2018). Molecular docking and dynamic simulation of AZD3293 and solanezumab effects against BACE1 to treat Alzheimer's disease. *Frontiers in Computational Neuroscience*, *12*. https://doi.org/10.3389/fncom.2018.00034

Huang, C. W., Lui, C. C., Chang, W. N., Lu, C. H., Wang, Y. L., & Chang, C. C. (2009). Elevated basal cortisol level predicts lower hippocampal volume and cognitive decline in Alzheimer's disease. *Journal of Clinical Neuroscience*, *16*(10), 1283–1286. https://doi.org/10.1016/j.jocn.2008.12.026

Jack, C. R., Albert, M. S., Knopman, D. S., McKhann, G. M., Sperling, R. A., Carrillo, M. C., Thies, B., & Phelps, C. H. (2011). Introduction to the recommendations from the national institute on aging-Alzheimer's association workgroups on diagnostic guidelines for Alzheimer's disease. *Alzheimer's and Dementia*, *7*(3), 257–262. https://doi.org/10.1016/j.jalz.2011.03.004

Jones, G., Willett', P., & Glen, R. C. (1995). Molecular recognition of receptor sites using a genetic algorithm with a description of desolvation. *Journal of Molecular Biology*, *245*.

Justice, N. J., Huang, L., Tian, J. Bin, Cole, A., Pruski, M., Hunt, A. J., Flores, R., Zhu, M. X., Arenkiel, B. R., & Zheng, H. (2015). Posttraumatic stress disorder-like induction elevates β-amyloid levels, which directly activates corticotropin-releasing factor neurons to exacerbate stress responses. *Journal of Neuroscience*, *35*(6), 2612–2623. https://doi.org/10.1523/JNEUROSCI.3333-14.2015

Karantzoulis, S., & Galvin, J. E. (2011). Distinguishing Alzheimer's disease from other major forms of dementia. *Expert Review of Neurotherapeutics*, *11*(11), 1579–1591. https://doi.org/10.1586/ern.11.155

Katchalski-Katzirtt, E., Shariv, I., Eisenstein, M., Friesem, A. A., Ii, C. A., & Vaksert, I. A. (1992). Molecular surface recognition: Determination of geometric fit between proteins and their ligands by correlation techniques (protein-protein interaction/surface complementarity/macromolecular complex prediction/ molecular docking). *Biophysics*, *89*.

Kelley, B. J., & Petersen, R. C. (2007). Alzheimer's disease and mild cognitive impairment. *Neurologic Clinics*, *25*(3), 577–609. https://doi.org/10.1016/j.ncl.2007.03.008

Khan, M. M., Sanusi, F., Ahsan, A., Ahmad, U., Akhtar, J., & Badrudeen, M. M. (2016). Alzheimer's disease: A review. *World Journal of Pharmacy and Pharmaceutical Sciences*, 5, 649–666.

Knibb, J. A., Kipps, C. M., Hodges, J. R., & Hodges, J. R. (2006). Frontotemporal dementia. *Current Opinion in Neurology*, *19*.

Kong, G. K. W., Adams, J. J., Harris, H. H., Boas, J. F., Curtain, C. C., Galatis, D., Masters, C. L., Barnham, K. J., McKinstry, W. J., Cappai, R., & Parker, M. W. (2007). Structural studies of the Alzheimer's amyloid precursor protein copper-binding domain reveal how it binds copper ions. *Journal of Molecular Biology*, *367*(1), 148–161. https://doi.org/10.1016/j.jmb.2006.12.041

Kramer, B., Rarey, M., & Lengauer, T. (n.d.). Evaluation of the FLEXX incremental construction algorithm for protein-ligand docking. *Proteins*, *37*(2), 228–241.

Leng, Y., Blackwell, T., Stone, K. L., Hoang, T. D., Redline, S., & Yaffe, K. (2016). Periodic limb movements in sleep are associated with greater cognitive decline in older men without dementia. *Sleep*, *39*(10), 1807–1810. https://doi.org/10.5665/sleep.6158

Li, J. G., Chiu, J., Ramanjulu, M., Blass, B. E., & Praticò, D. (2020). A pharmacological chaperone improves memory by reducing Aβ and tau neuropathology in a mouse model with plaques and tangles. *Molecular Neurodegeneration*, *15*(1). https://doi.org/10.1186/s13024-019-0350-4

Li, N. M., Liu, K. F., Qiu, Y. J., Zhang, H. H., Nakanishi, H., & Qing, H. (2019). Mutations of beta-amyloid precursor protein alter the consequence of Alzheimer's disease pathogenesis. *Neural Regeneration Research*, *14*(4), 658–665. https://doi.org/10.4103/1673-5374.247469

Lührs, T., Ritter, C., Adrian, M., Riek-Loher, D., Bohrmann, B., Döbeli, H., Schubert, D., Riek, R., & Hoffmann-La Roche, F. (2005).3D structure of Alzheimer's amyloid-(1–42) fibrils. *Proceedings of the National Academy of Sciences of the United States of America*, *29*(48). www.pnas.orgcgidoi10.1073pnas.0506723102

Makarian, M., Gonzalez, M., Salvador, S. M., Lorzadeh, S., Hudson, P. K., & Pecic, S. (2022). Synthesis, kinetic evaluation and molecular docking studies of donepezil-based acetylcholinesterase inhibitors. *Journal of Molecular Structure*, *1247*. https://doi.org/10.1016/j.molstruc.2021.131425

McKeith, I. G., Dickson, D. W., Lowe, J., Emre, M., O'Brien, J. T., Feldman, H., Cummings, J., Duda, J. E., Lippa, C., Perry, E. K., Aarsland, D., Arai, H., Ballard, C. G., Boeve, B., Burn, D. J., Costa, D., Del Ser, T., Dubois, B., Galasko, D., . . . & Yamada, M. (2005). Diagnosis and management of dementia with Lewy bodies: Third report of the DLB consortium. *Neurology*, *65*(12), 1863–1872). Lippincott Williams and Wilkins. https://doi.org/10.1212/01.wnl.0000187889.17253.b1

McKhann, G. M., Knopman, D. S., Chertkow, H., Hyman, B. T., Jack, C. R., Kawas, C. H., Klunk, W. E., Koroshetz, W. J., Manly, J. J., Mayeux, R., Mohs, R. C., Morris, J. C., Rossor, M. N., Scheltens, P., Carrillo, M. C., Thies, B., Weintraub, S., & Phelps, C. H. (2011). The diagnosis of dementia due to Alzheimer's disease: Recommendations from the National Institute on Aging-Alzheimer's Association workgroups on diagnostic guidelines for Alzheimer's disease. *Alzheimer's and Dementia*, *7*(3), 263–269. https://doi.org/10.1016/j.jalz.2011.03.005

Meyer, J. C., & Schellack, N. (2016). Overview of Alzheimer's disease and its management providing an overview on antipsychotics: Schizophrenia a psychiatric challenge? A 2017 update view project point prevalence survey of antimicrobial utilisation in selected public sector hospitals in the gauteng province of South Africa view project. *South African Pharmaceutical Journal*, *83*(9). www.researchgate.net/publication/316214529

Miller, M. D., Kearsley, S. K., Underwood, D. J., & Sheridan, R. E. (1994). FLOG: A system to select "quasi-flexible" ligands complementary to a receptor of known three-dimensional structure. *Journal of Computer-Aided Molecular Design*, *8*.

Neagoie, C., Vedrenne, E., Buron, F., Mérour, J. Y., Rosca, S., Bourg, S., Lozach, O., Meijer, L., Baldeyrou, B., Lansiaux, A., & Routier, S. (2012). Synthesis of chromeno[3,4-b]indoles as Lamellarin D analogues: A novel DYRK1A inhibitor class. *European Journal of Medicinal Chemistry*, *49*, 379–396. https://doi.org/10.1016/j.ejmech.2012.01.040

Otzen, D. E., Kristensen, O., & Oliveberg, M. (2000.). Designed protein tetramer zipped together with a hydrophobic Alzheimer homology: A structural clue to amyloid assembly. *Proceedings of the National Academy of Sciences of the United States of America*, *97*(18), 9907–9912. www.rcsb.org

Ray, B., Gaskins, D. L., Sajdyk, T. J., Spence, J. P., Fitz, S. D., Shekhar, A., & Lahiri, D. K. (2011). Restraint stress and repeated corticotrophin-releasing factor receptor activation in the amygdala both increase amyloid-β precursor protein and amyloid-β peptide but have divergent effects on brain-derived neurotrophic factor and pre-synaptic proteins in the prefrontal cortex of rats. *Neuroscience*, *184*, 139–150. https://doi.org/10.1016/j.neuroscience.2011.03.067

Roman Laskowski, B. A., Macarthur, M. W., & Thornton, J. M. (1983). Computer programs PROCHECK: A program to check the stereochemicai quality of protein structures. *Physica Status Solidi B*, *13*.

Shahroz, M. M., Sharma, H. K., Altamimi, A. S. A., Alamri, M. A., Ali, A., Ali, A., Alqahtani, S., Altharawi, A., Alabbas, A. B., Alossaimi, M. A., Riadi, Y., Firoz, A., & Afzal, O. (2022). Novel and potential small molecule scaffolds as DYRK1A inhibitors by integrated molecular docking-based virtual screening and dynamics simulation study. *Molecules*, *27*(4). https://doi.org/10.3390/molecules27041159

Sims, R., & Williams, J. (2016a). Defining the genetic architecture of Alzheimer's disease: Where next. *Neurodegenerative Diseases*, *16*(1–2), 6–11. https://doi.org/10.1159/000440841

Sims, R., & Williams, J. (2016b). Defining the genetic architecture of Alzheimer's disease: Where next. *Neurodegenerative Diseases*, *16*(1–2), 6–11. https://doi.org/10.1159/000440841

Skoog, I., & Gustafson, D. (2006). Update on hypertension and Alzheimer's disease. *Neurological Research*, *28*(6), 605–611. https://doi.org/10.1179/016164106X130506

Sommerlad, A., Ruegger, J., Singh-Manoux, A., Lewis, G., & Livingston, G. (2018). Marriage and risk of dementia: Systematic review and meta-analysis of observational studies. *Journal of Neurology, Neurosurgery and Psychiatry*, *89*(3), 231–238. https://doi.org/10.1136/jnnp-2017-316274

Spiess, M., Schwartz, A. L., Lodish, H. F., Steer, C. J., Kempner, E. S., Ashwell, G., Stoorvogel, W., Geuze, H. J., Griffith, J. M., Strous, G. J., Siidhof, T. C., Goldstein, J. L., Brown, M. S., Russell, D. W., Suter, U., Bastos, R., Hofstetter, H., Takahashi, T., Nakada, H., . . . & Garoff, H. (1990). X-ray crystal structure of the protease inhibitor domain of Alzheimer's amyloid, & protein precursor+,*. *Biochemical and Biophysical Research Communications*, *29*.

Sticht, H., Bayer, P., Willbold, D., Dames, S., Hilbich, C., Beyreuther, K., Frank, R. W., & Rösch, P. (1995). Structure of amyloid A4-(1–40)-peptide of Alzheimer's disease. *European Journal of Biochemistry*, *233*(1), 293–298. https://doi.org/10.1111/j.1432-1033.1995.293_1.x

Svedružić, Ž. M., Vrbnjak, K., Martinović, M., & Miletić, V. (2021). Structural analysis of the simultaneous activation and inhibition of γ-secretase activity in the development of drugs for Alzheimer's disease. *Pharmaceutics*, *13*(4). https://doi.org/10.3390/pharmaceutics13040514

Sweeney, M. D., Montagne, A., Sagare, A. P., Nation, D. A., Schneider, L. S., Chui, H. C., Harrington, M. G., Pa, J., Law, M., Wang, D. J. J., Jacobs, R. E., Doubal, F. N., Ramirez, J., Black, S. E., Nedergaard, M., Benveniste, H., Dichgans, M., Iadecola, C., Love, S., . . . & Zlokovic, B. V. (2019). Vascular dysfunction—The disregarded partner of Alzheimer's disease. *Alzheimer's and Dementia*, *15*(1), 158–167. Elsevier Inc. https://doi.org/10.1016/j.jalz.2018.07.222

Tapiola, T., Alafuzoff, I., Herukka, S.-K., Parkkinen, L., Hartikainen, P., Soininen, H., & Pirttilä, T. (n.d.). Cerebrospinal fluid-amyloid 42 and tau proteins as biomarkers of Alzheimer-type pathologic changes in the brain. *Archives of Neurology*, *66*(3), 382–389. http://archneur.jamanetwork.com/

Taylor, J. S., & Burnett, R. M. (2000). DARWIN: A program for docking flexible molecules. *Proteins*, *41*(2), 173–191. www.interscience.wiley.com/jpages/0887-3585/suppmat/

Toledo, J. B., Arnold, M., Kastenmüller, G., Chang, R., Baillie, R. A., Han, X., Thambisetty, M., Tenenbaum, J. D., Suhre, K., Thompson, J. W., John-Williams, L. S., MahmoudianDehkordi, S., Rotroff, D. M., Jack, J. R., Motsinger-Reif, A., Risacher, S. L., Blach, C., Lucas, J. E., Massaro, T., . . . & Kaddurah-Daouk, R. (2017). Metabolic network failures in Alzheimer's disease: A biochemical road map. *Alzheimer's and Dementia*, *13*(9), 965–984. https://doi.org/10.1016/j.jalz.2017.01.020

Toukan, K., & Rahman, A. (n.d.). Molecular-dynamics study of atomic motions in water. *Physical Review B*, *31*.

Weintraub, S., Wicklund, A. H., & Salmon, D. P. (2012). The neuropsychological profile of Alzheimer disease. *Cold Spring Harbor Perspectives in Medicine*, *2*(4). Cold Spring Harbor Laboratory Press. https://doi.org/10.1101/cshperspect.a006171

Xiang, X., Werner, G., Bohrmann, B., Liesz, A., Mazaheri, F., Capell, A., Feederle, R., Knuesel, I., Kleinberger, G., & Haass, C. (2016). TREM2 deficiency reduces the efficacy of immunotherapeutic amyloid clearance. *EMBO Molecular Medicine*, *8*(9), 992–1004. https://doi.org/10.15252/emmm.201606370

Xie, L., Kang, H., Xu, Q., Chen, M. J., Liao, Y., Thiyagarajan, M., O'Donnell, J., Christensen, D. J., Nicholson, C., Iliff, J. J., Takano, T., Deane, R., & Nedergaard, M. (2013). Sleep drives metabolite clearance from the adult brain. *Science, 342*(6156), 373–377. https://doi.org/10.1126/science.1241224

Xie, Y., Su, N., Yang, J., Tan, Q., Huang, S., Jin, M., Ni, Z., Zhang, B., Zhang, D., Luo, F., Chen, H., Sun, X., Feng, J. Q., Qi, H., & Chen, L. (2020). FGF/FGFR signaling in health and disease. *Signal Transduction and Targeted Therapy, 5*(1). Springer Nature. https://doi.org/10.1038/s41392-020-00222-7

Zhang, S., Iwata, K., Lachenmann, M. J., Peng, J. W., Li, S., Stimson, E. R., Lu, Y. A., Felix, A. M., Maggio, J. E., & Lee, J. P. (2000). The Alzheimer's peptide Aβ adopts a collapsed coil structure in water. *Journal of Structural Biology, 130*(2–3), 130–141. https://doi.org/10.1006/jsbi.2000.4288

Zhang, Z. H., Yu, L. J., Hui, X. C., Wu, Z. Z., Yin, K. L., Yang, H., & Xu, Y. (2014). Hydroxy-safflor yellow A attenuates Aβ1–42-induced inflammation by modulating the JAK2/STAT3/NF-κB pathway. *Brain Research, 1563*, 72–80. https://doi.org/10.1016/j.brainres.2014.03.036

Zvěřová, M., Fišar, Z., Jirák, R., Kitzlerová, E., Hroudová, J., & Raboch, J. (2013). Plasma cortisol in Alzheimer's disease with or without depressive symptoms. *Medical Science Monitor, 19*(1), 681–689. https://doi.org/10.12659/MSM.889110

10 Virtual Screening
Pioneering the Future of Drug Discovery in Alzheimer's Disease

*N. Yellasubbaiah, B. Sivakumar and B. Nagasudha**

10.1 INTRODUCTION

Although Alzheimer's disease (AD) is a multifactorial disease (Rosini, Simoni et al. 2016; Alvariño, Alonso et al. 2018), it is defined by increased generation and/or accumulation of amyloidogenic peptides (particularly A) produced by amyloid precursor protein (APP) proteolysis (Li, Zhao et al. 2019). The presence of senile plaques in the cerebral cortex is thought to stimulate inflammatory and neurotoxic processes, resulting in NO, cytokines, and reactive oxygen species (ROS) (Sivandzade, Bhalerao et al. 2019). In AD, this process contributes to neurodegeneration and neuronal cell death.

Depending on their concentration, ROS can have a favourable or negative impact on cellular activity (Zekry, Epperson et al. 2003; Manoharan, Guillemin et al. 2016). ROS in low concentrations can alter cellular activities via redox-dependent signalling and transcription factors (Tang and Le 2016; Kumar, Ganeshpurkar et al. 2018), whereas high levels of ROS can impair critical cellular functioning by causing damage to proteins, lipids, and DNA (Asiimwe, Yeo et al. 2016). Normal biological functions necessitate a delicate balance between the generation and elimination of ROS. Homeostasis imbalances can result in oxidative stress and the development of pathogenic illnesses (Islam 2017).

Stress causes deposition, tau hyperphosphorylation, and cognitive impairment. The reduction of endogenous antioxidant systems with age promotes the development of AD. As a result, oxidative stress is important to the aetiology of Alzheimer's disease (Santos, Correia et al. 2012; Chhetri, King et al. 2018). Current AD treatments include donepezil, galantamine, and rivastigmine, which are acetylcholinesterase inhibitors, and meanwhile, which is a noncompetitive inhibitor of N-methyl-D-aspartate (Bachurin, Bovina et al. 2017). These inhibitors work by blocking cholinergic and glutamate receptors. This is because AD (Kwong and Molkentin 2015) is linked to oxidative glutamate toxicity (Albrecht, Lewerenz et al. 2010), which is an excitatory neurotransmitter in the central nervous system (CNS).

Excess glutamate inhibits cysteine uptake by the xc system, which in turn limits glutathione synthesis (GSH), leading in the accumulation of reactive oxygen species (Albrecht, Lewerenz et al. 2010). Neurochemical dysfunction of cholinergic neurons in the central nervous system (CNS) can also contribute to AD aetiology (Albrecht, Lewerenz et al. 2010). Although these medications are the most effective pharmacological treatments for AD, they have a relatively small average overall effect and have no influence on the course of the underlying neurodegenerative process (Briggs, Kennelly et al. 2016). This is most likely related to the fact that AD is multifactorial and is linked to the activation or inactivation of a number of homeostatic enzymes.

Given that oxidative stress is important to AD pathology, oxidative defence mechanisms appear to be critical targets for creating new and potential AD treatments. One of the most effective anti-oxidative stress defence systems is the Kelch-like ECH-associated protein 1 (Keapl)/Nrf2/ARE pathway (Lu, Zhang et al. 2019). Furthermore, cyclooxygenase-2 (COX-2) has been linked to the expression of anti-inflammatory mediators, neuroprotection, and ROS regulation, and thus

DOI: 10.1201/9781003412069-10

represents a promising AD target (Chaudhary, Maurya et al. 2018). Natural products have substantial therapeutic potential for AD because they include substances known to have antioxidant activity, particularly polyphenol compounds (Liu, Xu et al. 2019). Lignans are polyphenol compounds that are chemically classified as dimers formed by the oxidative homocoupling of cinnamic alcohols or the coupling with cinnamic acids, according to Barbosa Filho in Simes (1999) (Peixoto, Nunes et al. 2015). Drug design is an important approach in medicinal chemistry that increasingly needs the use of new technologies to provide improved practicability and speed of results. For example, we routinely use in silico research to understand the properties of a ligand and its receptor (Teich, Sakurai et al. 2016).

10.1.1 N-Terminal Kinases of c-Jun (JNKs)

JNKs are a protein kinase family that consists of three genes (JNK1, JNK2, and JNK3) (Gourmaud, Paquet et al. 2015). JNK1 and JNK2 are found everywhere, but JNK3 is found mostly in the brain. To activate JNKs, extracellular stimuli such as UV radiation, cytokines, and A peptides phosphorylate (pJNK) and activate mitogen-activated protein (MAP) kinase (MAPK2)(Antoniou, Falconi et al. 2011). Furthermore, investigations have demonstrated that JNK in AD patients can be activated by stress and triggered by detrimental external stimuli via the kinase cascade and oxidative stress (Chaudhary, Maurya et al. 2018). JNKs are involved in a variety of biological processes, including inflammation, gene expression regulation, cell proliferation, and apoptosis. Because of its phosphorylation of APP, which increases A formation, JNK3 has been related to the aetiology of AD (Antoniou, Falconi et al. 2011). Because of its importance in neurodegeneration, JNK pathway signalling has been a target for developing pharmacological and potential therapeutic methods (Yarza, Vela et al. 2016).

The coordinated interaction of the scaffold proteins that constitute the JNK activation complex is required for JNK pathway activation (Flemming 2012). This complex is capable of signal amplification, substrate selectivity, and signal cascade coordination. JNK can be activated by a variety of stimuli, including JNK interaction protein 1a (JIP1a) and JIP1b (also known as IB1), JIP2, JIP3 (originally known as JSAP1), JNK-associated leucine zipper protein (JLP), and other SRC homology 3 (SH3) domain-containing proteins. JNK phosphorylation is mediated by c-Jun, which interacts with JunB, JunD, and c-Fos and activates transcription factor to form transcription factor activator protein 1 (AP-1) (Wang, Li et al. 2017). AP-1 regulates the maturation of the cellular stress response and modulates the signals that ultimately result in the activation of caspases and proteins associated with apoptosis. JNK-3 levels were found to be higher in the brains of living AD patients compared with controls, and kinase inhibitors, including JNK-3, can mitigate the effects of A-induced neuronal death (Tell, Hilbrich et al. 2016).

10.1.2 Phosphodiesterases (PDEs)

PDEs are a family of 11 enzyme subtypes (PDE1-PDE11) that regulate the hydrolysis rates of cAMP and cGMP (Fiorito, Vendome et al. 2017). PDE variants play diverse roles in a range of physiological and pathological processes. Although the majority of PDE isoforms (PDE1, PDE2, PDE3, PDE4, PDE5A, PDE7A, PDE7B, PDE8B, PDE9A, PD10A, and PDE11A) are expressed in the brain, their levels of expression vary between locations. PDE5 and PDE1 are found in the cerebellum but only in Purkinje neurons; PDE1B is found in subsets of Purkinje cells; PDE6 is restricted to the retina and pineal gland; PDE3B is expressed in proopiomelanocortin and neuropeptide neurons; PDE1 exhibits distribution patterns in the hippocampus, cerebral cortex, thalamus, and striatum; and PDE2A is widely distributed PDEs can influence neuronal cell survival, and when they fail, they can contribute to neurodegenerative illnesses such as Alzheimer's disease (Prickaerts, Heckman et al. 2017).

PDE5 has anti-inflammatory and neuroprotective effects by increasing NOS expression and cGMP accumulation and activating the protein kinase G (PKG) signalling pathway, which is important in the development of several neurodegenerative diseases, including AD, Parkinson's disease (PD), and multiple sclerosis (MS). PDE5 hydrolyses cGMP during AD progression. This critical intracellular messenger stimulates PKG, resulting in the initiation of several intracellular signals. The cyclic regulation of AMP and cGMP is crucial in numerous memory-related processes because these molecules are critical secondary brain messengers that are specifically associated with memory recovery (Nabavi, Talarek et al. 2019). The balance between their production and degradation, which PDEs govern, maintains their levels (Gulisano, Tropea et al. 2018).

PDE5 hydrolyzes cGMP exclusively. PDE5 inhibitors thereby increase cGMP concentrations in neurons. Increased PDE5 expression and activity, as well as the accumulation of A peptide, which blocks the NO/cGMP pathway activation, have been associated with age-related reductions in cGMP levels. Numerous studies have shown that PDE5 inhibitors reduce the risk of Alzheimer's disease by boosting NO/cGMP signalling. PDE5 inhibitors can cause vasodilation in the brain, hence boosting or maintaining the activation of neuroprotective signalling pathways (Través, Pardo et al. 2014). As a result, raising cGMP levels by blocking PDE5 is an additional strategy for improving the learning and memory skills of Alzheimer's patients.

10.1.3 Protein Phosphatase 1B Tyrosine (PTP1B)

PTP1B is a member of the nontransmembrane phosphotyrosine phosphatase family and a regulator of multiple CNS processes, many of which are pertinent to AD therapy (Song, Jung et al. 2016). In AD, elevated PTPB1 activity is associated with insulin deficiency and altered signalling pathways (Vieira, Lyra e Silva et al. 2017). In addition, endoplasmic reticulum neuroinflammation and stress, which are both associated with amyloidosis, can induce increased PTP1B activity.

The neuroinflammatory response is characterised by the activation of innate immune cells in the brain (microglia), the infiltration of macrophages, and the release of inflammatory mediators such as NO, cytokines, and chemokines, which are associated with the progression of neurodegenerative disorders (Song, Jung et al. 2016). Cognitive decline and neuronal mortality have been linked to inflammatory processes and amyloid aggregation. PTP1B inhibits numerous signalling pathways that activate GSK3 and are associated with neurodegeneration when it is active. Trodusquemina is a highly selective PTP1B inhibitor used in clinical trials for the treatment of diabetes and obesity, as well as the specific inhibition of PTP1B in neurons. Trodusquemina was sufficient to enhance spatial learning and memory deficits in hAPP-J20 mice and prevent neuronal death. Another study found that inflammatory stimuli modulate PTP1B expression and that PTP1B enhances microglial activation and functions as a key positive regulator of neuroinflammation. Thus, inhibiting PTP1B represents a novel therapeutic strategy for neuroinflammatory and neurodegenerative disorders (Ricke, Cruz et al. 2020).

10.1.4

NADPH oxidase, also known as nicotinamide adenine dinucleotide phosphate oxidase (NOX), is the most thoroughly studied ROS-generating system. Members of the NOX family are transmembrane proteins that use electrons from cytosolic NADPH to produce superoxide anion (Sorce, Stocker et al. 2017). Seven isoforms, NOX1, NOX2, NOX3, NOX4, NOX5, DUOX1, and DUOX2, produce active enzyme complexes. These membrane proteins are only known to catalyse the production of superoxide anion from hydrogen peroxide.

Hydrogen peroxide can cause direct cellular injury by oxidising deoxyribonucleic acid and lipids. Hydrogen peroxide readily permeates cell membranes. Multiple neurodegenerative causes can upregulate certain NOX isoforms. Free radical production is the principal function of NOX. According to research, genetic and pharmacological inhibition of NOX enzymes may mitigate the neurotoxic effects of brain traumas and neurodegenerative diseases, resulting in a neuroprotective effect (Ma, Wang et al. 2017).

In particular, the reported lack of benefits associated with various antioxidant techniques may be due to the inefficacy of antioxidant molecules in vivo or the concurrent diminution of oxidant regulatory functions (Sorce, Stocker et al. 2017). The translocation of p47phox and p67phox in the AD brain was reported by Shimohama et al., suggesting that NOX is activated in the AD brain. Oligomeric A- (oA-)-induced microglial proliferation and the production of proinflammatory factors including reactive oxygen species (ROS), nitric oxide (NO), tumour necrosis factor (TNF)-, and interleukin (IL-1) indicate that NOX inhibitors exert neuroprotective effects against AD due to their anti-inflammatory properties (Shimohama, Tanino et al. 2000).

10.1.5

NADPH quinone oxidoreductase 1 (NQO1) is a flavin adenine dinucleotide- (FAD-) dependent cytoplasmic flavoprotein that catalyses the reduction of two electrons from quinones, quinonimines, and nitroaromatic naphthoquinones and is substituted by glutathione, dichlorophenolindophenol dyes, and NADPH as an electron donor. Therefore, NQO1 plays a central role in monitoring cellular redox status and protecting against oxidative stress induced by a variety of metabolic situations (Ross and Siegel 2017), including the metabolism of quinones and other xenobiotics, through the following mechanisms: (i) functioning as a 2-electron donor to provide a derivation that competes with the formation of ROS reduced coenzyme Q; and (iii) regulating the stress-activated kinase pathway (Raina, Templeton et al. 1999). Inactivation of the detoxifying enzyme NQO1 is linked to the development of Alzheimer's disease.

NQO1 activity can be modified by genetic predispositions such as the C690T NQO1 polymorphism, advanced age, cigarette smoking, and a variety of medications. The early expression of NQO1 in astrocytes may indicate the activation of a neuronal cell antioxidant defence system at the onset of disease. In contrast, the delayed expression of NQO1 may indicate the activation of this system as a last-ditch effort to prevent neuronal cell demise (SantaCruz, Yazlovitskaya et al. 2004). The relationship between oxidative medicine and cellular longevity 3 NQO1's antioxidant activity is essential; however, additional research is necessary to determine whether it should be targeted in the treatment of AD.

10.1.6

Erythroid factor 2 (Nrf2) is linked to nuclear factor 2. Nrf2 is a transcription factor that promotes adaptation and survival under stress by modulating the gene production of multiple networks of cytoprotective proteins, including anti-inflammatory and antioxidant proteins as well as proteins that repair or remove damaged macromolecules (SantaCruz, Yazlovitskaya et al. 2004). Maintaining cellular redox balance and regulating mitochondrial ROS production are impossible without Nrf2. Nrf2 influences variations in mitochondrial membrane potential (m), ATP synthesis, and lipid peroxidation, and Nrf2 activation under stress conditions or by growth factors can neutralize increases in mitochondrial ROS production, promoting neuroprotection (Kensler, Wakabayashi et al. 2007).

Nrf2 is a crucial regulator of the body's antioxidant response and is essential for stimulating the expression of genes encoding antioxidant proteins and enzymes, as well as metabolic detoxification phase II enzymes, which are essential for cell protection and survival. HO-1, superoxide dismutase (SOD), catalase (CAT), NADPH, NQO1, GST, GSH reductase (GR), GSH peroxidase (GPx), thioredoxin (Trx), and glutamate-cysteine ligase (GCL) are Nrf2 substrates (Kaspar, Niture et al. 2009). In addition to mediating antioxidant and detoxification mechanisms, Nrf2 regulates the expression of 200 genes implicated in additional cellular functions, including inflammatory response, metabolic regulation, cell proliferation, senescence, and mitochondrial function. Researchers have investigated the role of Nrf2 in the mechanisms of apoptosis and neuroprotection associated with Alzheimer's disease and traumatic brain injury, as well as the reduction of ERO expression (Wright, Loo et al. 2011).

10.1.7 Sodium-Glucose Transport Protein (SGLT)

There are two essential families of glucose transporters: sodium-dependent glucose cotransporters (SGLTs) and facilitative glucose transporters (GLUTs). Five main SGLT isoforms, SGLT1, SGLT2, SGLT3, SGLT4, and SGLT5, have been identified; however, SGLT1 and SGLT2 play a significant role in the cellular processes of AD (**Shah, DeSilva et al. 2012**). The SLC5A1 gene produces the SGLT1 isoform, which transports glucose via a secondary active transport mechanism employing the Na+ gradient generated by the Na+/K+ ATPase pump. This receptor is predominantly expressed in the colon, trachea, heart, testes, prostate, brain, and kidneys.

SGLT1 is a metabotropic receptor linked to transmembrane G proteins with a secondary structure consisting of 664 amino acid residues organised into 14 transmembrane helices, with the NH2 and COOH terminals facing the extracellular side of the plasma membrane. Asn248 is the only site on the receptor for N-glycosylation. The kidneys, brain, liver, thyroid, and heart all contain the SGLT2 isoform, which is encoded by the SLC5A2 gene. The SGLT2 structure is extremely similar to that of the SGLT1 receptor, and it appears to be involved in the pathways for diabetes and renal disease (**Wright, Loo et al. 2011**). Due to its association with cellular mediators of vascular injury, SGLT1 has been implicated in AD (**Li, Agrawal et al. 2019**). Its activation is associated with a reduction in epidermal growth factor receptor levels, and its expression is associated with food and the regulation of insulin release by inhibiting amylase and glucosidase (**Lee, Heo et al. 2007**).

10.1.8 Consider DJ-1

DJ-1 is a sensor of oxidative stress and removes peroxide through self-oxidation (Mitsumoto and Nakagawa 2001). This receptor is also linked to the aetiology of cancer and may serve as a potential tumour marker (Yu, Pan et al. 2017). DJ-1 participates in multiple signalling pathways, such as the regulation of mitochondrial quality and the response to oxidative stress. It has been demonstrated that cells with high DJ-1 levels are resistant to oxidative stress and neurotoxins such as 6-OHDA, whereas cells with low DJ-1 levels are susceptible to oxidative injury (Inden, Kitamura et al. 2011). Dolgacheva and colleagues stated that the DJ-1 receptor exhibits anti-Parkinson's disease activity; the strategies discussed included protecting dopaminergic neurons from neurodegeneration in PD. According to scientists, the DJ-1 receptor in its natural state can serve as a sensor for oxidative stress and an antioxidant. DJ-1 regulates transcription, protects mitochondria from oxidative stress, and increases the quantities of uncoupling proteins (UCP) 4 and UCP 5, which are responsible for mitochondrial decoupling and the subsequent decrease in mitochondrial membrane potential. DJ-1 inhibits the synthesis of EROS and functions on redox factors, including NFB, which acts on anti-inflammatory factors (Xue, Jiang et al. 2017).

10.1.9 Cycloxygenase (COX)

Prostaglandins (PGs) are produced by prostaglandin endoperoxide through the enzymes synthase and cyclooxygenase (COX), which play essential roles in the aetiology and inflammation of autoimmune disorders. COX has two isoforms: COX-1, which is persistently expressed in the majority of tissues and organs, and COX-2, an inflammation-inducible enzyme that is essential during the inflammatory process and in autoimmune disease (Del Mundo, Castillo et al. 2020). COX-2 serves an important role in both ageing and skin cancer. PGE2 is an essential COX pathway product (Sandoughi, Saravani et al. 2020).

COX-2, also referred to as prostaglandin H synthase 2 (PGHS-2), catalyses the conversion of arachidonic acid and oxygen to prostaglandins (PGs), which are essential lipid mediators involved in a number of physiological and pathological processes. COX-2 is typically expressed at a modest level in standard physiological conditions. Nonetheless, this gene increases significantly in response to inflammation (He, Hua et al. 2020). COX1 is a constitutive enzyme necessary for sustaining a basic level of PGs in order to maintain physiological homeostasis, such as the integrity of the gastrointestinal tract (Jorda, Aldasoro et al. 2020). COX-1 and COX2 are responsible for the production of prostaglandins, prostacyclins, and thromboxanes. COX-1 and COX-2 have exceedingly similar active site structures and sequence identities (Gedawy, Kassab et al. 2020). Occasionally, neurodegenerative diseases such as AD are treated with nonsteroidal anti-inflammatory drugs (NSAIDs) that target COX-1 and COX-2 (Buzharevski, Paskaš et al. 2020).

10.1.10 NOS Is an Enzyme That Creates Nitric Oxide

NOS consists of three enzymes (eNOS, nNOS, and iNOS) that produce nitric oxide (NO) from L-arginine (Li, Wang et al. 2020). NO is a free radical gas involved in a number of biological processes, including the regulation of blood flow, blood pressure, and oxygen supply (Wilmes, Scheiper et al. 2020). NOS are endothelial NOS (eNOS or NOS1), inducible NOS (iNOS or NOS2), and neuronal NOS (nNOS or NOS3)(Kseibati, Shehatou et al. 2020). iNOS expression is solely stimulated by cytokines, TNF, infections, chronic inflammation, cancer, interferon, and hypoxia. Unlike the other two isoforms, iNOS inducement generates substantial levels of NO (Soskić, Dobutović et al. 2011). The pervasive production of iNOS in the CNS is detrimental and is frequently detected in neurological diseases like multiple sclerosis, stroke, and PD (Wang, Sun et al. 2020).

According to studies, the number of iNOS-positive neurons in the brains of AD patients increases markedly and is linked to neuronal damage. Based on its expression levels and biological activity, eNOS directly influences the NO production rate and serves as the rate-limiting enzyme for this process. In addition to affecting the structure and function of the vascular endothelium, eNOS activity also influences the maintenance of vascular and endothelial homeostasis (Chen, Li et al. 2020). NO is produced as a neurotransmitter by nNOS in both the central and peripheral nervous systems. nNOS is the neuronal enzyme responsible for NO production; however, not all neurons express nNOS. Nevertheless, excessive activation of nNOS can lead to neuronal mortality due to the production of toxic NO (Murciano-Calles, Coello et al. 2020).

10.1.11 Lipoxygenases (LOXs)

These enzymes catalyse the stereoselective addition of oxygen to arachidonic acid, docosahexaenoic acid, and other polyunsaturated fatty acids (Czapski, Czubowicz et al. 2016). The location of oxygen insertion in a substrate, with the exception of LOX-3, determines the naming of LOXs (Walther, Ivanov et al. 2001). There have been identified 5 types of LOXs in mammals: 5-, 8-, 12-, and 15-LOX, as well as LOX-3 (Kuhn and Thiele 1999). According to Chu et al., despite the fact that 5-LOX is predominantly recognised as a modulator of oxidation and inflammation, this pathway can directly affect the pathophysiology of AD.

The 5-LOX-secretase pathway regulates the synthesis of A peptides and other molecular maladies, including neuroinflammation, synaptic integrity, and cognitive function, which can aid in the development of novel treatments for AD and other neurodegenerative disorders. High 5-LOX levels in the nuclear envelope are associated with the release of leukotrienes, which attract inflammatory cells (Joshi and Praticò 2015). 5-LOX is widely distributed in the CNS and has been shown to be positively regulated in the postmortem brains of AD patients, playing a functional role in the disease's aetiology, with its activation impacting synapses and memory impairment (Firuzi, Zhuo et al. 2008).

According to Di Meco et al., 5-LOX is an essential enzyme for AD because it is involved in inflammatory responses and is expressed at a higher level in the hippocampi of AD patients than in those of healthy individuals (AlFadly, Elzahhar et al. 2019). Additionally, researchers discovered that several enzymes are directly and indirectly involved through oxidative stress mechanisms and that their activation and inactivation can contribute to neuroprotection or disease progression following a study was to identify potential anti-AD molecules through the virtual screening of ligands (Valencia, Herrera-Arozamena et al. 2018).

10.2 VIRTUAL SCREENING TOOLS AND METHODS

10.2.1 The Gathering and Organisation of Data

This study examined numerous enzymes with known biological activity and three-dimensional structures. Compounds with known activity against the enzymes JNK-3 (CHEMBL2637), PTP1B (CHEMBL335), NFR2 (CHEMBL1075094), NOX1 (CHEMBL1287628), PDE5 (CHEMBL1828), COX-2 (CHEMBL230), and iNOS (CHEMBL4EM1) were selected. These molecules were incorporated into predictive models (www.ebi.ac.uk/chembl/) (Fourches, Muratov et al. 2010). The information about the institutions is shown in Table 10.1. Based on their pIC50 log IC50 mol/l values, the compounds were categorized. IC_{50} indicates the concentration required to inhibit 50% of the target. However, activation data for the enzyme Nrf2 were utilized because activating this protein would have the desired effect. In addition, according to the methods described by Fourches et al., 159 CHEMBL ligands were evaluated via virtual screening for potential activity against enzymes associated with the progression of AD (Fourches, Muratov et al. 2010). The Chemaxon Standardizer version 18.17.0 produced three-dimensional structures (www.chemaxon.org).

10.2.2 Quantitative Structure-Activity Modelling (QSAR)

Knime 3.5.3 (KNIME 3.5.3, Konstanza Information Miner Copyright, 2018, www.knime.org) was utilized to execute the studies and produce Table 10.1: Set of Compounds from the ChEMBL databases for each selected enzyme. Database Active molecules Inactive compounds Total JNK-3 580 (pIC50 6:0) 642 (pIC50 6:0) 1.222 PTP1B 1.446 (pIC50 5:0) 1.354 (pIC50 5:0) 2,800 NFR2 163 (active) 85 (inactive). 248 NOX1 85 (pIC50 4:75) 60 (pIC50 4:75) 145 PDE5 873 (pIC50 7:0) 869 (pIC50 7:0) 1742 COX2 2.018 (pIC50 5:50) 1.702 (pIC50 5:50) 3.720 for iNOS 396 (pIC50 = 5:50) 367 (pIC50 5:50) Oxidative Medicine and Cellular Ageing Five of the computational models. Due to the success of our previous studies, we decided to conduct a 3D QSAR analysis for each enzyme bank. All compounds with resolved chemical structures were extracted as specific data files (SDF) and imported into Dragon 7.0 software to generate descriptors. The banks of molecules and their calculated descriptors were imported from the Dragon software, and the data were separated using the "Stratified sample" option of the "Partitioning" tool, which divided the data into training and testing sets, representing 80% and 20% of all compounds, respectively. The categories were selected at random, while the proportions of active and inactive compounds in both databases were maintained. The random forest method was used to construct predictive models utilizing WEKA nodes (Hall, Frank et al. 2009). RF was configured with the following parameters for all models: 250 total forests and a single seed for generating random numbers. Cross-validation was performed to evaluate the predictive ability of the developed models. The sensitivity (true-positive rate or active speed), specificity (true-negative rate or inactive speed), and accuracy of the external performances of the chosen models were evaluated (general predictability). Additionally, the sensitivity and specificity of the receiver operating characteristics curve were utilized because they more accurately reflect actual performance than accuracy.

The models were also evaluated with the Matthews correlation coefficient (MCC), which provides a global assessment of the model based on the confusion matrix data. The MCC is a correlation coefficient between the observed and predictive binary classifications, with values between −1 and +1, where +1 corresponds to a perfect prediction, 0 to a random prediction, and −1 to a total disagreement between the forecast and the observation (Matthews 1975). MCC is calculated with the following formula: MCC = VP × VN FP × FN VP + FP VP + FN VN + FP VN + FN, where VP represents genuine positives, VN genuine negatives, FP fraudulent positives, and FN fraudulent negatives. By analysing the substances in the test sets, the application domain (APD) was employed to assess the accuracy of the predictions.

The APD is a chemical space that incorporates the model's descriptors and the modelled response, enabling the calculation of uncertainty when predicting the activity of a compound from the training set used to build the model. By comparing predicted and observed values, this method is essential for validating the accuracy of QSAR models (Roy, Kar et al. 2015). APD is computed using the following formula: APD = d + Z, 2, where D and Z represent the Euclidean distances and the mean and standard deviation of the compounds in the training set, respectively. In this study, the empirical cutoff value Z was established at 0.5.

10.2.3 Docking Molecules

Molecular docking was performed using the Molegro Virtual Docker v6.0.1 (MVD) programme (Bitencourt-Ferreira and de Azevedo 2019), and 6 anchoring targets were selected (Table 10.2). The 3D structures of the enzymes used in this study were obtained from the Protein Data Bank (PDB) using the following codes: PDB ID 4Y46 for JNK-3, PDB ID 4Y14 for PTPB1, PDB ID 6FY4 for NQO1, PDB ID 3B2R for PDE5, PDB ID 5KIR for COX-2, and PDB ID 4NOS for iNOS. We did not attach the enzymes Nrf2 and NOX1 because PDB did not contain 3D human structures. Initially, all water molecules were eliminated from the crystalline structure, and the root-mean-square deviation (RMSD) was calculated from the poses to assess the reliability of the fit. The RMSD facilitates an effective connection mode close to the experimental structure if its value is less than 2.0. As a scoring function, the MolDock score was utilized to predict the optimal interactions between the ligand and the receptor. The enzyme and ligands were then introduced into the anchor assistant to determine the stability of the system based on the contacts identified with the active site of the enzyme (Bernstein, Koetzle et al. 1977).

10.2.4 The Forecasting of ADMET Properties

ADME properties were determined using the Swiss ADME open-access web tool (www.swissame.ch), which provides a set of rapid prediction models for evaluating physicochemical, pharmacokinetic, and pharmacological aspects. OSIRIS Property Explorer (www.organic-chemistry.org/prog/peo/) was utilized to estimate toxicity based on the parameters of mutagenicity, tumorigenicity, reproductive effects, and irritability. Membrane permeability, intestinal absorption, and the presence of a substrate or inhibitor for P glycoprotein were crucial for absorption. Therefore, we analysed compounds with a log P consensus of 4.15 or less and no more than 2 violations of Lipinski's criteria. Additionally, substances do not serve as substrates for the permeability glycoprotein enzyme (P-gp). We used variables such as the blood–brain barrier (log BBB) and CNS permeability to evaluate the distribution. We used CYP substrate or inhibition models (CYP1A2, CYP2C19, CYP2C9, CYP2D6, and CYP3A4) to predict metabolism (Daina, Michielin et al. 2017).

ENDNOTE

* Corresponding author

REFERENCES

Albrecht, P., J. Lewerenz, S. Dittmer, R. Noack, P. Maher and A. Methner (2010). "Mechanisms of oxidative glutamate toxicity: The glutamate/cystine antiporter system xc as a neuroprotective drug target." *CNS & Neurological Disorders-Drug Targets (Formerly Current Drug Targets-CNS & Neurological Disorders)* **9**(3): 373–382.

AlFadly, E. D., P. A. Elzahhar, A. Tramarin, S. Elkazaz, H. Shaltout, M. M. Abu-Serie, J. Janockova, O. Soukup, D. A. Ghareeb and A. F. El-Yazbi (2019). "Tackling neuroinflammation and cholinergic deficit in Alzheimer's disease: Multi-target inhibitors of cholinesterases, cyclooxygenase-2 and 15-lipoxygenase." *European Journal of Medicinal Chemistry* **167**: 161–186.

Alvariño, R., E. Alonso, R. Lacret, D. Oves-Costales, O. Genilloud, F. Reyes, A. Alfonso and L. M. Botana (2018). "Streptocyclinones A and B ameliorate Alzheimer's disease pathological processes in vitro." *Neuropharmacology* **141**: 283–295.

Antoniou, X., M. Falconi, D. Di Marino and T. Borsello (2011). "JNK3 as a therapeutic target for neurodegenerative diseases." *Journal of Alzheimer's Disease* **24**(4): 633–642.

Asiimwe, N., S. G. Yeo, M.-S. Kim, J. Jung and N. Y. Jeong (2016). "Nitric oxide: Exploring the contextual link with Alzheimer's disease." *Oxidative Medicine and Cellular Longevity* **2016**.

Bachurin, S. O., E. V. Bovina and A. A. Ustyugov (2017). "Drugs in clinical trials for Alzheimer's disease: The major trends." *Medicinal Research Reviews* **37**(5): 1186–1225.

Bernstein, F. C., T. F. Koetzle, G. J. Williams, E. F. Meyer Jr, M. D. Brice, J. R. Rodgers, O. Kennard, T. Shimanouchi and M. Tasumi (1977). "The Protein Data Bank: A computer-based archival file for macromolecular structures." *Journal of Molecular Biology* **112**(3): 535–542.

Bitencourt-Ferreira, G. and W. F. de Azevedo (2019). "Molegro virtual docker for docking." *Docking Screens for Drug Discovery*: 149–167.

Briggs, R., S. P. Kennelly and D. O'Neill (2016). "Drug treatments in Alzheimer's disease." *Clinical Medicine* **16**(3): 247.

Buzharevski, A., S. Paskaš, M.-B. Sárosi, M. Laube, P. Lönnecke, W. Neumann, B. Murganić, S. Mijatović, D. Maksimović-Ivanić and J. Pietzsch (2020). "Carboranyl derivatives of rofecoxib with cytostatic activity against human melanoma and colon cancer cells." *Scientific Reports* **10**(1): 4827.

Chaudhary, A., P. K. Maurya, B. S. Yadav, S. Singh and A. Mani (2018). "Current therapeutic targets for Alzheimer's disease." *Journal of Biomedicine* **3**: 74–84.

Chhetri, J., A. E. King and N. Gueven (2018). "Alzheimer's disease and NQO1: Is there a link?" *Current Alzheimer Research* **15**(1): 56–66.

Czapski, G. A., K. Czubowicz, J. B. Strosznajder and R. P. Strosznajder (2016). "The lipoxygenases: Their regulation and implication in Alzheimer's disease." *Neurochemical Research* **41**: 243–257.

Daina, A., O. Michielin and V. Zoete (2017). "SwissADME: A free web tool to evaluate pharmacokinetics, drug-likeness and medicinal chemistry friendliness of small molecules." *Scientific Reports* **7**(1): 42717.

Del Mundo, C. R., A. L. Castillo, S. S. A. An and M. A. Tan (2020). "In vivo COX-2 modulation and metabolite profiling of Pandanus tectorius leaves extracts." *3 Biotech* **10**: 1–10.

Fiorito, J., J. Vendome, F. Saeed, A. Staniszewski, H. Zhang, S. Yan, S.-X. Deng, O. Arancio and D. W. Landry (2017). "Identification of a novel 1, 2, 3, 4-tetrahydrobenzo [b][1, 6] naphthyridine analogue as a potent phosphodiesterase 5 inhibitor with improved aqueous solubility for the treatment of Alzheimer's disease." *Journal of Medicinal Chemistry* **60**(21): 8858–8875.

Firuzi, O., J. Zhuo, C. M. Chinnici, T. Wisniewski and D. Praticò (2008). "5-Lipoxygenase gene disruption reduces amyloid-β pathology in a mouse model of Alzheimer's disease." *The FASEB Journal: Official Publication of the Federation of American Societies for Experimental Biology* **22**(4): 1169.

Flemming, A. (2012). "JNK3 as new target in AD?" *Nature Reviews Drug Discovery* **11**(11): 829–829.

Fourches, D., E. Muratov and A. Tropsha (2010). "Trust, but verify: On the importance of chemical structure curation in cheminformatics and QSAR modelling research." *Journal of Chemical Information and Modelling* **50**(7): 1189.

Gedawy, E. M., A. E. Kassab and A. M. El Kerdawy (2020). "Design, synthesis and biological evaluation of novel pyrazole sulfonamide derivatives as dual COX-2/5-LOX inhibitors." *European Journal of Medicinal Chemistry* **189**: 112066.

Gourmaud, S., C. Paquet, J. Dumurgier, C. Pace, C. Bouras, F. Gray, J.-L. Laplanche, E. F. Meurs, F. Mouton-Liger and J. Hugon (2015). "Increased levels of cerebrospinal fluid JNK3 associated with amyloid pathology: Links to cognitive decline." *Journal of Psychiatry and Neuroscience* **40**(3): 151–161.

Gulisano, W., M. R. Tropea, O. Arancio, A. Palmeri and D. Puzzo (2018). "Sub-efficacious doses of phosphodiesterase 4 and 5 inhibitors improve memory in a mouse model of Alzheimer's disease." *Neuropharmacology* **138**: 151–159.

Hall, M., E. Frank, G. Holmes, B. Pfahringer, P. Reutemann and I. H. Witten (2009). "The WEKA data mining software: An update." *ACM SIGKDD Explorations Newsletter* **11**(1): 10–18.

He, R., K. Hua, S. Zhang, Y. Wan, H. Gong, B. Ma, R. Luo, R. Zhou and H. Jin (2020). "COX-2 mediated crosstalk between Wnt/β-catenin and the NF-κB signaling pathway during inflammatory responses induced by Haemophilus parasuis in PK-15 and NPTr cells." *Developmental & Comparative Immunology* **105**: 103588.

Inden, M., Y. Kitamura, K. Takahashi, K. Takata, N. Ito, R. Niwa, R. Funayama, K. Nishimura, T. Taniguchi and T. Honda (2011). "Protection against dopaminergic neurodegeneration in Parkinson's disease—model animals by a modulator of the oxidized form of DJ-1, a wild-type of familial Parkinson's disease—linked PARK7." *Journal of Pharmacological Sciences* **117**(3): 189–203.

Islam, M. T. (2017). "Oxidative stress and mitochondrial dysfunction-linked neurodegenerative disorders." *Neurological Research* **39**(1): 73–82.

Jorda, A., M. Aldasoro, C. Aldasoro, S. Guerra-Ojeda, A. Iradi, J. M. Vila, J. Campos-Campos and S. L. Valles (2020). "Action of low doses of aspirin in inflammation and oxidative stress induced by aβ1–42 on astrocytes in primary culture." *International Journal of Medical Sciences* **17**(6): 834.

Joshi, Y. B. and D. Praticò (2015). "The 5-lipoxygenase pathway: Oxidative and inflammatory contributions to the Alzheimer's disease phenotype." *Frontiers in Cellular Neuroscience* **8**: 436.

Kaspar, J. W., S. K. Niture and A. K. Jaiswal (2009). "Nrf2: INrf2 (Keapl) signaling in oxidative stress." *Free Radical Biology and Medicine* **47**(9): 1304–1309.

Kensler, T. W., N. Wakabayashi and S. Biswal (2007). "Cell survival responses to environmental stresses via the Keapl-Nrf2-ARE pathway." *Annual Review of Pharmacology and Toxicology* **47**: 89–116.

Kseibati, M. O., G. S. Shehatou, M. H. Sharawy, A. E. Eladl and H. A. Salem (2020). "Nicorandil ameliorates bleomycin-induced pulmonary fibrosis in rats through modulating eNOS, iNOS, TXNIP and HIF-1α levels." *Life Sciences* **246**: 117423.

Kuhn, H. and B. J. Thiele (1999). "The diversity of the lipoxygenase family: Many sequence data but little information on biological significance." *FEBS Letters* **449**(1): 7–11.

Kumar, D., A. Ganeshpurkar, D. Kumar, G. Modi, S. K. Gupta and S. K. Singh (2018). "Secretase inhibitors for the treatment of Alzheimer's disease: Long road ahead." *European Journal of Medicinal Chemistry* **148**: 436–452.

Kwong, J. Q. and J. D. Molkentin (2015). "Physiological and pathological roles of the mitochondrial permeability transition pore in the heart." *Cell Metabolism* **21**(2): 206–214.

Lee, Y. J., J. S. Heo, H. N. Suh, M. Y. Lee and H. J. Han (2007). "Interleukin-6 stimulates α-MG uptake in renal proximal tubule cells: Involvement of STAT3, PI3K/Akt, MAPKs, and NF-κB." *American Journal of Physiology-Renal Physiology* **293**(4): F1036–F1046.

Li, J., H. Wang, J. Li, Y. Liu and H. Ding (2020). "LC-MS analysis of Myrica rubra extract and its hypotensive effects via the inhibition of GLUT 1 and activation of the NO/Akt/eNOS signaling pathway." *RSC Advances* **10**(9): 5371–5384.

Li, S., X. Zhao, P. Lazarovici and W. Zheng (2019). "Artemether activation of AMPK/GSK3β (ser9)/Nrf2 signaling confers neuroprotection towards β-amyloid-induced neurotoxicity in 3xTg Alzheimer's mouse model." *Oxidative Medicine and Cellular Longevity* **2019**.

Li, Z., V. Agrawal, M. Ramratnam, R. K. Sharma, S. D'Auria, A. Sincoular, M. Jakubiak, M. L. Music, W. J. Kutschke and X. N. Huang (2019). "Cardiac sodium-dependent glucose cotransporter 1 is a novel mediator of ischaemia/reperfusion injury." *Cardiovascular Research* **115**(11): 1646–1658.

Liu, L., H. Xu, S. Ding, D. Wang, G. Song and X. Huang (2019). "Phosphodiesterase 5 inhibitors as novel agents for the treatment of Alzheimer's disease." *Brain Research Bulletin* **153**: 223–231.

Lu, M.-C., X. Zhang, F. Wu, S.-J. Tan, J. Zhao, Q.-D. You and Z.-Y. Jiang (2019). "Discovery of a potent Kelch-like ECH-associated protein 1-nuclear factor erythroid 2-related factor 2 (Keapl—Nrf2) protein—protein interaction inhibitor with natural proline structure as a cytoprotective agent against acetaminophen-induced hepatotoxicity." *Journal of Medicinal Chemistry* **62**(14): 6796–6813.

Manoharan, S., G. J. Guillemin, R. S. Abiramasundari, M. M. Essa, M. Akbar and M. D. Akbar (2016). "The role of reactive oxygen species in the pathogenesis of Alzheimer's disease, Parkinson's disease, and Huntington's disease: A mini review." *Oxidative Medicine and Cellular Longevity* **2016**.

Matthews, B. W. (1975). "Comparison of the predicted and observed secondary structure of T4 phage lysozyme." *Biochimica et Biophysica Acta (BBA)-Protein Structure* **405**(2): 442–451.

Mitsumoto, A. and Y. Nakagawa (2001). "DJ-1 is an indicator for endogenous reactive oxygen species elicited by endotoxin." *Free Radical Research* **35**(6): 885–893.

Murciano-Calles, J., A. Coello, A. Cámara-Artigas and J. C. Martinez (2020). "PDZ/PDZ interaction between PSD-95 and nNOS neuronal proteins: A thermodynamic analysis of the PSD95-PDZ2/nNOS-PDZ interaction." *Journal of Molecular Recognition* **33**(4): e2826.

Nabavi, S. M., S. Talarek, J. Listos, S. F. Nabavi, K. P. Devi, M. R. de Oliveira, D. Tewari, S. Argüelles, S. Mehrzadi and A. Hosseinzadeh (2019). "Phosphodiesterase inhibitors say NO to Alzheimer's disease." *Food and Chemical Toxicology* **134**: 110822.

Peixoto, C. A., A. K. S. Nunes and A. Garcia-Osta (2015). "Phosphodiesterase-5 inhibitors: Action on the signaling pathways of neuroinflammation, neurodegeneration, and cognition." *Mediators of Inflammation* **2015**.

Prickaerts, J., P. R. Heckman and A. Blokland (2017). "Investigational phosphodiesterase inhibitors in phase I and phase II clinical trials for Alzheimer's disease." *Expert Opinion on Investigational Drugs* **26**(9): 1033–1048.

Raina, A. K., D. J. Templeton, J. C. Deak, G. Perry and M. A. Smith (1999). "Quinone reductase (NQO1), a sensitive redox indicator, is increased in Alzheimer's disease." *Redox Report* **4**(1–2): 23–27.

Ricke, K. M., S. A. Cruz, Z. Qin, K. Farrokhi, F. Sharmin, L. Zhang, M. A. Zasloff, A. F. Stewart and H.-H. Chen (2020). "Neuronal protein tyrosine phosphatase 1B hastens amyloid β-associated Alzheimer's disease in mice." *Journal of Neuroscience* **40**(7): 1581–1593.

Rosini, M., E. Simoni, R. Caporaso and A. Minarini (2016). "Multitarget strategies in Alzheimer's disease: Benefits and challenges on the road to therapeutics." *Future Medicinal Chemistry* **8**(6): 697–711.

Ross, D. and D. Siegel (2017). "Functions of NQO1 in cellular protection and CoQ10 metabolism and its potential role as a redox sensitive molecular switch." *Frontiers in Physiology* **8**: 595.

Roy, K., S. Kar and P. Ambure (2015). "On a simple approach for determining applicability domain of QSAR models." *Chemometrics and Intelligent Laboratory Systems* **145**: 22–29.

Sandoughi, M., M. Saravani, M. Rokni, M. Nora, M. Mehrabani and A. Dehghan (2020). "Association between COX-2 and 15-PGDH polymorphisms and SLE susceptibility." *International Journal of Rheumatic Diseases* **23**(5): 627–632.

SantaCruz, K. S., E. Yazlovitskaya, J. Collins, J. Johnson and C. DeCarli (2004). "Regional NAD (P) H: Quinone oxidoreductase activity in Alzheimer's disease." *Neurobiology of Aging* **25**(1): 63–69.

Santos, R. X., S. C. Correia, X. Zhu, H.-G. Lee, R. B. Petersen, A. Nunomura, M. A. Smith, G. Perry and P. I. Moreira (2012). "Nuclear and mitochondrial DNA oxidation in Alzheimer's disease." *Free Radical Research* **46**(4): 565–576.

Shah, K., S. DeSilva and T. Abbruscato (2012). "The role of glucose transporters in brain disease: Diabetes and Alzheimer's disease." *International Journal of Molecular Sciences* **13**(10): 12629–12655.

Shimohama, S., H. Tanino, N. Kawakami, N. Okamura, H. Kodama, T. Yamaguchi, T. Hayakawa, A. Nunomura, S. Chiba and G. Perry (2000). "Activation of NADPH oxidase in Alzheimer's disease brains." *Biochemical and Biophysical Research Communications* **273**(1): 5–9.

Sivandzade, F., A. Bhalerao and L. Cucullo (2019). "Cerebrovascular and neurological disorders: Protective role of NRF2." *International Journal of Molecular Sciences* **20**(14): 3433.

Song, G. J., M. Jung, J.-H. Kim, H. Park, M. H. Rahman, S. Zhang, Z.-Y. Zhang, D. H. Park, H. Kook and I.-K. Lee (2016). "A novel role for protein tyrosine phosphatase 1B as a positive regulator of neuroinflammation." *Journal of Neuroinflammation* **13**: 1–14.

Sorce, S., R. Stocker, T. Seredenina, R. Holmdahl, A. Aguzzi, A. Chio, A. Depaulis, F. Heitz, P. Olofsson and T. Olsson (2017). "NADPH oxidases as drug targets and biomarkers in neurodegenerative diseases: What is the evidence?" *Free Radical Biology and Medicine* **112**: 387–396.

Soskić, S. S., B. D. Dobutović, E. M. Sudar, M. M. Obradović, D. M. Nikolić, J. D. Djordjevic, D. J. Radak, D. P. Mikhailidis and E. R. Isenović (2011). "Regulation of inducible nitric oxide synthase (iNOS) and its potential role in insulin resistance, diabetes and heart failure." *The Open Cardiovascular Medicine Journal* **5**: 153.

Tang, Y. and W. Le (2016). "Differential roles of M1 and M2 microglia in neurodegenerative diseases." *Molecular Neurobiology* **53**: 1181–1194.

Teich, A. F., M. Sakurai, M. Patel, C. Holman, F. Saeed, J. Fiorito and O. Arancio (2016). "PDE5 exists in human neurons and is a viable therapeutic target for neurologic disease." *Journal of Alzheimer's Disease* **52**(1): 295–302.

Tell, V., I. Hilbrich, M. Holzer, F. Totzke, C. Schachtele, I. Slynko, W. Sippl and A. Hilgeroth (2016). "Drug development of small-molecule inhibitors of AD-relevant kinases as novel perspective multitargeted approach." *Current Alzheimer Research* **13**(12): 1330–1336.

Través, P. G., V. Pardo, M. Pimentel-Santillana, Á. González-Rodríguez, M. Mojena, D. Rico, Y. Montenegro, C. Calés, P. Martín-Sanz and A. Valverde (2014). "Pivotal role of protein tyrosine phosphatase 1B

(PTP1B) in the macrophage response to pro-inflammatory and anti-inflammatory challenge." *Cell Death & Disease* **5**(3): e1125–e1125.

Valencia, M. E., C. Herrera-Arozamena, L. de Andrés, C. Pérez, J. A. Morales-García, A. Pérez-Castillo, E. Ramos, A. Romero, D. Viña and M. Yáñez (2018). "Neurogenic and neuroprotective donepezil-flavonoid hybrids with sigma-1 affinity and inhibition of key enzymes in Alzheimer's disease." *European Journal of Medicinal Chemistry* **156**: 534–553.

Vieira, M. N., N. M. Lyra e Silva, S. T. Ferreira and F. G. De Felice (2017). "Protein tyrosine phosphatase 1B (PTP1B): A potential target for Alzheimer's therapy?" *Frontiers in Aging Neuroscience* **9**: 7.

Walther, M., I. Ivanov, G. Myagkova and H. Kuhn (2001). "Alterations of lipoxygenase specificity by targeted substrate modification and site-directed mutagenesis." *Chemistry & Biology* **8**(8): 779–790.

Wang, Q., H. Sun, X. Qi and M. Zhou (2020). "eNOS rs2070744 polymorphism might influence predisposition to hemorrhagic cerebral vascular diseases in East Asians: A meta-analysis." *Brain and Behavior* **10**(5): e01538.

Wang, Y.-L., J.-f. Li, Y.-t. Wang, C.-y. Xu, L.-L. Hua, X.-p. Yang, S. Geng, S.-S. Wang, Z. Wang and H.-L. Yin (2017). "Curcumin reduces hippocampal neuron apoptosis and JNK-3 phosphorylation in rats with Aβ-induced Alzheimer's disease: Protecting spatial learning and memory." *Journal of Neurorestoratology* **5**(1): 117–123.

Wilmes, V., S. Scheiper, W. Roehr, C. Niess, S. Kippenberger, K. Steinhorst, M. A. Verhoff and S. Kauferstein (2020). "Increased inducible nitric oxide synthase (iNOS) expression in human myocardial infarction." *International Journal of Legal Medicine* **134**: 575–581.

Wright, E. M., D. D. Loo and B. A. Hirayama (2011). "Biology of human sodium glucose transporters." *Physiological Reviews* **91**(2): 733–794.

Xue, R., J. Jiang, B. Dong, W. Tan, Y. Sun, J. Zhao, Y. Chen, Y. Dong and C. Liu (2017). "DJ-1 activates autophagy in the repression of cardiac hypertrophy." *Archives of Biochemistry and Biophysics* **633**: 124–132.

Yarza, R., S. Vela, M. Solas and M. J. Ramirez (2016). "c-Jun N-terminal kinase (JNK) signaling as a therapeutic target for Alzheimer's disease." *Frontiers in Pharmacology* **6**: 321.

Yu, D., H. Pan, R. Zhang, Y. Li and X. Nie (2017). "Nucleus DJ-1/Park7 acts as a favorable prognostic factor and involves mucin secretion in invasive breast carcinoma in Chinese population." *International Journal of Clinical and Experimental Medicine* **10**(4): 6558–6567.

Zekry, D., T. K. Epperson and K. H. Krause (2003). "A role for NOX NADPH oxidases in Alzheimer's disease and other types of dementia?" *IUBMB Life* **55**(6): 307–313.

Index

Note: Page numbers in *italics* indicate a figure and page numbers in **bold** indicate a table on the corresponding page.

O

P